A WELL-ORDERED THING

A WELL-ORDERED THING

DMITRII MENDELEEV and the
SHADOW OF THE PERIODIC TABLE

MICHAEL D. GORDIN

BASIC
BOOKS

A Member of the Perseus Books Group
New York

Published by Basic Books,
A Member of the Perseus Books Group

Books published by Basic Books are available at special discounts for bulk purchases in the United States by corporations, institutions, and other organizations. For more information, please contact the Special Markets Department at the Perseus Books Group, 11 Cambridge Center, Cambridge MA 02142, or call (617) 252–5298, (800) 255–1514 or e-mail special.markets@perseusbooks.com.

Designed by Lisa Kreinbrink

Library of Congress Cataloging-in-Publication Data

Gordin, Michael D.
    A well-ordered thing : Dmitrii Mendeleev and the shadow of the periodic table / Michael D. Gordin.
        p. cm.
Includes bibliographical references and index.
    ISBN 0-465-02775-X
    1. Mendeleyev, Dmitry Ivanovich, 1834-1907. 2. Chemists—Russia (Federation)—Biography. 3. Periodic law. I. Title.

QD22.M43G67 2004
540'.92—dc22

04 05 06 07/10 9 8 7 6 5 4 3 2 1

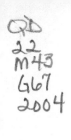
For my family

CONTENTS

LIST OF FIGURES

NOTE TO THE READER

One of the most confusing aspects of Mendeleev (pronounced Men-de-LAY-ev) is spelling his name in the Latin alphabet. Given the tremendous variety of systems used today to transcribe Cyrillic letters in England and the United States—not to mention the rivals in Germany and France, and those of the nineteenth century versus the twentieth—none of which were used invariably by the man himself or his peers, for the sake of consistency I have transliterated all Russian terms using a modification of the Library of Congress standard, except in the cases of well-known names such as Dostoevsky, or when those names belong to first-generation settlers in Russia from Western European nations (and alphabets). Soft signs at the ends of proper nouns have been suppressed in the text on occasion for the sake of readability.

A second difficulty is with dates. Until after the Bolshevik Revolution of 1917, all dates in Russia followed the old-style Julian calendar, which lagged 12 days behind the new-style Gregorian calendar in the nineteenth century, 13 days in the twentieth. I use old-style dates consistently without comment. New-style dates are either juxtaposed in parentheses or indicated by (N.S.), as in 3 (15) August 1868 or 15 August 1868 (N.S.).

Finally, there is the problem of translation. Mendeleev's style is peculiarly Russian, frequently employing idiomatic expressions and relying heavily on unique features of Russian syntax. I have attempted to translate taking these features and rhythms into account, although the result may sometimes appear unusual to the English reader.

Chemical symbols for various elements are kept to a minimum in the text, and they are defined upon first usage. The only exceptions appear within images of periodic systems. The reader can turn to one of the fully explicated systems, such as that in Chapter 8, for the full names of these elements. The most commonly used symbols include H (hydrogen), C (carbon), O (oxygen), N (nitrogen), Na (sodium), Al (aluminum), and Cl (chlorine).

The following abbreviations are used in the notes and bibliography:

ADIM: Arkhiv-Muzei D. I. Mendeleeva (D. I. Mendeleev Archive-Museum), Mendeleevskaia liniia, d. 2, St. Petersburg State University, St. Petersburg, Russia.

MS: D. I. Mendeleev, *Sochineniia,* 25 v. (Leningrad: Izd. AN SSSR, 1934–1956).

PD: Pushkinskii Dom (Pushkin House, Institute of Russian Literature), Makarova nab., d. 4, St. Petersburg, Russia.

PFARAN: Peterburgskii Filial Arkhiva Rossiiskoi Akademii Nauk (Petersburg Division of the Archive of the Russian Academy of Sciences), Universitetskaia nab., d. 1, St. Petersburg, Russia.

RGIA: Rossiiskii Gosudarstvennyi Istoricheskii Arkhiv (Russian State Historical Archive), Angliiskaia nab., d. 4, St. Petersburg, Russia.

TIIEiT: Trudy Instituta Istorii Estestvoznaniia i Tekhniki.

VGPMV: Vremennik Glavnoi Palaty Mer i Vesov.

VIET: Voprosy Istorii Estestvoznaniia i Tekhniki.

ZhRFKhO: Zhurnal Russkogo Fiziko-Khimicheskogo Obshchestva.

With the exception of ADIM, all archival documents are cited according to collection (*fond,* abbreviated f.), directory (*opis'*, abbreviated op.), file (*delo,* abbreviated d.), and page (*list,* abbreviated l.). Papers from ADIM are cited according to practices originally established by Dmitrii I. Mendeleev.

If passion less, and reason more,
My wayward nature checked and led,
If some great change of empire bore
The seat of rule from heart to head; . . .
If I could trim my muse's wing
Control her flight, abate her rage,
And teach her, a well-ordered thing,
To coo and warble in a cage . . .

—GEORGE HENRY BOKER (1882)

Rough draft of the first periodic system, dated 17 February 1869, from Smirnov, *Mendeleev*, insert 1.

PREFACE

The periodic table of chemical elements may be the most widely recognized talisman of modern science. The general contours of the periodic table—its squares piled into two peaks on the left and right, a long trough in the middle, an island of two rows on the bottom, all coded with obscure letter symbols and a series of numbers—are familiar by sight even to those with the most cursory high-school science education. When one becomes more familiar with the ordering of the hundred-odd chemical substances within the table, the symmetries seem so obvious, the sequences so natural, that most people find it hard to imagine a time when this object did not exist, when it had to be brought into being by individuals much like ourselves, striving to make sense of the disparate phenomena of the world. Of course, the periodic table had to be created *somewhere*—everything that we know about the world first appeared in a specific place at a specific time. The periodic system actually presents one of the more complicated cases, emerging independently during the 1860s in England, France, the United States, Germany, and Russia.

The most developed form of the periodic system of chemical elements, the one canonized as the standard across the world today, emerged from the last of these places: Russia. In fact, this form of the system, born in the northern Imperial capital of St. Petersburg in the late 1860s, was so suggestive that its formulator—a young chemistry professor at the local university—risked using the blank spaces in its framework to predict three yet undiscovered elements to supplement the sixty-three ones then recognized. In the face of all the competing periodic systems from Western Europe, no one in chemistry had yet hazarded so audacious a prediction. Even more amazing, the 35-year-old Petersburger's predictions were confirmed within fifteen years. This periodic table (or periodic system) of chemical elements was widely proclaimed as the periodic *law*, one of the cornerstones of the modern physical sciences. Shortly after its inception, the polychromatic icon of the periodic system appeared in chemistry classrooms and laboratories across the world, a position it will almost certainly retain far into the future.

But why Russia? How did this object, so apparently universal, appear in a distant corner of Europe (barely even in Europe, according to contemporary opinions) in a place so manifestly *particular?* One way of answering this question is to pose another: What kind of problem was the periodic system an attempt to solve? In the most basic sense, it was a chemical problem. The system juxtaposes diverse elements and attempts to detect regularities in their differences by proposing a fundamental similarity: a change in chemical properties predicated upon atomic weight. It is a system that attempts to comprise all our knowledge of the chemical elements into one ordering—despite the fact that a few elements refused to be tamed so readily, insisted on being misfits.

But the problem was not solely chemical. The periodic system was developed in Russia by an individual who was deeply immersed in a culture obsessed with systematizing such misfits, a man trying to bring order to a Russian society that was apparently disintegrating. Whatever its correlation with the natural world, the periodic system as we now know it is an artifact of the same culture that produced the novels of Dostoevsky and the pageantry of the Tsars. In order to understand the building of this part of modern chemistry, one must come to terms with the attempts to create a modern Russia. Both stemmed from a set of prior, more personal pressures.

The object of these pressures was a St. Petersburg chemist named Dmitrii Ivanovich Mendeleev. It seems that wherever one turns in late Imperial Russia—the period spanning from the Emancipation of the serfs in 1861 to the first Russian Revolution of 1905—one encounters this man. The same individual who composed the periodic system also helped design the highly protectionist Russian tariff of 1891, battled local Spiritualists, created a smokeless gunpowder, attempted Arctic exploration, consulted on oil development in Baku, investigated iron and coal deposits, published art criticism, flew in balloons, introduced the metric system, and much more. Far from the popular image of the chemist dutifully absorbed in experimentation, Mendeleev was (with a few important exceptions) not a laboratory chemist. The periodic law was only one of his efforts in building modern systems, although far and away his most successful one. This law remains a special case of a much broader phenomenon that was fundamentally rooted in the culture of the late Imperial period, and of nineteenth-century Europe in general. Mendeleev spent as much, if not more, of his time and energy pursuing attempts to transform Imperial Russia into a more stable, yet also more liberal, autocracy as he did on chemistry. With his close relationship to several government ministers and access to the Tsar, Mendeleev made a prodigious impact on the economy and politics of Imperial Russia. To follow Mendeleev through the Imperial capital is to unfold the tapestry of life in Imperial Russia. The image that emerges is more textured than the life of any one individual; it is, rather, a reflection of the systematizing spirit that engulfed modernizing Europe in the nineteenth century and that has left so many traces on the world today.

Mendeleev himself remains trapped in the shadow of the periodic table. Strictly speaking, he did not formulate the periodic table as it stands today: He was not aware of about a third of the elements we now know of; he organized elements in terms of "atomic weight," not atomic number (the quantity of protons in an atom of the element) as we do today; and he was remarkably inconstant in his feelings toward any particular layout of the periodic system (there are literally hundreds of topologically distinct forms today, both two-dimensional and not, and Mendeleev flirted with dozens on his own). Mendeleev championed his arrangement of elements as a periodic *system,* meant to be only a representation of a more fundamental *law.* The table, especially in its most common current form, was only one expression among many. Yet people today think of Mendeleev—when they think of him at all—in terms of the periodic table they know, and color in their history according to that assumption. The Mendeleev presented in the following pages roams free of this present-day understanding of the periodic table, but even during his lifetime that shadow began to cohere and cast him in its pall. He should be viewed as he saw himself: in the light of his contemporary world.

What follows is not a traditional biography. Here is no comprehensive account of every aspect of the adult Mendeleev's life, and we encounter precious little of his childhood. Instead, one confronts a complex culture by following the course of one man—admittedly a highly ambitious, intelligent, and well-connected one. Precisely because Mendeleev conceived of his systems as unifying and totalizing, and therefore applicable to every facet of Imperial life, observing him and his successes—and, more often, his failures—reveals a great deal about the dilemmas that faced a nation negotiating the uncertain path between tradition and revolution. Similarly, Mendeleev's highly diverse scientific activities belie any simplistic categorizations into individual sciences such as "chemistry" or "physics," and instead show a polymath who perceived all knowledge, both of the natural and the social world, to be fundamentally of a piece, able to be treated with the same intellectual equipment. This equipment was drawn from the dominant available source: the culture of Imperial Petersburg.

Mendeleev as a historical figure has left little lasting impact. Russian schoolchildren in chemistry class still learn some stories—mostly apocryphal—about his life, and occasionally his bearded visage with its straggly hair peeps out from an insert in Western chemistry textbooks, but much of the legacy he worked so hard to build has been lost. Except, of course, for the ubiquitous periodic system. This is a story about Russia and science, but it is also a very European tale, one broadly strewn across the sciences and the humanities.

We usually do not examine the origins of the tenets of our knowledge, at least not ones as stable as the periodic table. When we do, however, we often come upon a picture both strange and oddly familiar. Imperial Russia seemed that way to Mendeleev as well, as he returned from abroad in 1861 on the eve of momentous transformations at home. His universe would never look the same again.

CHAPTER 1

INTRODUCTION: AUTOCRACY AND MR. MENDELEEV

Mendeleev in 1869, from Trirogova-Mendeleev,
Mendeleev i ego sem'ia, facing page 9.

CHAPTER 1

INTRODUCTION: AUTOCRACY AND MR. MENDELEEV

Oh, what a marvelous affirmation of evolutionary theory! Oh, what a
great chain extends from a dog to Mendeleev the chemist!
                                        —MIKHAIL BULGAKOV[1]

THE BEST PLACE to begin this very Russian story is in Germany. On
three days during the first week of September 1860 in the southern town of
Karlsruhe, chemists from across Europe assembled to discuss weighty issues—
or, more accurately, the issue of weight, which was threatening to overload
their science with inconsistency and contradiction. German organic chemist
August Kekulé conceived of the gathering with the hope of resolving crucial
disagreements over the conventions of chemistry, such as the calculation of
atomic weights, and even what terms like "molecule" and "atom" meant. The
appeal soliciting attendance was sent out in July over the signatures of some of
the most prominent names in chemistry.[2] A young chemistry postdoctoral re-
searcher from St. Petersburg—then conveniently living in nearby Heidelberg—could not pass up the opportunity to attend such an event and meet the
luminaries of his field. His name was Dmitrii Ivanovich Mendeleev.

The Karlsruhe Congress was a significant event in the history of nineteenth-
century chemistry for several reasons, none of which depends upon
Mendeleev's attendance. Karlsruhe represents the first time that chemists from
across Europe gathered in one place to resolve central scientific issues, and thus
was an important stage in the professionalization of chemistry as an interna-

tional science.[3] Second, Amedeo Avogadro's 1811 hypothesis to standardize atomic weights was revived at the Congress by Italian chemist Stanislao Cannizzaro, a move that bridged a chasm of widespread confusion and seeded the consensus about the notion of atomic weights that remains the basis of chemistry to this day. Young Mendeleev, for example, who was twenty-six at the time, would for the rest of his life recall Cannizzaro's innovations as central to the formation of his periodic system of chemical elements.[4]

The memory of the Karlsruhe Congress also had much more personal consequences for Mendeleev as he reflected on his imminent return to St. Petersburg, capital of an empire then on the brink of substantial reforms. The model of Karlsruhe offered an opportunity to think about organizing expertise to resolve conceptual disputes calmly; the experience proved so important that he felt he had to share it with the Russian public. Mendeleev wrote a letter to his Russian mentor A. A. Voskresenskii on 7 September, which the latter (at Mendeleev's request) published in Petersburg's chief daily newspaper. "The chemical congress which just finished in Karlsruhe is such a remarkable event in the history of our science that I consider it an obligation to describe to you—even in a few words—the sessions of the congress and the results it achieved," he exulted.[5] Mendeleev was a young chemist passionate about his science, and he was also an ambitious man craving a place in the limelight of St. Petersburg culture. By enlightening the public about Karlsruhe, he sought a grand entrance as a public intellectual.

The Congress thus raises three issues that provide a convenient entry into our story. At the most basic level, the Congress changed Mendeleev's understanding of several chemical concepts in a way that would resonate throughout chemistry. Beyond that, the Congress placed him in contact with other specialists, providing him with a rational model for the coordination of civil servants. In the Russia of the Great Reforms to which Mendeleev returned, such models were posited as deliberate and pointed contrasts with a culture of officialdom legendary for its arbitrariness and indeterminacy. Karlsruhe held the potential to redeem Petersburg. Finally, Karlsruhe changed the way Mendeleev thought of himself as a Petersburg intellectual. Although he was certainly not the only chemist in the city (and far from the most prominent), he boldly chronicled his own experiences, communicating to Petersburgers the meaning of chemistry and its consequences for everyday life. We will follow Mendeleev on his triple path—as chemist, bureaucratic expert, and public figure—from this opening gambit at Karlsruhe until his death amid revolution and turmoil.

This is the story of two systematic misfits: Dmitrii Mendeleev and the Russian Empire. The central figure is the former, but the central object of inquiry is the latter; through Mendeleev and his vocation in chemistry, the turbulent culture of late Imperial Russia is laid bare. The periodic law, Mendeleev's chief claim to fame, was at once a symptom of underlying pressures in the Russian environment and within chemistry. Both Russian history and the history of science converge around the notion of a "systematic misfit": the tension between attempting comprehensive, orderly systems, constructed for stability and clarity, and the awkward application of those systems to the real world. To the extent that a system can predict future behavior or events, it provides stability; on the other hand, such regularity makes it vulnerable to misfits that refuse to comply with its rigor. This is not the fault of Russians, or chemistry, or Mendeleev, but is merely a consequence of the inevitable messiness of the natural and social worlds we live in. When one encounters such a misfit—in the periodic system, in economics, in private life—one has three choices: ignore the misfit; attempt to rebuild the system around the misfit; or, like the mythical Procrustes who lopped off the legs of travelers to fit them into his bed, jam the misfit into the confines of the original system. Each approach, with varying degrees of hope and violence, appears in Mendeleev's story, tracing a path through the cultural politics of the late nineteenth century that ranges from the machinations of empires to the vibration of atoms.

LIBERALISM IN THE NAME OF AUTOCRACY

Mendeleev was excited by the Karlsruhe Congress not just because it resolved some thorny confusions within chemistry. That was hardly a reason to compose a newspaper article for the chemically illiterate public. Imagine today a write-up of a scientific meeting becoming national news, and you will appreciate the peculiarity. Mendeleev wanted Russian readers to hear another message. After describing Cannizzaro's reform of atomic weights, Mendeleev offered special praise of the unanimity with which the chemists had validated the measure:

> The result [of voting on Cannizzaro's suggestions] was unexpectedly unanimous and important. Having adopted the distinction between atom and molecule, chemists of all countries adopted the basis of the unitary system. . . . To this story let me add that in all the discussions there was not one malicious word between both parties. All this, it seems to me, is a complete guarantee of the quick success of these new foundations in the future. Not one among 150 chemists agreed to vote against these foundations.[6]

In Mendeleev's view, proper decision-making proceeded in a courteous, communal, reasoned, and consensual environment. All in all, a perfect model for fundamental reform.

This reaction to Karlsruhe was fundamentally *conservative*, in a very specific sense. It is difficult to characterize precisely Mendeleev's political position because it does not fall into the easy categories of "reactionary," "liberal," or "radical" that usually organize our understanding of past politics. Mendeleev was one of many Russians who borrowed very heavily from liberal rhetoric while pursuing ends such as autocracy or Russian chauvinism that mesh poorly with nineteenth-century conceptions of liberalism (the latter being a doctrine, distinct from today's credo of the same name, that emphasized free trade, property rights, and individual autonomy from the state). A liberal in the name of autocracy, Mendeleev supported the rule of law only insofar as it was the best way, in his view, to preserve traditions essential to Russian stability—traditions embodied in the institution of the autocracy. By contrast, his Russian contemporaries who identified themselves with liberalism were liberals in the name of Russia. For them, liberalism linked Russia to the legal and political traditions of European progress. For Mendeleev, these liberals were deluded or misinformed—or simply dangerous—and he had no patience for them.

Viewing Mendeleev as a conservative opens up Russian culture in surprisingly novel ways. In light of the momentous events of 1917, historians have understandably emphasized radicals in late Imperial Russia to the exclusion of multiple competing movements. While we now work with an incredibly rich taxonomy of trends among radicals—populists, legal Marxists, Bolsheviks, Socialists-Revolutionaries, anarchists, nihilists, Mensheviks, Empirio-Critics, and so on—every position on the Right has been lumped together under the general banner of Reaction. In trying to understand Tsarist autocracy through the eyes of those wanting to overthrow it, we have lost sight of vital distinctions among the guardians of the established order. To be sure, there *were* extreme reactionaries who wanted to halt all change in Russia, but there was also an equally important group of conservatives who actively lobbied for gradual reform in order to maintain, wherever possible, the aspects of Tsarism they considered worthy of preservation. They recognized that the world was changing, and that Russia had to change with it, or perish. These conservatives consistently attempted to exploit particular features of autocracy in order to tame its most dangerous opponents.

Liberalism in the name of autocracy was a specific Russian variant of European conservatism. Not exactly an ideology, it was more of an attitude toward history and the state. Following nineteenth-century historian Nikolai Karamzin (and his precursor Edmund Burke), conservatives believed that tradition, the residue of historical epochs as revealed in national institutions, was a valuable force for stability.[7] (Reactionaries, by contrast, held to tradition for its own sake.) When adherence to *all* traditions threatened the stability of society, conservatives embraced gradual reform to adapt to change *from within the framework* of historical traditions. The prerogative of selecting among various traditions was tremendously liberating to conservatives, who could reject even the most venerated of Russian social institutions, such as the nobility, in order to uphold autocracy. Autocracy itself was nonnegotiable: It was the most characteristic Russian national tradition, and it served as the instrument of gradualist reform. Thus the functions, values, and structures the state had accumulated through historical accretion were the features that made those institutions worth preserving. This political position bears more than an accidental relation to the common understanding of the scientific method.

In theory (and often in practice), the Tsar's authority was absolute and unconstrained.[8] But this did not imply political stagnation. To the contrary, in the context of rapid economic modernization and social dislocation, the autocrat often proved willing and able to reform the state surprisingly widely within the bounds of his own theoretically unbounded authority. In fact, the Tsar was the *only* individual in Russia with the authority to change any aspect of the system, however minor. Repeatedly during Mendeleev's life, he witnessed different Tsars issue transformative, even progressive decrees through fiat. It was a power he respected and coveted. Mendeleev and other conservatives valorized the Tsar's reforming powers as free of the inefficiency of parliamentary compromise and senseless debate. Rationalism (suitably conformed to tradition, of course) dictated Russia's path, and conservative experts were the arbiters of the rational. They did not want to eliminate the Tsar's autocratic power; they wanted *access* to it. In attacking the person of the Tsar, terrorist radicals pointed to the same feature of the topography of power: for change to happen, the top had to permit it—or be eliminated.[9]

The immediate mechanism for reform was the much-maligned Imperial bureaucracy, which in Mendeleev's lifetime experienced the rise of a new stratum of civil servants—the *raznochintsy* (literally, "people of various ranks").[10] *Raznochintsy* originated overwhelmingly from families affiliated with the increasingly professionalized civil service of the early nineteenth century, and thus were neither landed nobility nor serfs. They tended to be educated and socialized within bureaucratic culture. Mendeleev was a *raznochinets* by virtue of his father's position as an educational administrator. But Mendeleev was something more than just his father's son (if anything, he was his mother's son): He was a highly trained chemist. He was not *just* a civil servant: He was a civil servant with specialized skills. Such professionals assumed a new importance in Petersburg after the defeat of Russian forces in the Crimean War (ended in 1856 after the death of Nicholas I and the accession of his son, Alexander II). The state moved from being a world closed to advice to an administration starving for information and expertise. Besides a cohort of influential reforming bureaucrats, a sizable portion of the now active cultural elite consisted of professionals—lawyers, physicians, engineers, economists, even chemists.[11]

The crucial point here is that each of these different professionals offered a different form of *expertise* to the reforming state. Mendeleev, for one, was quite aware of the distinctions between different types of expertise that the state might wish to consult—such as the difference between legal and scientific expertise. In late October 1870, after a disappointing experience serving as an expert witness in a court case (on the determination of poisons), he wrote his first and only legal article at the very moment he was finalizing his periodic law:

> Of course the expert is not the judge, defense attorney, or prosecutor, but nevertheless, if he is called then one must give him the right to express his opinion on those subjects he was called to judge; without this the expert's role and the utility expected from his specialized knowledge is significantly diminished to the detriment of the truth sought in court. It is necessary not to forget here that the expert is also subject to oath like witnesses, and true statements are demanded from him, but [the officers of the court] don't give him the opportunity to speak truly.[12]

Mendeleev, the *raznochintsy,* and the rest of the professionals each wanted the opportunity to speak truly. What they considered to be true and which opportunities they were willing to exploit, however, varied widely depending on their political and intellectual commitments.

Yet both Russian liberals and Russian conservatives from among this professional stratum joined forces in support of the program of reforms enacted by Tsar Alexander II beginning in 1861, the so-called Great Reforms.[13] These Reforms not only structured the political and social landscape of Imperial Russia long after most of them were scaled back or scuttled, they profoundly shaped how individuals like Mendeleev viewed the notion of reform in general. Mendeleev and like-minded conservatives continually attempted to replay the Great Reforms scenario—and at times it became a farce. An understanding of the Great Reforms from the conservatives' point of view is thus necessary to understanding Mendeleev's bureaucratic work and, moreover, his chemistry.

The Great Reforms were a set of seven measures: Emancipation (1861); the university statute (1863); rural councils (*zemstva*) (1864); the European-style judicial system (1864); censorship reform (1865); municipal autonomy (1870); and the universal draft (1874). Alexander II viewed the Reforms as necessary to provide for a more stable military and fiscal structure, thus reinforcing the bulwarks of autocracy while presenting them (particularly Emancipation) as acts of unbounded love for his people.[14] This view of the Great Reforms had a profound resonance among a wide group of intellectuals and elites. The bureaucrats (most of them *raznochintsy*) who developed the reforms considered Russia's fundamental problem to be a surfeit of arbitrary power (*proizvol*), and they wished to transform Russia from a realm of subjects to a polity of citizens subject to the rule of law (*zakonnost'*). To be sure, the Reforms did not curtail the absolute power of the Tsar, but rather they hoped to restrict the domains in which he deployed it arbitrarily. He remained an absolute autocrat, but, like Scotus's God, he ordained a rule-bound order. For Mendeleev, the Great Reforms were not a "revolution from above," they were an *anti*-revolutionary force rooted in tradition. At the base of these Reforms there persisted what we, adapting Isaiah Berlin, might call the "conservatives' dilemma": How does one reconcile a rational basis for modernization (the liberal project centered on law) with attention to national traditions that do not necessarily have anything rational at their base (the conservative project emphasizing stability)? How could conservatives eliminate the deleterious consequences of *proizvol* while relying on the mechanism of *proizvol* themselves?[15]

Although many of the reforming measures were watered down in practice, the sense of a tremendous rupture in the Russian state and in what it *meant* to be a subject of the Tsar spread widely. Members of the Petersburg elite addressed such issues in a stunning variety of ways. Their common concern was often not so much whether a *civil* society existed, but about how to *create* an *ordered* society.[16] Diverse groups on both the left and the right faced an equally acute problem of order. Liberals in the name of Russia sought solutions to problems of order in parliamentarism and constitutionalism. Radicals sought them in socialism, populism, and revolution. Reactionaries looked to Orthodoxy, a revivified autocracy, and Great Russian nationalism.[17] Conservatives, or liberals in the name of autocracy, more than anyone else remained wedded to state-sponsored reform: Committed to the principles of the Great Reforms *and* those of autocracy, they compromised wherever they could. *How* you compromised was what mattered.

MAKING SENSE OF THE MAN

Chemists provided useful expertise for the state in the form of consultation on sewage, oil, coal, dyestuffs, pharmaceuticals, and other chemical products, and Mendeleev engaged in all of these activities. But his involvement highlights other ways in which scientific expertise could contribute to the *cultural* redefinition of the Russian public. In the nineteenth century, many scientists considered chemistry to be a model of a "unified" culture because its subject matter was both accessible and dramatic. In Russia in particular, chemistry was seen as the exemplar of almost all the sciences, and beginning in the middle of the century, chemists acquired high public visibility, in large part due to Mendeleev's good offices.[18] For example, when lawyers wanted an expert witness on poisons to testify in a murder trial, Mendeleev volunteered. When public lectures in chemistry were authorized for general education, Mendeleev stepped forward. When cheese needed to be inspected, kerosene street lamps evaluated, alcohol measured and taxed—in short, almost every and any public chemical venture in the Imperial capital—Mendeleev made it clear that he was more than useful: He was essential.

Past studies of this phenomenon have tended to emphasize the "public face" of chemistry, as if chemists always made a hard and fast distinction between the forms of their science they showed to the public and the "real" science performed in the laboratory. For Mendeleev, not only was chemistry performed before the public of a piece with his more "standard" chemical investigations, but it formed an inextricable part of his politics. Arguing against Spiritualism, advising on tariffs, and formulating the periodic law were related activities of creating order in a society sorely in need of it. Mendeleev's world always remained rooted in chemistry, for it was there that he sought to produce a coherent solution to the crisis of modernization faced by the Russian elite.

All of which suggests why Mendeleev presents an ideal case for investigating the place of science, particularly chemistry, in Russian culture, and in turn the place of Russian culture within chemistry. He spent roughly thirty years teaching at St. Petersburg University at a time when the role of the natural sciences in the university curriculum underwent dramatic changes. He served as a consultant for the Ministry of Finances and the Naval Ministry, as director of the Chief Bureau of Weights and Measures, and as close advisor to Tsar Alexander III, Sergei Witte, and other central figures in the late Tsarist state. This was a period in Russian history when an extremely small coterie of individuals comprised the cultural elite of Petersburg, and Mendeleev's contacts extended from novelists to painters to engineers, leaving his imprint on almost all areas of the humanities and the sciences. His dominant role in the Russian Physico-Chemical Society meant that he had his finger on the pulse of those sciences in Russia; he was their ambassador to Western Europe, and he was the West's representative in the scientific periphery that was Petersburg.

Petersburg, the capital of Imperial Russia and Mendeleev's home throughout his entire adult life, served as both the scene of his extraordinarily rich scope of activity and the mechanism—through its copious bureaucracy—by which he carried out his designs. For many of the young professionals and bureaucrats who were implicated in the Great Reforms, service in the capital presented the surest path for advancement in the Empire.[19] Mendeleev was no exception. He would become a public celebrity, alternately fêted and criticized in the emerging organs of the popular press, themselves products of the Great Reforms. A public figure in at least three senses, Mendeleev was a subject of public discussion, a public servant, and a prominent interpreter of chemistry; he was the embodiment of public knowledge. The central warrant for his claims was his discovery of a law of nature—the periodic law—and the predictions he made from them. By virtue of prediction, Mendeleev could in turn argue for an economics and a political structure that would make individual agents more predictable and pliant before a modernizing autocracy. Laws of society became metaphors for laws of nature.[20]

Mendeleev and his periodic law have exerted a surprisingly persistent pull on Western intellectuals, as seen, for example, in two rather different works written by former chemists. The first is Oliver Sacks's recent memoir of his childhood experimentation in chemistry. Sacks uses his memories of England before and during the Second World War as a framework for a narrative of the history of chemistry. In this account, chemistry progresses through history until it arrives at the synthesis of the periodic law—and its hero, a Russian savant named Dmitrii Mendeleev.[21] The second, Primo Levi's breathtaking *The Periodic Table,* is a collection of autobiographical essays, vignettes, and short stories, all organized around the elements of the periodic system in a variety of ways. In some the element in question is the subject of chemical research; in others, the element stands as a metaphor for personal qualities; in yet others, the elements are personified.[22] This wonderful text makes sparse reference, though, to Mendeleev, the formulator of the system that serves Levi so well. And while it is a fact that Mendeleev the man is more often than not subsumed in historical reference by his great work, he deserves to be rescued, as it were, from the shadow of the periodic table. In the narrative that follows, Mendeleev and his law are sometimes the subjects of study, sometimes metaphors (consciously or unconsciously deployed), sometimes things to be built upon, at other times notions to be defended—but always pliant images that reflect the varieties of historical experience in late Imperial Russia.[23]

That material is shaped here into a narrative that differs from conventional biography. Rather than beginning with Mendeleev's birth in January 1834 and ending at his death in January 1907, I concentrate on Mendeleev and the Russian Empire from Emancipation to the Revolution of 1905, the epoch of Mendeleev's greatest chemical achievements and of Russia's greatest hope for a reformed liberal state. I have selected seven major episodes from Mendeleev's life not because they were objectively the "most important" (whatever that would mean), but because each emphasizes a different feature of the cultural life of both Imperial Petersburg and nineteenth-century science.[24] This cultural biographical study aims to illuminate both the history of chemistry and the history of the Russian Empire. In the person of Mendeleev, both staked their claims together.

ELEMENTS OF THE SYSTEM

Mendeleev and his first wife, Feozva Nikitchna, in 1862, from Dobrotin et al., *Letopis' zhizni I deiatel'nosti D. I. Mendeleeva*, 95.

ELEMENTS OF THE SYSTEM: BUILDING PERIODICITY AND A SCIENTIFIC PETERSBURG

At present one can consider it universally acknowledged that among the phenomena of inanimate nature there is no arbitrary will; here the unshakable connections between phenomena rule with complete authority—relations which we call laws. In the invariance of these relations we are even inclined to see the characteristic sign which differentiates the inanimate from the living.

—A. N. SHCHUKAREV[1]

PICTURE A HISTORIAN SEARCHING for the origins of the periodic law. Knowing that it emerged in the late 1860s, he begins to scour the major chemical journals in English, French, and German. Eventually, this search pays off, and our historian finds a lengthy article published in 1871 where it would be expected: in the most prominent of German chemical journals, the *Annalen der Chemie und Pharmacie*. A cursory glance at the footnotes, however, reveals that this is *not* the original publication: This periodic system of chemical elements has appeared before in a rather obscure St. Petersburg chemical journal, published in Russian. In fact, in only the second issue of this journal—restricted from a broader European readership for linguistic reasons—one finds a rather casual description of a chemical classification. This is hardly the universal law of nature our historian had set out to find. But the quest does not stop there, for in the body of this first article, dated April 1869, it appears that the author of this law first published his scientific findings in a *textbook*—an introductory textbook for first-year college students at that. This law of nature, therefore, which has become so ubiquitous that it appears in every classroom and textbook of chemistry, actually first emerged in a classroom and a textbook of chemistry.

That much has long been known. The formulator of the periodic system's most successful and widespread variant, D. I. Mendeleev, made no secret of its conceptual genesis during the writing of a chemical textbook. Yet the implications of taking this historical curiosity seriously—it is not every day that our most fundamental concepts of the world stem from a basic exercise in pedagogy—have scarcely been realized. Let us consider Mendeleev's path toward the periodic law as a *path*—a historical movement through time, with all the contingencies that implies. The periodic system was the product of twin pedagogical trajectories: Mendeleev's personal trajectory through the educational institutions of St. Petersburg in his attempt to solidify a scientific career; and an effort to introduce the totality of chemistry through a set of easily understood basic principles. How the classification of elements became a periodic system and then a law of nature was intimately tied with how Mendeleev became increasingly secure at St. Petersburg University.

One of the most striking aspects of Mendeleev's eventually successful endeavors to provide a stable framework for both inorganic chemistry and his personal career is how haphazard the whole process was. When he returned to Petersburg from his two years studying abroad at Heidelberg, he was neither famous nor on the track of the periodic law. Little more than a cold breeze met Mendeleev as he disembarked from the international platform of the Vitebsk station in St. Petersburg on 14 February 1861. Mendeleev had few close friends to greet him in this city where he was still a relative outsider, his Siberian origins not quite washed away by a decade of schooling in Petersburg and Heidelberg. He arrived at a most auspicious time: Within a few days, the centuries-long tradition of serfdom was to be abolished in the first and most prominent of the Great Reforms. On 16 February Mendeleev noted in his diary that he had "heard a lot about Emancipation" in the bathhouse.[2] The very air was charged.

Mendeleev, like his peers, bridled with anticipation. A young, bright newcomer, he arrived at precisely the moment when the Great Reforms provided astonishing upward mobility for professionals, especially those with technical expertise. The story of the creation of the periodic law is the story of Mendeleev finding his way in this culture of rapid transformation and developing local, stopgap solutions to pressing personal crises. Mendeleev would take the University and elevate it as a symbolic citadel for the priests of technical expertise and develop his hasty periodic system into a "law" that would undergird his evolving worldview. Similarly, the Great Reforms themselves were a series of ad hoc measures, designed to bolster the fiscal and military stability of the Empire, which were retrospectively recast by their principal agents into a unified picture of a reformed Russia. Mendeleev was loyal in his intellectual affections. Long after the Reforms were curtailed or repealed, Mendeleev would continue to consider them the only cultural model that had partially succeeded in modernizing Russia's economy and society.

Consider the personal transformation that took place in the 1860s. Mendeleev returned to Petersburg burdened by debt. He had to find an apartment, pay back a 1,000 ruble loan for the laboratory equipment he had purchased in Heidelberg, and locate resources for new research projects. Arriving at the middle of the academic year, he was unlikely to find a speedy appointment at one of the capital's many teaching establishments. In less than a month after his return, he had already contacted a publisher about translating J. R. Wagner's German text on chemical technology and had obtained a contract for his own proposed organic chemistry textbook.[3] From these modest beginnings, flash ahead to the end of the decade. In 1871 he was professor of general chemistry at St. Petersburg University, the most important chemistry chair in the country, and had expanded the chemistry faculty into one of the strongest in Europe. He had also developed a periodic system of chemical elements that he considered sufficiently "lawlike" to hazard the prediction of three undiscovered chemical elements. In addition, Mendeleev had published two highly successful textbooks, joined the ranks of the Ministry of Finances as advisor on alcohol taxation and agricultural reform, and served as a private consultant for the burgeoning Baku oil industry. His star was on the rise, and he knew it.

The ambitious and energetic Mendeleev did not, at age 35 in 1869, believe his arrangement of elements to be the apex of his career.[4] He did not even recognize his "periodic system" as a law. Why would he? He was no prophet—at least not until 1871. His confidence in that eventual prophecy is a tale of the emergence of periodicity out of the confluence of local concerns—professionalization, pedagogy, authorship—and how Mendeleev built up not just a "periodic law" out of his "periodic system," but a notion of the rightful place of chemical experts in Great Reforms Russia out of the model of St. Petersburg University.[5] These were refractions of the metaphor of Karlsruhe that had propelled him along his new chemical path. As Mendeleev became more convinced of the potential of his periodic system, he transformed it into a law by invoking the power of prediction, a tactic he would continually employ to legitimize both the notion of chemical expertise and his own status as the archetypal expert. Mendeleev was not just building his own career as a scientist in Imperial Russia, he was constructing what it *meant* to be a scientist in Imperial Russia. This process began with Mendeleev's path through Petersburg educational institutions and culminated in the codification of the periodic law. By the end of Mendeleev's first decade back in Petersburg, he had assembled what would become the elements of his utopia of chemical prophecy.

THE EDUCATION OF DMITRII MENDELEEV

Mendeleev came to St. Petersburg in 1850 as a last resort. After his graduation from the local *gymnasium* in Tobol'sk, Siberia, his mother brought him to European Russia to further his education. She first tried to enroll him at Moscow University, the nation's oldest and most prestigious institution, but was refused. The next option was to take young Dmitrii to St. Petersburg. When the University there did not take him, he eventually registered—through the help of a family friend—at the Chief Pedagogical Institute, his father's alma mater. The Institute that fostered Mendeleev from 1851 to 1855 was a transformed place since his father's days. Ivan Ivanovich Davydov, who directed the Institute from 1847 until 1858 (when it closed for good), shifted the school's focus from training teachers to independent research. The curriculum was built around the standard backbone of theology, logic and psychology, pure mathematics, mathematical and general geography, physics, general history with ancient geography, Russian, Greek, Latin, German, and French. The students then broke off into three faculties: Philosophical-Juridical, Physical-Mathematical, and Historical-Philological. Mendeleev was enrolled in the second of these. All of the 100–200 students received free education in return for devoting two years of teaching in secondary *gymnasia*.[6] The Institute was located on the grounds of St. Petersburg University, but was "closed" (in the parlance of the time), meaning that students lived and studied on campus, and were denied access to ordinary University students or the public at large. Rather than finding this environment stifling, as did many of his peers, Mendeleev thrived here, enjoying the attention of distinguished University faculty (who also taught at the Institute): mathematician M. V. Ostrogradskii, mineralogist S. S. Kutorga, physicist Heinrich F. E. Lenz, and chemist A. A. Voskresenskii.[7]

Mendeleev was encouraged to pursue his scientific interests, in particular by Voskresenskii, who was Mendeleev's chief mentor in the Petersburg academic world. Mendeleev's early work explored organic isomorphism, the phenomenon whereby two substances with different chemical composition express the same crystalline structure. Discovered by Eilhard Mitscherlich in 1822, this finding cast into disrepute the long-standing notion that crystalline structure was a unique reflection of underlying chemical composition. Mendeleev's 1856 candidate thesis, "Isomorphism in Connection with Other Relations of Crystalline Form to Content," reflected an early interest in connecting internal properties to external structure. Heavily influenced by French chemist Charles Gerhardt, Mendeleev conducted what was essentially a broad literature review and concluded that specific volume was the best means to examine the influence of composition on form. He continued to explore specific volumes with a Gerhardtian emphasis in his master's thesis, also published in 1856.[8] Mendeleev would later draw on certain aspects of this research when formulating his periodic system: a concern with the elements' *physical,* not chemical, properties, attention to classification, and a reconsideration of atomic-weight values.[9]

On 14 April 1859, after an unpleasant stint teaching secondary school in the Crimea—inadequate facilities, stifling weather, and abominable students—Mendeleev left for Heidelberg on a government-subsidized trip to further his studies in chemistry. Upon arrival, he obtained a spot in the laboratory of distinguished German chemist Robert Bunsen, but the fumes and the noise so annoyed him that he instead transformed his apartment into a "very cute laboratory" that even had its own gas supply.[10] Mendeleev almost immediately threw himself into chemical researches on capillarity (the effect whereby liquid is drawn up in a narrow tube against the pull of gravity). He conducted a broad array of experiments with a variety of organic liquids, which eventually led both to his doctoral thesis on alcohol solutions and to his claim to co-discovery of the "critical point" of liquids.

Mendeleev was socially active among Russian students and travelers, many of whom later remarked that his powerful personality formed the center of the Russian student community. As physiologist I. M. Sechenov recalled: "Mendeleev made himself, of course, the center of the circle; all the more since, despite his young years (he is years younger than I), he was already a trained chemist, and we were [merely] students."[11] Mendeleev's closest friends while he was abroad were Sechenov, who later cut a colorful and politically charged career in Russia, and fellow chemist A. P. Borodin, who would eventually divert some of his attention from chemistry to compose music, including the renowned opera *Prince Igor.* All were repelled by what they characterized as the bourgeois pretensions of the German students.[12] Despite this obstacle, the time in Heidelberg was extremely important for the young Mendeleev, cementing bonds between fellow Russian chemistry students and bringing him to Karlsruhe.

Before leaving Heidelberg, his Russian friends (and his German patron, Emil Erlenmeyer) threw the young chemist a farewell party that, according to Mendeleev's diary, touched him deeply. The contrast between the collegiality and ease of Heidelberg and the loneliness of the subsequent struggle in Petersburg could scarcely be more blatant. Poor and desperate for money in February 1861, he turned to publication. In a matter of months, he composed a textbook, *Organic Chemistry,* one of the last defenses of Gerhardt's style of organic chemistry—a theoretical framework that concentrated on the classification of families of compounds and maintained an agnostic stance toward the internal structure of molecules—which was being displaced by the rise of structure theory, which broke organic molecules down into their component parts and remains the basis of organic theory today.[13]

Mendeleev submitted the manuscript to the Petersburg Academy of Sciences in hope of winning their Demidov Prize for outstanding scholarly work. The committee, composed of two of his patrons, J. Fritzsche and N. N. Zinin, awarded Mendeleev the prize in early 1862; he used the money to marry Feozva N. Lesheva. As the Demidov citation pointed out, most textbooks were either an abbreviation of existing data or a catalog of limited facts: "Mr. Mendeleev's book *Organic Chemistry* presents us with the rare occurrence of an autonomous development of a science in a brief textbook; a development, in our opinion, which is very successful and in the greatest degree appropriate to the mission of the book as a textbook."[14] Reception among students was equally enthusiastic. "I remember with what interest we, still students, greeted the appearance in 1861 of his *Organic Chemistry*," N. A. Menshutkin, later professor of analytic chemistry at St. Petersburg University, recalled. "At that time this book was the only one in Russia, standing at the height of science, even distinguished in comparison to foreign works in its interest, clarity of exposition, and completely unique integrity."[15] Mendeleev now had the reputation on which he would build a career.

That career was almost entirely bounded by St. Petersburg University.[16] Upon returning to Petersburg, Mendeleev approached his undergraduate mentor, Aleksandr Voskresenskii.[17] Although Voskresenskii managed to find some job openings, Mendeleev was too occupied writing *Organic Chemistry* to take on heavy teaching commitments, and he worked on the book for the remainder of the summer, securing an adjunct position at the University for the fall. This University, founded only in the second decade of the nineteenth century, would become—for political, demographic, and intellectual reasons—the apex of the Russian university system by the end of the century, a transformation in which Mendeleev figured centrally. The importance of a university education in Imperial Russia changed significantly with the restructuring of promotions in the civil service. In 1809, legislation made university-level examinations mandatory for advancement in the bureaucratic ranks system, a connection cemented by the 1835 university statute.[18] As a result, any young man who wanted advancement in Russia needed to attend university, and attending the one in the Imperial capital was the surest way to ascend rapidly. (Women's higher education would not even emerge as a contested political issue until the late 1860s, to say nothing of their service in the bureaucracy.)

The Emancipation of the serfs in 1861 considerably changed the situation. Once the serfs were legally freed of their obligations to the landowners, it became possible for poorer students to flock to the universities. Even though, in the late 1850s, the proportion of noble and civil-servant sons increased in the university, so did the absolute number of students from other (lower) estates. In the next two decades, 40–60% of students received some sort of financial assistance, and an average of 2,000 a year received tuition exemptions.[19] These new realities were particularly apparent at St. Petersburg University, which in fall 1861 became the center of student turmoil, an event that would profoundly shape Mendeleev's vision of the role of the University in a modernizing Russia.

This unrest was the result of a misunderstanding, the consequence of a bureaucracy slow to adapt to the policies of Alexander II. During the spring and summer of 1861, new regulations substantially relaxed restrictions concerning student assembly and university policing. These changes, however, were kept secret until the very last minute. Rumors that stricter regulation was imminent sparked a walkout by students. The situation exploded into rampant street protests.[20] The professoriate was caught in the middle. Uneasy in their role as civil servants, unaccustomed to enforcing police orders, and pressed by their own liberal sympathies and the desire to gain popularity among the student body, professors walked a tightrope that put them in the bad graces of the Ministry of Popular Enlightenment and, when they wavered in their convictions, called forth the antipathy of their students.[21]

Mendeleev experienced this oscillation firsthand as a docent at the University. Amid fog-of-war rumors about the nature and extent of the protest, Mendeleev recorded in his diary a strong sympathy for the students in their desire for more openness from the regime. What most upset him, in turn, was the lack of proper procedures to guide the police. He had heard that the police had received authorization to shoot and beat students: "Horrible things. It is unbelievable that this went through the hands of the ministers and the sovereign in our times."[22] On 24 September the University was shut down until further notice. On 12 October, after some students were wounded in conflicts with police, Mendeleev was so incensed at the perceived violations of legality that he contemplated resigning.[23] The University remained closed into 1862; not until fall 1863 did it resume normal operations. Mendeleev, profoundly shocked by the semester of rebellion, became a fierce advocate for the government's eventual solution to the stalemate: the university statute of 1863.

This law was a complete revision of the standing 1835 statute. Minister of Popular Enlightenment A. V. Golovnin directed the negotiations over the new law, and succeeded not only in recruiting nineteen new professors per university (mostly in the natural sciences), but also in giving the institutions greater autonomy. The statute allowed the faculty to elect their own deans, gave them disciplinary jurisdiction over students and tenure, and provided more money to aid poor students. More than any other Imperial university statute, this came closest to meeting the demands of the professoriate. A vast majority of professors clung to this new arrangement not only as a solution to student unrest, but also as an embodiment of what it meant to be a member of the community of scholars.[24] Although it was eventually replaced in 1884, Mendeleev remained faithful until his death to the vague set of principles—autonomy, academic freedom, scholarship—of the 1863 law. His confidence in the statute was somewhat misplaced. The 1863 statute was intrinsically unstable since it was not based on a fundamental *policy* of governance, but was cobbled from a ramshackle set of regulations designed to bolster the professionalization of academics. The absence of a unified philosophy of how the universities should interact with the government meant that, ironically, the more successful the professionalization, the more suspicious the government became of the universities.[25] As Mendeleev would painfully come to realize at the end of his life, professors were civil servants just like any other bureaucrats, and had to behave accordingly.

But this would be a long time coming. Mendeleev's defense of the statute of 1863 became most pronounced after 1867, when he obtained tenure as professor of general chemistry at St. Petersburg University and could fully appreciate the benefits of professorial autonomy. For the six years between the dislocations of student unrest in 1861 and his final ensconcement at the University, he circulated among a variety of local institutions. The Technological Institute in Petersburg, administered by the Ministry of Finances, hired him as extraordinary (untenured) professor of chemistry on 19 December 1863. He had a relatively light teaching load (compared to the previous generation of Russian chemists) consisting of three lectures a week on organic chemistry for sophomores, one lecture a week on analytic chemistry for upperclassmen, and the supervision of laboratory exercises.[26] He became an ordinary (tenured) professor of chemistry there in 1864. The next year he was elected extraordinary professor of technical chemistry at St. Petersburg University, and he held both posts simultaneously, neglecting the Technological Institute even more after he was promoted to the chemistry professorship at the University in October 1867. He only resigned his post at the Institute in 1871, even though his petition to have his time-intensive laboratory teaching load eliminated or split with another professor was granted.[27] St. Petersburg University provided the framework in which he approached his classification of chemical elements.

PRINCIPLES OF CHEMISTRY AND THE PERIODIC SYSTEM

The periodic law emerged out of the periodic system of elements, the tabular classification that Mendeleev composed in early 1869 at St. Petersburg University. He created the periodic system to address a specific set of demands required in the composition of a new inorganic chemistry textbook—pedagogical problems of classification and organization.[28] The

Karlsruhe Congress had made the problem of creating a consistent general chemistry textbook more acute. The reform of atomic weights meant that all prior textbooks needed to be heavily revised, supplemented by the array of new elements discovered in this decade due to the innovation of spectroscopy. But hidden within this population explosion in the elemental world was the seed of its own solution, for without those consistent atomic weights, the patterns of periodicity would have remained hidden. It is striking, in fact, that the six competing versions of the periodic system, including Mendeleev's, emerged following the assimilation of Cannizzaro's resurrection of Avogadro's hypothesis. Karlsruhe set the stage for the periodic system, and the periodic law returned the favor by furthering the post-Karlsruhe regime of atomic weights.[29]

St. Petersburg University proved to be a fruitful setting for Mendeleev. When he took over his mentor Voskresenskii's post as professor of chemistry in October 1867, he assumed the large inorganic chemistry (or "general chemistry," as Mendeleev liked to call it) lecture course that was required of all students in the natural sciences faculty. In order to teach such a course, he had to find an appropriate textbook. With a few exceptions—including two important texts on organic chemistry (one being Mendeleev's own)—Russian chemical textbooks in this period were adapted translations from Western European texts. With the rapid advances in chemistry, however, any new translation would be almost certainly out of date as soon as it appeared.[30] When Mendeleev began teaching at the University, there were 63 known elements, each identified by atomic weights newly determined by Avogadro's hypothesis. He had to develop some system of classification. The two basic methods for dividing the elements—into metals and metalloids (nonmetals) or by using the new concept of valency—seemed unhelpful to Mendeleev. He chose to write his own textbook instead and work out the challenges of classification himself.

Textbooks are a much-maligned genre in today's science, seen as merely second-rate reiterations of "real" science. This grossly undervalues both the historical and pedagogical functions of these texts. A brief mental juxtaposition with the now ubiquitous short scientific article should make this plain. One could not possibly train chemists using solely a barrage of scientific articles—or at least not nearly as efficiently as with a textbook. Not only does a textbook stand as a codification of what is considered "universal" knowledge within a field at a given moment, but the application of these textbooks to teach a younger generation of scientists reinforces that very universality. Particularly in the field of chemistry, which at both the beginning and middle of the nineteenth century underwent tremendous transformations even in the definitions of central terms (affinity, valency, atom, element, atomic weight, molecule), textbooks were used not only to codify what was standard knowledge, but to *create* the very set of standard concepts.[31]

So did introductory lectures, and the freshman inorganic course presented quite a challenge. This was a year-long large lecture course with integrated laboratory demonstrations. Mendeleev's responsibilities lay entirely in lecturing (laboratory duties were handled by assistants), although in the first few years this obligation was all-encompassing. The immensely successful text he wrote to guide himself and the students, *Principles of Chemistry* (*Osnovy khimii*), was divided into two volumes, each with two parts. The two parts of volume 1 were largely written in 1868, and concluded in the first month of 1869.[32]

Rather than structuring the first volume of his textbook around a classification of the elements, Mendeleev described chemistry in terms of the *practices* by which one acquired knowledge of the chemical world. Early in volume 1, which was entirely written before the inception of the periodic system, Mendeleev's definition of chemistry illustrated the text's structure:

> [Chemistry] is a natural science which describes homogeneous bodies, studies the molecular phenomena by which these bodies undergo transformations into new homogeneous bodies, and as an exact science it strives . . . to attribute weight and measure to all bodies and phenomena, and to recognize the exact numerical laws which govern the variety of its studied forms.[33]

Notice that Mendeleev did not introduce elements, atoms, or any theory of chemical combination. Instead, volume 1 is littered with definitions, plans for basic chemical experiments, and natural-historical information. The reader finds no direct hints of the forthcoming periodic law. Volume 1 is an empirical introduction to chemical practices and the inductive aspects of chemistry; volume 2 is a series of deductions from chemical theories, most saliently from the periodic system.

The theory that one would expect to be most connected to periodicity was also the one Mendeleev was most loath to take literally: atomism. Physical atomism—the belief that atoms are discrete physical bodies, which we now take for granted—was heavily contested in chemistry in the nineteenth century, and the periodic law eventually served as one of the strongest arguments in its favor. It does not follow, however, that Mendeleev must have been thinking in terms of physical atomism when he conceived his system.[34] In his practical work Mendeleev, of course, used the notion that substances combine in defined ratios with each other ("chemical atomism")—it was practically impossible to be a chemist without doing so—but he had long maintained a conflicted attitude to the physical interpretation of atomic theory. In his 1856 candidate thesis, he explained that, while the atomic hypothesis was a useful explanation, it "does not possess even now a part of that tangible visualizability, that experimental reliability, which has been achieved, for example, by the wave hypothesis [of light], not even to mention Copernicus's theory, which one can no longer call a hypothesis."[35] In an 1864 lecture, Mendeleev argued that since definite compounds pointed toward atomic theory and indefinite compounds (like solutions) pointed away from it, "one should not seek in chemistry the foundations for the creation of the atomic system."[36] Even as late as 1903, Mendeleev accepted atomism only as a pedagogically "superior generalization."[37]

Mendeleev's skepticism toward atomism sharply emphasizes the difference between the present-day interpretation of the periodic system and Mendeleev's views of 1869. Today's periodic system is widely understood as revealing periodic properties caused by the gradual filling of electron shells in individual atoms. Elements with one free electron in the outer shell will have similar propensities to combine in certain ratios, and thus have similar chemical properties. The primary ordering of today's system—atomic number—measures the number of protons in the nucleus of an atom, which in turn determines the electrons and thus the chemical properties.[38] This entire concept is structured around *atoms*. For Mendeleev, any atoms that might exist had *absolutely no* substructure, and he resisted the notion of electrons (discovered in 1897) until his death. (He never even heard of protons.) Mendeleev's system had no notion of atomic number, and everything was ordered by atomic weights—or, as Mendeleev would prefer, "elemental weights." This raises the crucial concept that underlay the entirety of the periodic system, and what would serve as the chief warrant for Mendeleev's elevation of the convenient classification to a law: the abstraction of an "element." There is, strictly speaking, no such thing as an element in nature; what exist instead are "simple substances," a concept initially developed by Antoine Lavoisier. That is to say, no one (even after the advent of scanning-tunneling microscopes) has ever seen "carbon"; instead, they have seen diamond, or graphite, or other forms (and, today, carbon *atoms*). Oxygen is observable in nature as the oxygen molecule or ozone. We *infer* the notion of an "element" as the metaphysical basis that relates the various forms, much as Mendeleev later inferred the periodic law as the metaphysical basis to explain the diversity of "elements."[39]

This distinction would only come to Mendeleev halfway through writing his *Principles of Chemistry*. Instead, chemical practice and not chemical theory provided his initial organizing principle, which begins to transition into the origins of the periodic system at Chapter 20, which addressed table salt. Up to this point, Mendeleev had only treated four elements in any detail: oxygen, carbon, nitrogen, and hydrogen—the so-called "organogens." Mendeleev began this chapter as usual by purifying the central substance, sodium chloride, from sources such as seawater. A discussion of sodium and chlorine followed in the next few chapters, and finally the halogens appeared, the family of elements (bromine, iodine, fluorine) that were clearly related to chlorine. Thus ends volume 1, and the alkali metals (the sodium family) form the first chapter of volume 2.

Mendeleev faced a serious predicament at this point, in late January 1869. His textbook was pedagogically sound so far, and he had just sent volume 1 to the publishers, but had dealt with only 8 elements, relegating 55, fully seven-eighths of known elements, to the second volume.[40] Clearly, Mendeleev had to come up with a less rambling organizational method or he would never finish in the contractually agreed-upon time and space. Mendeleev had traversed the material of volume 1 several times in earlier chemical lectures, but he had yet to settle on a mechanism to solve the organization of the remaining elements.[41] Now, with a contract hanging over his head, he had to devise a more consistent solution. As he recalled in April 1869:

> Having undertaken the compilation of a guidebook to chemistry, called "Principles of Chemistry," I had to set up simple bodies in some kind of system so that their distribution was not governed by accidents, as if by instinctive guesses, but by some definite exact principle. Above we saw the almost complete absence of numerical relations in the establishment of a system of simple bodies; but any system based on exactly observed numbers, of course, will already in this fashion deserve preference over other systems which do not have numerical foundations, in which there remains little place for arbitrariness (*proizvolu*).[42]

Mendeleev's earlier system of pedagogically useful organization—using laboratory practices to explain the common substances (water, ammonia, table salt) in which they are found—could no longer sustain the burden of exposition. He needed a new system that would still be pedagogically useful, and he hit on the idea of using a numerical marker for each element. Atomic weight seemed the most likely candidate for a system that would (a) account for all remaining elements; (b) do so in limited space; and (c) maintain some pedagogical merit. His solution, the periodic system, remains one of the most useful teaching tools in chemistry.

Early in February 1869, while Mendeleev was writing Chapter 1 and Chapter 2 of volume 2 on sodium and the alkali metals, he listed these elements in order of increasing atomic weight and compared them with the halogens, similarly arranged.[43] By Chapter 4, on the alkaline earths (the calcium family), Mendeleev was entirely converted to the idea of organizing all of the elements according to a numerical system. He no longer enumerated the elements according to substances in which they could be found; instead, he began on the first page of this chapter to show that the arithmetical difference between rows followed a similar pattern in all three groups: halogens, alkali metals, and alkaline earths.[44] In addition, these alkaline earth elements, with a valency of 2, succeeded the alkali metals, with a valency of 1. While Mendeleev remained resistant to aspects of valency theory, his system followed the progression of combining power across the elements. Note that atomic weight was not yet of *dominant* importance. Atomic weight was used as a secondary quality that showed the hierarchical ordering within families. As volume 2 proceeded, Mendeleev would begin to emphasize atomic weights so much that they were listed even in chapter titles, and elements were always introduced along with their atomic weight.

It is extremely difficult to reconstruct the process by which Mendeleev came to his periodic organization of elements in terms of their atomic weights. He did not simply list them in order of increasing weight, but observed the periodic repetition of chemical properties, thus correlating two parameters. The problem from the historian's perspective is that, while Mendeleev kept almost every document and draft that crossed his hands *after* he believed he would become famous, he did not do so *before* the periodic law. As a result, we have just *four* relevant documents that precede the first publication on the periodic law—and one of these is a fair copy of another. Thus we are forced to consider volume 1 of *Principles* in conjunction with these documents and come to some informed speculations.

There are two basic ways Mendeleev could have moved from a recognition of the importance of atomic weight as a good classifying tool to a draft of a periodic system: Either he wrote out the elements by order of atomic weight in rows and noticed periodic repetition; or he assembled several "natural groups" of elements, like the halogens and the alkali metals, and noticed a pattern of increasing atomic weight. Most analyses of Mendeleev stand dogmatically on either the "row" or "group" version.[45] Mendeleev's only direct statement on this matter, however, shows a middle way. He wrote in April 1869 that he "gathered the bodies with the lowest atomic weights and placed them by order of their increase in atomic weight."[46] This produced what he called his "first try," marking elements with their atomic weights:

Li = 7;	Be = 9.4;	B = 11;	C = 12;	N = 14;	O = 16;	F = 19
Na = 23;	Mg = 24;	Al = 27.4;	Si = 28;	P = 31;	S = 32;	Cl = 35.5
K = 39;	Ca = 40;	—	Ti = 50;	V = 51[47]		

I will return shortly to the "—" underneath aluminum (Al). For the moment, however, consider Mendeleev's list. These are most of what Mendeleev called the "typical elements"—the set of light elements up to chlorine that provide a neat encapsulation of the periodic system (all of the major groups are included, and the differences in the properties of each group are expressed most starkly).[48] The list also emphasizes elements treated in volume 1 and the first chapter of volume 2 of *Principles:* the "organogens" (minus hydrogen), the halogens, and the alkali metals. Here Mendeleev built "groups" and "rows" simultaneously. He took the lightest elements and listed them by rising atomic weight, building a row; but each of the "typical elements" in the top row *encoded as typical* the properties of the elements below it, precisely because the contrast between the properties of, say, beryllium (Be) and boron (B) are sharper than between two heavier members of their respective groups. These elements stood in for their groups, and Mendeleev could see both patterns at once.

This realization happened sometime early in 1869. After considerable work to make a system that contained all of the elements, he sent a draft of a single sheet to the printers on 17 February 1869. This draft was printed in both Russian (150 copies) and French (50 copies), and the sheet, entitled "An Attempt at a System of Elements, Based on Their Atomic Weight and Chemical Affinity," was sent off to various chemists (Figure 2.1). The fact that he had more printed up in Russian indicates that his primary audience at this point was local and not international.

This "Attempt" (*Opyt*) was not the final version of the periodic system—it contains many errors, and Mendeleev spent the better part of the next two years reinventing it. To transform it into a recognizable form similar to today's representation, one must rotate it clockwise by 90° and reflect it across the vertical axis. Even then, the alkali metals and the halogens are next to each other, which is counterintuitive if you organize the elements according to any physical property—atomic volume, electronegativity, electron shells, and so on. Mendeleev's "Attempt" convinced him of the importance of atomic weights as a parameter for classification and of some natural correlation embedded under the surface. He would keep tinkering with the system until he found a chemical property that monotonically separated each group: the degrees of oxidation in a saturated chemical compound of the element.[49] The quality of a first draft is evident in his title of the "Attempt." In a rough draft, Mendeleev crossed out in both French and Russian the word "classification" (*classification, raspredelenie*) and replaced it with "system" (*système, sistema*), once he became convinced that his organization was not arbitrary (see frontispiece to the Preface). But in the French title, he forgot to change the gender of the indefinite article from feminine to masculine (*une* to *un*). The version he sent out thus bears the traces of Mendeleev's gradual process of construction.[50]

This system, then, emerged out of need—the need for a pedagogical "classification" that answered specific necessities in presenting material to beginning chemistry students. The pedagogic utility of Mendeleev's periodic system would later be universally recognized, even by its critics. Mendeleev often invoked the system's pedagogical origins: "I note also that the outlining for beginners of the facts of chemistry and their generalization benefits very much

			Ti $=$ 50	Zr $=$ 90	? $=$ 180.
			V $=$ 51	Nb $=$ 94	Ta $=$ 182.
			Cr $=$ 52	Mo $=$ 96	W $=$ 186.
			Mn $=$ 55	Rh $=$ 104,4	Pt $=$ 197,4
			Fe $=$ 56	Ru $=$ 104,4	Ir $=$ 198.
		Ni $=$ Co $=$ 59	Pl $=$ 106,6	Os $=$ 199.
			Cu $=$ 63,4	Ag $=$ 108	Hg $=$ 200.
H $=$ 1					
	Be $=$ 9,4	Mg $=$ 24	Zn $=$ 65,2	Cd $=$ 112	
	B $=$ 11	Al $=$ 27,4	? $=$ 68	Ur $=$ 116	Au $=$ 197?
	C $=$ 12	Si $=$ 28	? $=$ 70	Sn $=$ 118	
	N $=$ 14	P $=$ 31	As $=$ 75	Sb $=$ 122	Bi $=$ 210?
	O $=$ 16	S $=$ 32	Se $=$ 79,4	Te $=$ 128?	
	F $=$ 19	Cl $=$ 35,5	Br $=$ 80	J $=$ 127	
Li $=$ 7	Na $=$ 23	K $=$ 39	Rb $=$ 85,4	Cs $=$ 133	Tl $=$ 204.
		Ca $=$ 40	Sr $=$ 87,6	Ba $=$ 137	Pb $=$ 207.
		? $=$ 45	Ce $=$ 92		
		? Er $=$ 56	La $=$ 94		
		? Yt $=$ 60	Di $=$ 95		
		? In $=$ 75,6	Th $=$ 118?		

FIGURE 2.1 The first published form of Mendeleev's periodic system, dated 17 February 1869 and entitled "An Attempt at a System of Elements, Based on Their Atomic Weight and Chemical Affinity." Mendeleev had 50 of these images printed up under a French title and 150 under a Russian one, which he mailed to various chemists. In order to transform this image into a modern periodic system, it must first be rotated clockwise 90°, reflected, and then the halogens (the row beginning with F = 19) need to be placed at the opposite extreme from the alkali metals (the row beginning with Li = 7). Notice that spaces are left with question marks for elements that Mendeleev suspected existed. Source: Mendeleev, *Periodicheskii zakon. Klassiki nauki,* 9.

from the use of the periodic law, as I became convinced not only in lectures in the last two years, but also during the preparation of a course of inorganic chemistry now already finished and published by me (in Russian). At the foundation of its presentation I placed the periodic law."[51] This does not mean that the first half of the textbook was written without pedagogical goals in mind. In fact, Mendeleev found the pedagogical format of volume 1 to be so important that he never revised the fundamental structure of the book throughout its eight editions, although he later made the periodic law more prominent.[52] This old model centered on chemical *practices* derived from his *Organic Chemistry* textbook, a heritage he was loath to disown even as he confronted the lawlike status of periodicity.

SYSTEM INTO LAW: MAKING PERIODICITY NATURAL

It is unlikely that Mendeleev understood the generality of his system when he first developed it in February 1869. Had he been cognizant of the implications of the periodic system, he would most likely not have relegated the initial presentation of it to the Russian Chemical Society in March 1869 to his friend Nikolai Menshutkin while he went off to inspect cheese-making cooperatives for the Imperial Free Economic Society. (Mendeleev was at the time well known as a consultant on agricultural matters, and small-scale cheese production by independent artisans intrigued him as a possible model for organizing industry. His positive report on artisanal cheese was less well received than the paper he had delegated to his friend.)

By late 1871, however, in the last of Mendeleev's research articles on the periodic law, he was quite sure that he had isolated a new law of chemistry.[53] Much as the creation of the "Attempt" was rooted in one of Mendeleev's local contexts—the classrooms of St. Petersburg University—the tale of how Mendeleev came to understand the periodic *system* as a periodic *law* can only be told outside the University, as Mendeleev addressed himself to the community of chemical practitioners. Through the Russian Chemical Society's journal, Mendeleev targeted a larger audience with each elaboration of the regularities of his system, as he became increasingly bold about the possibilities of his elemental arrangement.

Mendeleev's first scientific article on his findings was published in April 1869, two months after the mailing of the "Attempt." He began this piece with an enumeration of different schemes to order the elements, most of which were based on arbitrary distinctions that could not possibly reflect the order of the world. He concluded that "at the present time there is not a single general principle which withstands criticism, is able to serve as a basis for a judgment on the relative properties of elements, and which allows one to array them in a more or less strict system."[54] But Mendeleev was interested in more than just a "system." In this same article he used the word "law" to refer to periodicity for the first time:

All the comparisons which I made in this direction bring me to the conclusion that *the magnitude of atomic weight determines the nature of the element* as much as the weight of a molecule determines the properties and many reactions of a complex substance. If this conviction is supported by further application of the established principle for the study of elements, then we will approach an epoch of understanding the essential distinction and reason for the affinity of elementary bodies.

I propose that the law (*zakon*) I have established does not go at cross-purposes with the general direction of the natural sciences, and that until now its proof has not appeared, although there were already hints of it. From now on, it seems to me, a new interest will develop in the determination of atomic weights, in the discovery of new simple substances, and in the seeking out of new analogies between them.

I introduce for this one of many systems of elements, founded on their atomic weight. It serves only as an attempt (*opytom*), an endeavor (*popytkoi*) to express the result which it is possible to achieve in this matter. I myself see that this endeavor is incomplete. . . .[55]

Notice, importantly, that this "law" is not the periodic law we now know, or the one that Mendeleev would endorse within two years. Here he claimed that the weight of atoms determined their properties—which also happens to be false—not that there was a periodic dependence of properties with increasing weight. In fact, Mendeleev was rather loose with the term "law" (*zakon*), citing as laws such generalizations as Auguste Laurent's even number rule and P. L. Dulong and A. T. Petit's rule on specific heats, neither of which would pass Mendeleev's later criteria.[56]

By August of 1869 Mendeleev appeared to have developed a stricter conception of what it took to be a law of nature. In an article published that month on the variation of atomic volumes over the periodic system, he shied away from the word "law" and called it a "regularity" (*pravil'nost*).[57] This retreat was motivated by his realization of the persistence of exceptions to inflexible ordering by physical properties. Yet, by October 1869 Mendeleev found that ordering the elements by the quantity of oxygen in their oxides revealed how "natural" his system was, and how it evolved from the alkali metals to the halogens.[58] That is, taking R to be a generic element, the alkali metals combined as R_2O, the alkaline earths as RO (or R_2O_2), all the way to the halogens, which combined as R_2O_7. This provided a neat ordering of groups from 1 to 7—later codified as I to VII—based on the subscript attached to oxygen.[59]

Over the next year, Mendeleev conducted broad-ranging investigations into aspects of his scheme, trying to account for the problems that beset the rare earths, indium, and other irregularities. By November 1870, he was utterly convinced of both the "naturalness" and the lawlike character of his periodic system. That month, he published a Russian article in which he predicted the discovery of new elements, proposed changes in the atomic weights of current elements, and formed the framework for his more detailed German article the following year that would eventually create his European reputation. This confidence is foreshadowed in the title: "The Natural System of Elements and Its Application for the Indication of the Properties of Undiscovered Elements." In this piece Mendeleev first uses "law" in the strict sense as referring to periodicity: "I propose also that the law of periodicity (i.e. the periodic dependence in the change of properties of the elements on their atomic weight) gives us a new means to determine the magnitudes of the atomic weight of elements, because here already in two examples, namely with indium and cerium, the propositions which were drawn from the foundation of the law of periodicity were affirmed."[60]

In this article (and in its German successor), Mendeleev recapitulated the process by which he came to the periodic law. First he surveyed current systems; then he created his own conventional system; then he tested it on items about which we have stable knowledge (such as the typical elements); then he tried it on less stable elements and corrected their properties (such as doubling the atomic weight of uranium); and next on extremely doubtful objects (indium and cerium). Building incrementally on these foundations, he moved to the prediction of new elements. He called the system "*a natural system of elements*," because "not in a single instance does one meet any essential obstacles for the application of this system for the study of the properties of elements and their compounds. . . ."[61] This cautious transition from convention to broader and broader claims about less and less stable knowledge is a recurrent pattern for Mendeleev that transcended the boundaries between science, politics, and culture.

So far, there has been nothing to distinguish this process of reasoning from that which produced the less rigorous "regularities." After he had already moved the reader to extremely doubtful elements and showed the application of periodicity, he now moved to *completely* doubtful elements, that is, those that were unknown:

> With the pointing out of the periodic and atomological dependence between the atomic weight and the properties of all elements, it appears possible not only to point to the absence of certain of them [elements], but also to determine with greater certainty and likelihood for success the properties of these still unknown elements; one can point to their atomic weight, density in free form or in the form of an oxide, the acidity or basicity of their degrees of oxidation, the possibility of reduction and the formation of double salts, to decide with this the properties of metalloorganic and chlorine compounds of the given element—there is even the possibility to describe the properties of certain compounds of the still undiscovered elements with very great detail. I decide to do this for the sake of having the possibility, when with time one of these substances I predict will be discovered, of finally assuring myself and other chemists of the justification of the propositions which lay at the foundation of the system I propose. For me personally these propositions were finally solidified from the moment when these propositions, which were based on the periodic law (*zakonnosti*), which lies at the basis of all this research, were justified for indium.[62]

Mendeleev's landmark 1871 article supports this thinking with a great deal more detail. This article was written in July and translated into German by Felix Wreden, and eventually appeared in November.[63] Mendeleev was now convinced that his system was superior to those of his predecessors, and even hinted that there might be a mathematical function underlying the pattern produced by the atomic weights. It was, after all, from the mathematical concept of a periodic function (like a sine or cosine wave) that Mendeleev had borrowed the term "periodicity" in the first place, a term not used by any of the other proponents of systems of elements.[64] At the basis of this relation was the importance of prediction:

> That is the essence of the law of periodicity. Each natural law (*estestvennyi zakon*), however, only acquires scientific significance when there is the possibility of drawing from it practical, if one can put it that way, consequences, that is, those logical conclusions which explain what is not yet explained, point to phenomena not yet known, and especially when it gives the possibility to make such predictions which can be confirmed by experiment. Then the utility of the law becomes obvious and one has the possibility to test its validity.

It was at this point that he declared that the system "has a significance not just pedagogical, not only easing the study of various facts, bringing them into order and connection, but it also has a purely scientific significance, discovering analogies and pointing through them to new paths for the study of elements."[65] He had moved from pedagogy to pure science through prediction.

CLAIRVOYANCE: THE EKA-ELEMENTS

Clearly, the crux of Mendeleev's attitude toward what made the periodic system into a "law" was the role of prediction. It was this capacity for prediction that convinced him of the "naturalness" of his system of elements, and it was the discovery of new elements that would eventually astonish chemists internationally.[66] Understandably, the discoveries of the three predicted elements have received a great deal of attention from historians and chemists, but the process by which Mendeleev made his predictions has received almost none. However, to understand how the periodic law—and prediction as a central component—formed the underlying metaphor and warrant for Mendeleev's vision of restructuring Imperial Russia, then the primary question is what convinced *him*. For him, following mainstream philosophies of the scientific method in the nineteenth century, prediction was what made a science "scientific," as he expressed in his notes for a public lecture in the early 1870s: "A theory is a connection of the internal with an entire worldview: beginning as an hypothesis, it ends with the theoretical discovery of new phenomena, drawing everything from one proposition. This corresponds to the prediction of phenomena in their complete accuracy, the discovery of new unprecedented phenomena. Astronomy [and] physics are in this situation, chemistry still isn't."[67]

It is important to stress that prediction was *not* what Mendeleev was after when he first began constructing his periodic system, as the "Attempt" (Figure 2.1) demonstrates. He had been trying to assemble a teaching tool, and he used question marks as placeholders for elements that were needed to keep the system viable. The atomic weights offered were educated guesses, and would fluctuate as he moved beyond this first draft to a complete revision of the system. But the question marks are there all the same, and they (as well as the "—" in his "first try") indicate the moment at which Mendeleev began to think of his system as something scientific—as something that could predict.

The first explicit mention of prediction was in his April 1869 article, which he concluded with a list of eight advantages of his system over the competing classifications of the day. The sixth point read: "One should expect the discovery of many yet *unknown* simple bodies, for example, elements with affinity to Al and Si, with atomic weights 65–75."[68] But this is a weak prediction: It is only the *sixth* point in his list, and it is remarkably vague (he did not even specify the number of elements expected). In fact, it is apparent from his notes that he tried to fill the blanks at first with existing elements on the grounds of chemical consistency, to see if perhaps their atomic weights had been inaccurately measured.[69] By late August 1869, in his article on atomic volumes, he had abandoned this approach and some of his earlier vagueness: "Therefore it is possible to say, that the two elements which are not yet in the system and which should display affinity with aluminum and silicon and have atomic weight of about 70, will display an atomic volume around 10 or 15, i.e. will have specific weight of about 6 and, thus, will occupy exactly in all relations the average or will comprise the transition in properties from zinc to arsenic."[70]

But prediction was not emphasized as a primary function of the system for over a year, until Mendeleev's extensive Russian article of November 1870, and repeated more intensely in the German expansion of 1871. During this period, Mendeleev tinkered with aspects of the system, especially the problematic rare earths, such as indium and cerium.[71] Once again he displayed a characteristic of his prophetic work: If an idea was promising, he would retreat from his bolder claims in favor of comprehensively studying the minutiae of the project, making sure that all easily answerable questions were resolved before using these as a stable platform to leap into the undiscovered. He would employ this process again in Russian tariff policy twenty years later.

[31]	Группа I	Группа II	Группа III	Группа IV	Группа V	Группа VI	Группа VII	Группа VIII. Переход к группе I
Типическіе элементы	H = 1							
	Li = 7	Be = 9,4	B = 11	C = 12	N = 14	O = 16	F = 19	
Первый период — Ряд 1-й	Na = 23	Mg = 24	Al = 27,3	Si = 28	P = 31	S = 32	Cl = 35,5	
— 2-й	K = 39	Ca = 40	— = 44	Ti = 50?	V = 51	Cr = 52	Mn = 55	Fe = 56, Co = 59, Ni = 59, Cu = 63
Второй период — 3-й	(Cu = 63)	Zn = 65	— = 68	— = 72	As = 75	Se = 78	Br = 80	
— 4-й	Rb = 85	Sr = 87	(?Yt = 88?)	Zr = 90	Nb = 94	Mo = 96	— = 100	Ru = 104, Rh = 104, Pd = 104, Ag = 108
Третій период — 5-й	(Ag = 108)	Cd = 112	In = 113	Sn = 118	Sb = 122	Te = 128?	J = 127	
— 6-й	Cs = 133	Ba = 137	— = 137	Ce = 138?	—	—	—	—
Четвертый период — 7-й	—	—	—	—	—	—	—	
— 8-й	—	—	—	—	Ta = 182	W = 184	—	Os = 199?, Ir = 198? Pt = 197?, Au = 197
Пятый период — 9-й	(Au = 197)	Hg = 200	Tl = 204	Pb = 207	Bi = 208	—	—	
— 10-й	—	—	—	Th = 232	—	Ur = 240	—	
Высшая соляная окись	R^2O	H^2O^2 или RO	R^2O^3	R^2O^4 или RO^2	R^2O^5	R^2O^6 или RO^3	R^2O^7	R^2O^8 или RO^4
Высшее водородное соединеніе			$(RH^5?)$	RH^4	RH^3	RH^2	RH	—

FIGURE 2.2   Short-form periodic system from Mendeleev's November 1870 article. Mendeleev used this system to calculate the properties of his three eka-elements, which are located in groups III and IV. The "long periods" (represented here by the brackets at the left) have been collapsed into two individually numbered rows—after excluding the first two rows of "typical elements." Thus, an element in row 2 in the short-form system is in the fourth period of the long-form system. The staggering of the elements within columns allows the determination of secondary chemical analogies. The degrees of oxidation and hydrogenation are indicated at the bottom of the columns. Source: Mendeleev, *Periodicheskii zakon. Klassiki nauki*, 76.

By November 1870 Mendeleev was persuaded that the major difficulties posed by the rare earths and other problematic elements had been resolved (as with uranium or indium), or contained (as with the cerite metals) so that substantial revisions were unlikely. It was at this point that he publicly articulated the process of prediction. He began by displaying a revised system, what would come to be called the "short-form" periodic system (Figure 2.2). These systems are built as direct analogies from the list of typical elements—those elements with sharp characteristics that stand at the top of the twin peaks of today's long-form periodic systems. Short-form systems compress the "long" periods that contain the transition elements (the valleys) into a second "little period" that folds underneath the first, so those periods with sixteen elements are shown in two rows of eight. The advantage of this form from Mendeleev's perspective was that it expanded the analogies one could draw between an element and its neighbors by increasing the number of neighbors, as well as simplifying the progression of levels of oxidation indicated in the headings of the groups. (The electron-shell interpretation of the system has today completely eradicated the viability of Mendeleev's short form.)

Mendeleev began by noticing that there were not enough analogs of aluminum and boron. Most groups seemed to have six analogs down to the fifth row, whereas the third group had only four. That is, when you go to the third column (group), there were two spaces that seemed unoccupied after boron (B) and aluminum (Al) before one hit the next element. Put another way, that meant that immediately after potassium (K) and calcium (Ca) in the second row, there was a gap, and immediately after copper (Cu) and zinc (Zn) in the next row there was a similar gap. Mendeleev began with the first of these. Since these elements had atomic weights close to 40, and then the next one, titanium (Ti), was close to 50, he opted for an "average" 44. [This is not an exact average of the atomic weights of K, Ca, Ti, and V (vanadium), which would be 45—Mendeleev modified the mathematics to suit his intuition.][72] Since this element was in an even row, it should have more alkali properties than lighter elements in the same group (boron and aluminum), and its oxide R_2O_3 should be a more energetic base. Mendeleev developed this point through a strong analogy to titanium, comparing TiO_2 and its lighter analogs. As with titanium, this element's oxide should have a sharper basic character and thus it should form alkali compounds insoluble in water, although it ought to form stable acidic salts. He also went into detail on its chloride and atomic volume. While some of these predictions displayed remarkable virtuosity, many others were repetitive (such as the forms of the chloride, oxide, and hydride of the element, all of which are functions of the valency). Mendeleev chose to call the element eka-boron, "creating the name from the fact that it follows boron, as the first element of the even groups, and the prefix *eka* comes from the Sanskrit word for *one*. Eb=44."[73]

After treating eka-boron in great detail, he made similar arguments for eka-aluminum (El=68). This element was immediately above indium in the short-form table, and Mendeleev had recently successfully reclassified this element, which enabled him to give an extended account of its properties.[74] He devoted less space to it, however, than to eka-boron. Finally, he considered the "most interesting" of his elements to be eka-silicon (or eka-silicium, Es=72). Unlike the other two cases, Mendeleev actually suggested in which minerals chemists might begin to search for this new element.[75] He especially valued this element because it occupied the center of the short-form table: the fourth element in the fourth row. In that sense, it was the centerpiece that would tie the whole system together.[76]

Yet, in both the Russian and the German articles, Mendeleev set off eka-boron from the other two elements, not eka-silicon. In a contemporary Russian review, this separation was seen as marking eka-boron as the most important eka-element.[77] He could have placed his strongest prediction, eka-silicon, first, and then moved easily from heaviest to lightest. Instead, Mendeleev followed the logic I traced in the last section, beginning with the most stable knowledge and then moving to less and less reliable claims. The prediction of eka-boron may not have been the best, since it had only four elements near it (K, Ca, Ti, and V) that could serve as analogs, but those analogs were extremely well studied, and thus served as the best way to persuade a skeptic. In the 1871 article, Mendeleev again would place eka-boron first, and spend the same amount of space on both it and eka-silicon, leaving eka-aluminum with only a third of the attention.[78] He wanted to stress his predictions, but he did not want to sacrifice his credibility with the audience by moving to the most extreme case first.

Even though Mendeleev clearly believed that his ability to make such claims transformed his system into a law of nature, he was well aware that chemists— with little experience of such laws—might dismiss the possibility of his new elements. After all, there was no reason to expect that every gap in the system had to be filled, and it was perhaps easiest to disregard Mendeleev's predictions as just so much wishful thinking. After such a bold departure, Mendeleev immediately retreated and expressed the vague hope that *one* of these three would eventually be discovered. And then he retreated yet again, saying that even if these predictions did not work at all, at least he had managed to correct several atomic weights and determine the properties of poorly studied elements.[79] The image of a sage utterly confident in his predictions, experiment and community consensus be damned, is belied by the text.[80] He concluded his 1871 article:

> Not getting carried away with the immediately apparent advantages of such a system, one will have to, however, recognize its justification finally, at least, when the properties derived on its foundation for the yet unknown elements are justified by the actual discovery of them, because, one must confess, up till now chemistry has had no means to predict the existence of new simple substances, and they were only discovered via direct observation. . . . When the periodic dependence of properties on atomic weight and the atomological relations of elements will be able to be attributed to exact laws (*zakonam*), then we will approach even more the comprehension of the very essence of the distinction of elements among themselves and then, of course, chemistry will be already in a state to leave the hypothetical field of static concepts which dominate it now, and then the possibility will appear of delivering it to a dynamical direction, already so fruitfully employed in the study of the majority of physical phenomena.[81]

Even though Mendeleev's article appeared in German, his periodic system received little attention aside from brief priority disputes with John Newlands in England and Lothar Meyer in Germany. But the confirmations of both gallium (eka-aluminum) and scandium (eka-boron) would make Mendeleev a household name in scientific Europe.[82]

THE VINDICATION OF PROPHECY: THE EKA-DISCOVERIES

The first of these elements to be found was the very one to which Mendeleev had paid the least attention in his predictions—eka-aluminum—discovered in France in 1875 as gallium by Paul Émile (François) Lecoq de Boisbaudran.[83] Lecoq de Boisbaudran was trained as a physical chemist, and in the late 1860s he became one of the foremost practitioners of the relatively new technique of spectroscopy (heating a substance, observing its emitted light through a diffraction grating, and noting its characteristic spectral lines). Using this method he discovered not only gallium, but also samarium (1879), gadolinium (1886), dysprosium (1886), and europium (1892). His discovery of gallium—so named in a burst of patriotism after his native France (Gallia)—earned him the cross of the Legion of Honor in 1876. In 1879, the English awarded him the Davy Medal for his discovery, three years before Mendeleev and Lothar Meyer would share one for the periodic system.[84]

Lecoq de Boisbaudran made his discovery on the afternoon of Friday, 27 August 1875 (N.S.), when he noted a distinctive spectral line in a metal from Pierrefitte, a mine in the Pyrenees. Over the course of the next year, he published a series of articles that explicated the various properties of this new element.[85] It is clear that he had no prior knowledge of Mendeleev's predictions of eka-aluminum, but that does not imply that he was simply an empiricist blindly searching for new elements. Rather, he had some years earlier produced his own classification of the elements, based on spectral lines. Using those regularities, he made a prediction of the atomic weight of an analog of aluminum that was actually fairly close to Mendeleev's value (and closer to today's accepted value).[86] Clearly, Mendeleev was far from the only chemist interested in prediction.

Two features of the discovery of gallium make it distinctive among the eka-elements. It was the first, and the obvious similarity of this element with eka-aluminum drew substantial attention to Mendeleev's 1871 system. Second, this was the only case among the three where Mendeleev scoured the foreign literature for possible confirmations of his predictions, and publicly made the connection himself. In the cases of eka-boron and eka-silicon, intermediaries stepped in, although they extended full credit to Mendeleev.

At the 6 (18) November 1875 meeting of the Russian Chemical Society, Mendeleev observed that the properties of gallium looked a great deal like eka-aluminum, and he hoped that this would be further confirmed.[87] The article that truly made Mendeleev's name was published in the French *Comptes Rendus,* the same journal where Lecoq de Boisbaudran had announced his findings. Mendeleev published a short-form periodic system, which showed a space for "68?" in the center. He then recounted the cases in which he had corrected atomic weights and been confirmed before moving into a much more detailed account of his prediction of eka-aluminum than he had in either 1870 or 1871:

> The properties of eka-aluminum, following the periodic law, should be the following. Its atomic weight will be El=68; its oxide will have the formula El^2O^3; its salts will display the formula ElX^3. Thus, for example, the (unique?) chloride of eka-aluminum will be $ElCl^3$; it will give in analysis 39 out of 100 of metal and 61 of 100 of chlorine and will be more volatile than $ZnCl^2$. The sulfide El^2S^3, or oxysulfide $El^2(S,O)^3$, should be precipitated by hydrogen sulfide and will be insoluble in ammonium sulfide. The metal will be obtained easily by reduction; its density will be 5.9; therefore, its atomic volume will be 11.5, it will be almost fixed, and will melt at a rather low temperature. On contact with air, it won't oxidize; heated to red, it will decompose water. The pure and molten metal will be attacked by acids and bases only slowly. The oxide El^2O^3 will have specific weight around 5.5; it should be soluble in energetic acids, forming an amorphous hydrate insoluble in water, it will dissolve in acids and bases. The oxide of eka-aluminum will form neutral and basic salts $El^2(OH,X)^6$, but not acidic salts; the alum $ElK(SO^4)^212H^2O$ will be more soluble than the corresponding aluminum salt and less crystallizable. The basic properties of El^2O^3 being more pronounced than those of Al^2O^3 and less than those of ZnO, one must expect that it will be precipitated by carbonate of barite. The volatility, as well as the other properties of saline combinations of eka-aluminum, present the average between those of aluminum and those of indium, it is probable that the metal in question will be discovered by spectral analysis, as were indium and thallium.[88]

Many of the properties he lists are derivatives of the others, as simple cases of valency.

Lecoq de Boisbaudran had received a letter from Mendeleev almost immediately after publishing his first account of the discovery, and said he would not comment on Mendeleev's corrections of his data (Mendeleev questioned the density findings) until he did more work.[89] Interestingly, the simple substance of gallium was a liquid at room temperature, which was unpredictable from the periodic system. After further research, Lecoq de Boisbaudran found that the density of the metal was 5.935, which was strikingly close to Mendeleev's predicted value of 5.9, but not at all close to the average of 4.8 of indium and aluminum, which once again shows how much chemical intuition was built into Mendeleev's predictions to correct the simple averages.[90]

There was an understandable reluctance among contemporaries to accept the two other predictions on the basis of one, possibly lucky, guess. When the second eka-element was discovered in 1879, Mendeleev's case was much more than twice as strong; it seemed as if there were really some deep regularity reflected in his system. This element, scandium (eka-boron), was a rather complicated case, since it was more similar to the rare earths than either of Mendeleev's other two eka-elements, and these elements were very close to each other in both atomic weight and chemical properties, and thus proved hard to isolate. This is a large part of why Mendeleev chose to rely on calcium and titanium to make his predictions.[91] This element was discovered among various rare earths by L. F. Nilson of Sweden. In his original publication announcing this (once again) patriotically named element, Nilson made no mention of the correspondence with Mendeleev's eka-boron; Mendeleev, for his part, could not read Swedish and make the connection himself.[92] It was Nilson's countryman, Per Cleve, who did so.

Cleve wrote to Mendeleev on 19 August 1879: "I have the honor to inform you that your element eka-boron has been isolated. It is scandium, discovered by Nilson this spring. . . ."[93] Much more important was his article to the *Comptes Rendus,* where he drew out the similarities in detail. After chronicling the properties of scandium (Sc), he wrote: "What makes the discovery of scandium interesting is that its existence had been announced in advance. In his article on the periodic law, Mr. Mendeleev predicted the existence of a metal with atomic weight 44. He called it *eka-boron.* The characteristics of eka-boron correspond rather well with those of scandium." Cleve then produced what would later become a famous double table, as shown on page 41.

Such double tables would soon become standard presentations of the discovery. The correspondence is all the more remarkable in that it was impossible to confirm all Mendeleev's predictions until 1937, thirty years after his death, when scandium was finally isolated in pure form.[94] Nilson himself was delighted at the coincidence of properties, and believed that Cleve's observations, when combined with the case of gallium, had truly confirmed the periodic law.[95]

Yet Mendeleev's "most interesting" element, eka-silicon, the core of the periodic system, remained elusive. It is ironic that this was the last of Mendeleev's three eka-elements to be discovered, since Mendeleev had believed it would be discovered first. It would eventually become the most persuasive example of the power of Mendeleev's predictions; in his most extensive obituary, the comparison of eka-silicon and germanium was the only one discussed in detail, presented again in a Cleve-style dual table.[96] The process of the discovery of germanium was very similar to that of scandium.

Supposed characteristics of eka-boron	*Observed characteristics of scandium*
Atomic weight, 44	Atomic weight, 45
Eka-boron should have only one stable oxide, Eb^2O^3, a base more energetic than that of aluminum, with which it should have several characteristics in common. It should be less basic than magnesium.	Scandium only gives the oxide Sc^2O^3, a base more energetic than aluminum, but weaker than magnesium.
Just as yttrium must be a more energetic base, one can predict a great resemblance between yttrium and the oxide of eka-boron. If eka-boron is found mixed with yttrium, the separation should be difficult and based on delicate differences, for example, on differences of solubility, on differences in basic energy.	Scandium oxide is less basic than yttrium, and their separation is based on the differing stability of their nitrates under heat.
The oxide of eka-boron is insoluble in alkalis; it is doubtful that it will decompose ammoniac salt.	The hydrate of scandium is insoluble in alkalis; it does not decompose ammoniac salt.
The salts should be colorless and give, with KOH, Na^4CO^3 [*sic*] and HNa^2SO^4, etc., gelatinous precipitates.	The salts of scandium are colorless and give, with KOH, Na^2CO^3 and HNa^2SO^4, etc., gelatinous precipitates.
With potassium sulfate, it should form a double salt, having the composition of alum, but barely isomorphic with that salt.	The double sulfate of scandium and of ammonium is anhydrous, but it has the composition of alum.
Only a small number of salts of eka-boron should crystallize well.	The sulfate of scandium does not give distinct crystals, nor does its nitrate, its acetate, and its formiate.
Water should decompose the anhydrous chloride of eka-boron with the liberation of HCl.	The crystallized chloride decomposes and liberates HCl when heated.
The oxide should be infusible, and it should, after calcination, dissolve in acids with some difficulty.	The calcinated oxide is an infusible powder which dissolves with difficulty in acids.
The density of the oxide is around 3.5.	The density of the oxide is exactly 3.9.[97]

On 6 February 1886 (N.S.), German chemist Clemens Winkler announced his discovery of a new nonmetallic element in a mineral that had been found in the summer of 1885 near his Mining Academy in Freiberg, and—in a somewhat curious pattern—named this element after his native country.[98] (None of the three chemists knew of the connection with the other two elements when they discovered their own, which makes this coincidence entirely fortuitous.) On 25 February 1886 (N.S.), V. F. Richter, who had once been the Petersburg correspondent of the German Chemical Society (and had reported on the first announcements of Mendeleev's periodic system in 1869), wrote to Winkler of the correspondence with Mendeleev's prediction:

> Germanium, which name you should preserve since you are factually its father, is the element eka-silicon, Es-73, predicted by Mendeleev, the lowest homolog of tin, standing in the first large period between Ga (69.8) and As (79.9). . . . Eka-silicon is the element which we have awaited with great anticipation, and in any case the immediate study of germanium will be the most definitive *experimentum crucis* for the periodic system.[99]

Winkler was immediately enthusiastic about the connection. In a telling comment that would reinforce Mendeleev's own views about the physics-like predictive powers of his law, Winkler suggested renaming the element neptunium, because like the planet Neptune it was discovered by a prediction from interpolation. Newton's laws were famously confirmed by the independent ascription of perturbations in the orbit of Uranus to a hypothesized Neptune by John Couch Adams of England (1843) and Urbain-Jean-Joseph Le Verrier of France (1846), and Mendeleev would later draw on this physical analogy and the power of prediction to defend his periodic law. (There is an element neptunium today, but it occupies the space between uranium and plutonium, following an alternative astronomical analogy.) Winkler retreated from the analogy and resolved instead to retain the name of his country, which—while it drew attacks as overly nationalistic from some French chemists—received approval from Mendeleev.[100]

Eka-silicon was the only eka-element that Mendeleev seriously undertook to isolate in the laboratory. Even before finishing his theoretical work on the periodic law, he outlined a research program directed toward finding this element.[101] On 5 December 1870, he asked Karl F. Kessler, rector of St. Petersburg University, to obtain specific minerals from the Mining Institute (a few blocks away from the University): "Wanting to verify even a part of the conclusions I expressed with respect to this [periodic dependence], I am obliged to occupy myself with research of certain rare minerals, which I thus request you to turn to the Mining Institute and ask from them certain minerals, which they have in their reserves designated for scientific work." Mendeleev made a similar request to the Russian Technical Society, and received his supplies. He even refused a post at Moscow University on the grounds that he did not want to give up his current research on the rare earths.[102] Nevertheless, this particular effort was soon abandoned, and Mendeleev's attention would drift. He would not return again to active research on the periodic law.[103]

CONCLUSION: GATHERING THE ELEMENTS OF THE SYSTEM

The view of the periodic system as the pinnacle of Mendeleev's career—encouraged by the chemist himself—was a retrospective construction. Mendeleev was not concerned in 1869 with establishing a basic law of chemistry. He was concerned with writing a textbook for young chemists at St. Petersburg University. These very local concerns are exactly what become obscured when one detaches the man from his context. From 1871 on, Mendeleev himself would repeatedly abstract periodicity from its context at St. Petersburg University under the university statute of 1863 and the pressures of writing an introductory textbook, making it seem an emblem of pure science. This is the process that accounts for the surprise of our imagined historian at the beginning of this chapter upon encountering Mendeleev's very early efforts. Those early papers *are* the origin of the periodic law; it is how those sketches turned into an immutable law that requires explaining. The vision that Mendeleev would develop over his life was consciously built, just like the periodic system, out of diverse elements that were harmonized for the sake of internal consistency. Mendeleev's predictions themselves had naturalized periodicity by demonstrating the predictive power of his system. He then used this success to naturalize the other components of his Great Reforms model. They had all been created together, and they were all naturalized together.

This success of periodicity was bolstered by the paucity of critics. Rarely has a foundational scientific development been introduced with so little debate. That is not to say that the system was immediately accepted by practicing chemists, but it was not dismissed either. The early attention it received was not about its utility for pedagogy or its potential for chemistry, but mostly concerned priority disputes among the major competing systems.[104] Among the few criticisms, two are especially revealing for the way they resisted periodicity's redefinition of what it meant to do chemistry. The first came from one of Mendeleev's Petersburg mentors, Nikolai Zinin, who in 1871 told Mendeleev to "get back to work" and stop engaging in speculations. He wanted Mendeleev to return to empirical organic chemical research. Irritated, Mendeleev wrote to Zinin in 1871:

> I write you directly: what do you want, that I leave my area [of study], that I busy myself with the discovery of new bodies, that I worry about how often people are citing me? . . . I consider the elaboration of the facts of organic chemistry in our time as not leading to a goal as quickly as it did 15 years ago, and so I'm not going to busy myself with the petty facts of this sprig of chemistry. . . . I ask either don't censure and don't judge me, or say already what the errors are in my works, and not that I am not working. . . . I would look at who would have done as much as I have in my position, and I attribute your words to a lack of attention to my works, which suffer precisely from the fact that they do not comprise only the one-sided interest found in studies *today*. . . . [105]

He never sent this letter, more from a diplomatic calculation than a reconsideration of his position. From the moment Mendeleev came to believe that he could predict using periodicity, he was no longer interested in empirical fumbling. A law of nature demanded no less.

The other criticism came from G. N. Wyrouboff, a Russian émigré chemist working in France, at the very late date of 1896. This was long after the three eka-elements had been established and Mendeleev's position as the prophet of chemistry was quite secure. Wyrouboff, however, critiqued precisely the vulnerable core of Mendeleev's worldview, the notion of *law:*

> But M. Mendeleeff has aimed at producing something more and better than a mere *catalogue raisonné* of the elements. He converted his classification into the periodic system. It was a philosophic view, borrowing arguments from the kindred sciences, and imposing itself on us by its universality. He formulated, as the fundamental law of the physico-chemical sciences, the dictum that "all the properties of bodies are periodic functions of their atomic weights". . . .
>
> On reaching this point of its development, the conception of Prof. Mendeleeff becomes essentially injurious. Under the pretext of a law which has still to be demonstrated, it forbids us to throw light upon pure matters of observation, and forces us to remain in a vicious circle from which there is no escape. I think that it is time to show clearly that there is nothing which merits the name of *law* or *system.* [106]

The example Wyrouboff cited was the inversion of tellurium and iodine in the periodic system. Up to the present day, heavier tellurium has preceded the lighter iodine, breaking the order of monotonically increasing atomic weights in favor of chemical analogy. (Rethinking periodicity in terms of atomic number instead of weight removes the conceptual difficulties this raises.) The fault of Mendeleev's program was that periodicity was not *enough* of a law, not that it was too much of one. But Wyrouboff's criticism was too little, too late—and increasingly beside the point. Mendeleev's reformulation of chemistry and the notion of law had long ago saturated the field—and, just as importantly, the culture of late Imperial Russia.

THE IDEAL GAS LAWYER

Mendeleev, painted by I. N. Kramskoi, 1878, from Dobrotin et al., *Letopis' zhizni I deiatel'nosti D. I. Mendeleeva*, 526.

THE IDEAL GAS LAWYER: EXPANDING SCIENCE ON THE BANKS OF THE NEVA

The assistant looked at me with an amused, vaguely ironic expression: better not to do than to do, better to meditate than to act, better his astrophysics, the threshold of the Unknowable, than my chemistry, a mess compounded of stenches, explosions, and small futile mysteries.
—PRIMO LEVI[1]

IN 1871, MENDELEEV'S SUCCESSES lay far in the future. Gallium, the first predicted element to be discovered, did not make its appearance until 1876, and scandium and germanium had to wait a few years more. Mendeleev's bold predictions of 1871 had two glaring deficiencies: They were unsubstantiated, and they were not in the *chemical* tradition. The whole thing smacked of physics; and, as he abandoned his half-hearted attempts to uncover his missing elements, his wavering attention shifted to that science. As Mendeleev's successor at the Technological Institute in Petersburg, Friedrich Beilstein, wrote to Emil Erlenmeyer in Heidelberg in spring 1872:

Mendeleev, who in general was never a chemist from conviction, but rather from speculation, has now abandoned our unsophisticated science. He has prophesied the existence of all sorts of new elements & believes that he needs only to conceive of them in order to have them immediately in the bag. The so-and-sos don't want to, however, and thus he has become very morose. . . . He himself has announced with great fanfare that he is busy with a test of Mariotte's Law under high pressures. Through good connections & noise-making he has already gathered some capital & perhaps you will soon see him, since he is undertaking this year a (subsidized) vacation jaunt under the pretense of ordering some gear & a manometer. He is in general an odd chap. . . .[2]

In the eyes of his peers, Mendeleev had abandoned his chemical guesses—and chemistry altogether—in favor of subsidized research on gas laws, of all things. This new effort would dominate the 1870s, but, in contrast to the repeated successes of the periodic law, every aspect of this broadly conceived gas project would end in dramatic failure.

Where did this program come from? In 1871, while composing his major German article on the periodic law, Mendeleev began investigating the rare earth elements—those that today dangle under the periodic table as an isolated island—and began also to search for eka-silicon (the future germanium). In the inside cover of one of his working notebooks, he pasted two periodic tables from his *Principles of Chemistry,* one small and schematic, and one large with chemical information.[3] This was not just for ready reference as a stable, certain thing—the equivalent of the modern polychromatic icons that loom over today's high-school student—but part of a research program. In 1871, Mendeleev thought of the periodic system as a tool that not only formulated all of current chemistry, but also generated questions to ground a research school. Mendeleev's patience with this research agenda ran out quickly, however. As his more critical contemporaries correctly surmised, he was not one to devote all his energies to one topic when a shinier prize beckoned him. On 20 December 1871, in this same notebook, after 67 pages of work on the rare earths and scribbling several public lectures on the philosophy of science, Mendeleev sketched an instrument for measuring pressure. On this date, on this page, we can pinpoint the death of *all* research by Mendeleev on the periodic system. Gas expansion, not elemental discovery, became his goal.[4]

Science has never been just an endeavor of searching out knowledge of the natural world; it is also always a question of *organization,* both of concepts and of people, to ensure the stability of that knowledge. And so it is across the backdrop of 1870s Petersburg, and the perils of organization found therein, that Mendeleev (under the auspices of the Russian Technical Society) conducted an elaborate experiment on how to organize laboratory research on a large scale (large relative to the norms of anyone's personal experience in Russia, and unparalleled until the creation of Ivan Pavlov's "physiology factory" in the 1890s).[5] Mendeleev embedded the gas-expansion experiments in a highly specialized division of labor intended simultaneously to resolve technical scientific issues and the question of how to organize information and individuals in Great Reforms Petersburg. For Mendeleev in 1871, *scientific* identity became his central focus, and thus questions of autonomy, integrity of research, and proper organization were thrown into relief in a way that would have been inconceivable twenty years earlier. It is only after professional identity has been achieved that one can fret about its loss. Throughout this narrative, then, our ears must be tuned to signs of concern about autonomy and anxiety about "freedoms" that remained contingent during this transformative period of Russian history. In this chapter and the next, we will see Mendeleev repeatedly return to how expertise could be properly organized—and made useful— without compromising its identity as expertise.

What Mendeleev introduced in the Russian milieu was the potential for large-scale, organized scientific research. In the 1870s he was on the track of creating a "big science," Imperial-style. "Big science" has generally been reserved for the gigantomania of the well-funded experimental projects, especially in physics, of the post–World War II era.[6] But in truth, the concern of "control" over both experiment and laboratory and the anxieties of dealing with large funds are *relative* ones, and the epistemic and managerial anxieties and concrete difficulties of "big science" were by no means products of the 1940s, or even the twentieth century. The scattered periodic-system experimental researches were all desktop projects, as were Mendeleev's experiments on capillarity in his apartment laboratory in Heidelberg. The gas project, on the other hand, required a team of young assistants and laboratory researchers, each working on separate tasks in an elaborately designed division of labor intended to uncover the exact laws that governed gas expansion and compression. Such a project required considerable funds, and Mendeleev enrolled various high-level contacts in the War and Navy Ministries—even the uncle of Tsar Alexander II himself—by perpetuating the false impression that he was interested in high-pressure research with applications to ballistics. Mendeleev even turned this local Petersburg project, firmly ensconced in the University, into an international endeavor of exact calibration.

Mendeleev's venue for these efforts was the Imperial Russian Technical Society, which came to stand for Mendeleev as a symbol of how one could effectively organize science in a new age of big experiments. While his own Russian Chemical Society had been intended to develop an autonomous chemical community and provide a Russian outlet for domestic research, it had not served Mendeleev's further agenda of providing a pipeline of expertise directly to the state. The Russian Technical Society showed how this could be done: Through its patronage by members of the crucial ministries, it was able not only to funnel its recommendations to appropriate points in the intricate bureaucracy, but it could also obtain funding for new research. For most of the 1870s, Mendeleev valued the Russian Technical Society as an exemplar of how science should be properly organized—second only to the Olympus of the Russian Academy of Sciences. Both would lose their appeal before the decade was out.

TRUE BEDROCK: THE CULTURAL SIGNIFICANCE OF ETHER

One might expect that Mendeleev's gas project stemmed directly from his formulation of the periodic law. That is, having uncovered the phenomenon of periodicity, he undertook an exploration of mass as the property that made periodicity work.[7] This conjecture is wrong. Mendeleev's gas work was indeed an extension of his earlier research, just not of the periodic law. He had conducted gas work as far back as his 1856 master's thesis, and he had long been interested in gaseous phenomena.[8] Now Mendeleev extended this earlier interest with a lesson learned from the periodic system: He wanted to move from the periodic law to a *yet more* fundamental law of nature. For a physical scientist circa 1871, the most fundamental quarry was the luminiferous ether.

In Mendeleev's personal copy of the first edition of *Principles of Chemistry*, on the pull-out periodic system in volume 2, he left a strong indicator of his thinking. The short-form system (Figure 3.1) contains numerous scribbles in red about the properties of the rare earths that he had begun to investigate. It also includes in blue in the upper left corner above and to the left of hydrogen the following notation, written around 1871: "The ether is lighter than all of them by a million times."[9]

The ether Mendeleev sought did not even have an empty place waiting for it in the periodic system, yet he was convinced it existed, and he began his gas project as a conscious attempt to look for it. He was hardly unique in his emphasis on the ether as the unifying concept of all physical sciences; this was a hallmark of much theoretical research in the nineteenth century. It was to be Mendeleev's *idée fixe* to transform these theoretical speculations from England, France, and Germany into an experimental project. But even while doing so, he remained as attached to the metaphysical and emotional appeal of the ether as were his Western fellows. Throughout Mendeleev's life, his passion for unified law often centered on the ether as both motivating force and metaphorical touchstone.

However it was understood in its particulars, the ether was the substance that served as the substrate for, at the very least, light waves. Even before Aristotelian mechanics posited an ether to fill the heavens' plenum, there had been discussions of an ethereal substance that penetrated all things. The "modern" ether theories, however, almost all trace their heritage to Isaac Newton, specifically to Query 31 of his *Opticks*. Ether then began to be understood as a "subtle fluid" that carried a variety of effects: light, heat, electricity, chemical action, and gravity.[10] As natural philosophers of the late eighteenth and early nineteenth centuries began to reformulate the concept of the ether, they endowed it with general properties that altered the Newtonian conception. First the diversity of subtle fluids was curtailed: As physicists mechanically modeled the ether in the nineteenth century, they gave it an articulated mathematical

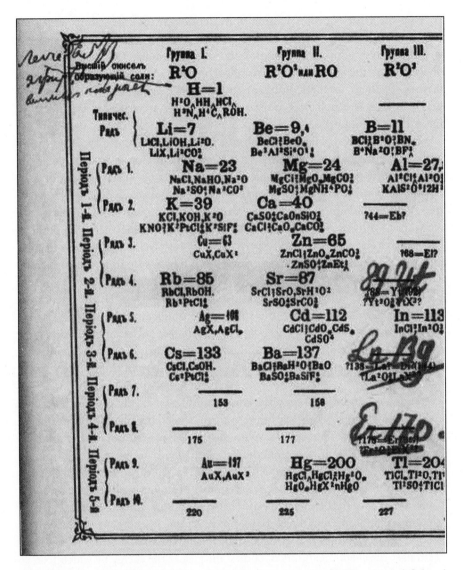

FIGURE 3.1 Periodic system from Mendeleev's gas notebook. The upper left-hand corner indicates Mendeleev's incipient interest in the ether. The other notations refer to rare-earth investigations. Source: Mendeleev, *Nauchnyi arkhiv: Periodicheskii zakon,* photocopy 29.

structure that reduced much of its earlier variety (magnetic ethers, chemical ethers, optical ethers, etc.) into aspects of one or two models. Many mechanical and mathematical models were created to derive the ether's mechanics, and many of these provided valuable scientific results that led to further depend-

ency on the ether concept.[11] Aside from the diverse internal reasons that promoted an essentially universal belief in the ether among scientific communities, even after its supposed abolition by Albert Einstein's 1905 Special Theory
of Relativity, the notion attracted many physical scientists for nontechnical
reasons. To them, aesthetic unity, programmatic integrity, or even spiritual coherence were as much part and parcel of science as were the equations, in
many cases even more so.[12]

Mendeleev himself had spent almost his entire academic life, beginning
with his undergraduate work in the mid-1850s, interested in aspects of the
ether. He first wrote on these issues in two abstracts of Western scientific articles for the *Journal of the Ministry of Popular Enlightenment (Zhurnal Ministerstva Narodnogo Prosveshcheniia)*. The first of these was written in 1856 on the
question of a lunar atmosphere, which eventually became the centerpiece of
Mendeleev's early theory of the ether. This may very well have been his first exposure to the literature on celestial gases and fluids. A second abstract on theories of fluorescence demonstrated a remarkable facility with theories of
chemical, heat, and light rays, including ample citations to the Western literature.[13] He delved further into the ether in the publications resulting from his
master's thesis on specific volumes (1856). Here, Mendeleev argued for a view
of matter as divided into "corporeal" and "ethereal" atoms, the interactions between which explained the phenomena of heat and light in chemical reactions:

> We suppose, and should suppose, atoms standing separately from each other
> or forming groups with space between them, sometimes filled with the ether
> or ethereal atoms. Thus specific volume shows the volume not only of a sim-
> ple atom, but also of its surrounding ethereal cloud. . . . Here, in our opin-
> ion, are the foundations by which one should explain the evolution of light
> and heat in chemical processes. Each process of union is carried out by the
> motion of corporeal atoms; they also bring to motion ethereal atoms which,
> upon insignificant vibrational speed, produce some phenomena of heat, and
> upon faster vibration rays of light, and, finally, chemically acting rays. The
> stronger the chemical affinity of bodies, the greater the energy by which [cor-
> poreal] atoms try for mutual attraction, the more forceful the vibrations they
> communicate to the ethereal atoms, i.e., the more heat is evolved upon com-
> bination.[14]

Mendeleev then saw the goal of the new science—meaning Charles Gerhardt's
organic chemistry—to be the unification of light, heat, and other forces.[15] The
gaseous ether was the next step.

Beyond this conceptual interest in the scientific possibilities afforded by a
material ether, Mendeleev had often worked in the experimental borderland
between chemistry and the physics of various states of matter. Mendeleev saw

his gas work as an effort that would straddle the border between physics and chemistry as defined in his time and place by providing results that would reconcile competing interpretations of gas behavior. He was quite familiar with Victor Regnault's classic researches into the elasticity of gases, a masterpiece of precision measurement, the replication and correction of which formed the basis of his own efforts.[16] When Mendeleev edited a translation of a textbook on analytic chemistry in 1866, he gathered a series of colleagues to expand chapters on various topics; he himself wrote the section on gases.[17] While examining the expansion of organic liquids and formulating his version of the "temperature of absolute ebullition" (1860)—what Thomas Andrews would label the "critical point" in 1869—Mendeleev considered the issue from the gas side as well. (The critical point is the temperature above which it is impossible to compress a gas into a liquid.) By 1870, in a priority dispute with Andrews, Mendeleev drew attention to the inadequacy of the standard gas laws: "It seems to me very likely that all gases heated above their absolute boiling temperature, under high pressures will behave like hydrogen. If this is so, then under these conditions one could not compress without limit, but upon compression they would approach a certain limiting volume which it is impossible to cross."[18] The absolute limits to compression would form the theoretical point of entry for Mendeleev's explicitly anti-theoretical project.

The connection between the ether and rarefied gases lay in interstellar space. If there really were a limit to gas elasticity at low pressures, Mendeleev reasoned, then our atmosphere, faced with less and less pressure as it extended from earth, would at some point cease expanding and leave room beyond for the ether:

> Then one should also admit that the expansive luminiferous ether of celestial space is composed of a substance as different from the gases [of air] as one chemically simple body is from another, that is, that they do not transition into each other. This is in agreement with what became known recently concerning the movement of Encke's comet. . . . I don't dare hope that I will solve such capital questions, but as an approach to their resolution I will still continue to work out observations on the compression of gases under the smallest possible measurable pressures.[19]

Thus, the ubiquity of the ether was tied to seminal questions in astrophysics. In one of the more visible astronomical anomalies of the nineteenth century, Encke's comet (1819) refused to adhere to standard applications of Newton's gravity. After all planetary perturbations were accounted for, the comet still displayed erratic, nongravitational behavior, which J. F. Encke attributed to a resistant ether. If true, these deviations would have significant consequences for the Newtonian worldview, suggesting that the solar system would slowly decay (what would later be termed "entropic death"). Russian physicist G. Levitskii held that the ether could be detected by observing such an "ether drag" on comets. Mendeleev, however, considered more persuasive the contrary numerical analyses of Pulkovo astronomer E. von Asten.[20] This did not, however, imply the ether's nonexistence:

> Mr. Asten's very clear researches, apparently impervious to further doubt, do not allow us any more to admit a universal resistive medium or do not allow us to ascribe resistance to the luminiferous ether, which everyone recognizes. . . . The environment filled with the infinitely expansive material of the luminiferous ether does not offer resistance; [resistance] appears only where there is gas. The boundaries of the atmospheres of heavenly bodies, and especially of those like the sun, probably still exist far from their surfaces.[21]

Mendeleev suggested that drag could be attributed to the interaction of Encke's comet with the atmosphere of the sun, thus using the gas laws to safeguard the existence of the ether. The stakes for him, then, were enormous—rivaled only by his ambition. He sought to isolate in a laboratory the all-pervading substance that linked the heavens and could unify the various sciences.

CONFINED SPACES: THE PROSECUTION OF THE GAS PROJECT

In the early nineteenth century, gas laws had been the cutting edge of experimental physical research, especially in France, which had assumed the mantle of leading nation in mathematics and physical experimentation. The combination of the discovery of uniform thermal gas expansion independent of the substance of the gas by John Dalton and Joseph Gay-Lussac in 1801–1802, and widespread knowledge of the adiabatic heating effect, led to extensive interest in gases as a unique state of matter.[22]

This interest eventually peaked in the work of Victor Regnault, who led a brilliant if fairly routine career through the institutions of French science. Educated at the École Polytechnique, he studied later under Justus von Liebig in his massive training laboratory in Giessen in the German states, and in 1836 returned to Paris to work with Gay-Lussac. By 1840, he had been admitted into the Académie des Sciences just shy of age 30. According to historian Robert Fox, he switched his research interests from organic chemistry to gases after taking "the fatal step" of accepting money for his experiments from the Ministry of Public Works, which hoped to work out gas laws for better steam engines. This led to what Fox considers "the most precise and yet the most unimaginative of compilations."[23] Such were the dangers of taking money for "big science" in the nineteenth century. Nevertheless, Regnault's precision experiments on the deviations of gas expansion from Boyle-Mariotte at pressures slightly above and below standard pressure (1 atmosphere = 760 millimeters of mercury) are considered exemplary of "proper" empiricism in research.[24]

Mendeleev's gas work was thus colored by twin concerns stemming from Regnault: How to redo the work of a man considered to have definitively determined the exact deviations from the gas laws; and, more implicitly, how to avoid the corruption a donor's money might bring to a project. Mendeleev, as it turned out, failed to answer either of these questions before beginning his research, and this failure would return to plague him.

Mendeleev, like Regnault, centered his work on the empirical verification of the accuracy of the two canonical gas laws. The first, the Boyle-Mariotte law, describes the inverse relationship between pressure and volume. That is, when a specific gas is subjected to a higher amount of pressure, it will compress to an equal degree; the product of pressure and volume is constant for each gas. The second, the Gay-Lussac equation, shows the direct proportionality of temperature and volume. When a given gas is heated, its volume will expand in direct correlation to the temperature increase; cool a gas, and it contracts to an equal degree—the quotient of volume and temperature is constant for each gas.[25] Mendeleev's goal was to determine deviations in the behavior of actual gases with respect to these equations. Faced with a consistent discrepancy, one might presume that an additional gas was present with particularly high elasticity. This ghost in the error bars was, presumably, the ether.

How might Mendeleev finance such a sweeping project? In principle, in the early 1870s, he could approach several organized scientific societies that might be able to underwrite such a venture. His first choice would likely have been the Russian Chemical Society, which was built in conjunction with Mendeleev's scientific career in Petersburg, and which would form a constant touchstone of all his efforts—scientific, political, cultural—until his death.[26] While there had been apothecaries, metallurgists, and physicians in Russia before the advent of Peter the Great, the institutions of organized Western science became possible only with the importation of Western European academics and tradesmen in the early eighteenth century. The sole organization of any note was the Academy of Sciences, founded in 1725. Those scientific societies that emerged early in the nineteenth century, such as the Pharmaceutical Society in Riga (1803) and the St. Petersburg Mineralogical Society (1817), were sparsely attended amateur or artisanal societies.[27] These organizations had neither the publication outlets, the regular meetings, nor the official recognition that would make them scientific societies (*obshchestva*) as generally understood, and they did not have the requisite numbers and common interests that could cohere into what Mendeleev would consider a community (*obshchenie*). Efforts to address these deficiencies in the 1850s collapsed for lack of support.[28]

The defeat of Russian forces in Crimea in 1856 sparked a series of reforms of Russian higher education and the disposition of technical expertise. The atmosphere of the Great Reforms helped raise hopes for all kinds of formal organization. The loosening of rigid restrictions against independent (meaning nonstate) organizations and free assembly led to the proliferation of voluntary associations, from charitable societies to learned gatherings, most under officially approved charters.[29] In the late 1850s, the state also decided to let Russian students travel abroad to complete their education and hopefully become more effective teachers. This state-led transition would eventually precipitate interest in a Russian chemical association. Take Mendeleev's case. Included in his Heidelberg circle at various times was the kernel of the next generation of Russian chemists. Their strong camaraderie contrasted with the lack of community that met them on their return home. Efforts to hold chemical "evenings" in Petersburg did not really bear fruit until after Mendeleev's return in 1861, when he took an active lead in their organization.[30]

The first rumblings for a chemical society began in the capital's daily newspapers. An anonymous note in the *Russian Invalid* on 17 August 1861, almost certainly written by Mendeleev—just months after his return from abroad— stated the case clearly:

A chemical society, in our opinion, is entirely possible in Petersburg. There live our most famous chemists, Messrs. Voskresenskii, Zinin, Mendeleev, Sokolov, Shishkov, Khodnev, and Engel'gardt—and in general in Petersburg many young people occupy themselves by studying chemistry. Why shouldn't our scientists gather around themselves an entire society?

We consider it unnecessary to discuss the utility of such a society. Under the society there could be a public laboratory, which there isn't in Petersburg at this time. The University laboratory is too small and serves only for University students. . . . It is too hard to get access to the Academy laboratory. . . . It is almost impossible to study physics in Petersburg. But we propose that it is possible to find funds, although of course with great difficulty; but is it necessary to prove to everyone the danger of sparse studies? The establishment of a physico-chemical society could enable the publication of a "Chemical Journal," in which a division could also be opened for physics.[31]

In this proselytizing for a chemical society, pedagogy (training chemists) was explicitly united with a society (organizing chemical expertise). The desire to combine physics with chemistry was partially a reflection of Mendeleev's interest in both fields, but it was also indicative of the inadequacy of support for a society of *just* chemists; physicists provided more leaven. In the end, however, the Russian Chemical Society was founded without the physicists Mendeleev had wanted—the Russian Physical Society was created as a sister organization to the Chemical Society in the early 1870s and the two were fused together only in 1878.

Not every chemist (or physical scientist) in the capital wanted to make the society formal, yet after a series of petitions to the Ministry of Popular Enlightenment, the supporters of a formal organization prevailed.[32] In January 1868, at the first Russian Congress of Natural Scientists and Physicians in St. Petersburg, part of a government effort to increase communication among Russian naturalists, the Chemical Division turned the event into a plea for a Chemical Society, a plea that was approved on 26 October.[33] Born of a Congress instigated by the government and armed with a charter approved by the Ministry, the organization was not exactly "independent." Nevertheless, it financed itself through subscriptions and was administered autonomously under its first president, N. N. Zinin.[34] The Chemical Society, in the spirit of the university statute of 1863, was a government attempt to let scholars manage their own affairs.

The Chemical Society, the scene of many of Mendeleev's earlier and later exploits, was financially unable to bankroll the gas project, so he turned instead to the Russian Technical Society. The Technical Society was not a decentralized voluntary organization. It was founded in 1866 as a forum for engineers and bureaucrats to articulate the future direction of technical policy and the expansion of technical education. The Technical Society, along with its partner, the "Russian Industrial Society" (literally designated by the mouthful "The Russian Society for the Further Development of Industry and Trade," founded in 1867), was nearly an adjunct of the Ministry of Finances, and was established as a means to obtain industrial consulting by taking advantage of local business, academic, and technical interests. The Society was provided with lavish funds by various ministries to finance projects that promised imminent technological application.[35] Mendeleev was a member of both the Technical and Industrial Societies, and would later (especially in the 1880s) try to use their quasi-official status to alter government policies. His administrative connection with both societies grew deeper in the 1880s; this gas project was his introduction to the politics of grant-getting in Imperial Russia. It was, in fact, *everyone's* introduction, since before Mendeleev no one outside the Academy of Sciences had ever proposed or received funding for a laboratory project on such a scale.

It is uncertain when Mendeleev approached members of the Technical Society, or whether they were the ones who approached him.[36] He certainly did dangle the prospect of practical benefits before them, most of which would stem from proposed high-pressure experiments. Mendeleev's real interest concerned the experimental identification of the ether, and this was a low-pressure phenomenon, but in order to get the money, he had to depict his low-pressure project as merely preparatory to the work on high pressures, which, he suggested, would generate significant practical results in terms of artillery shells and engines, and especially for experiments on gunpowder.[37] To be fair to Mendeleev, he did begin to work on high pressures after the publication of the first volume of *On the Elasticity of Gases* (*Ob uprugosti gazov,* 1875). Yet these experiments were never completed.[38]

The first inquiries made by the Society to Mendeleev concerning his gas work focused entirely on the problem of adequate *controls* for various phenomena—both experimental control of physical phenomena and administrative control of the division of labor in the laboratory—a problem that is at the heart of anxieties about big science.[39] Both experimental and administrative controls were integrated throughout Mendeleev's work. The chairman of the Russian Technical Society, Prince P. A. Kochubei, applied for funds from the War Ministry and the Navy Ministry.[40] The Navy, under the Tsar's uncle Grand Prince Konstantin Nikolaevich, had been a bastion of support for the Great Reforms in general and technical advancement in particular. Both ministries gave 5,000 rubles apiece to Mendeleev's project, and the Minister of Popular Enlightenment, D. A. Tolstoi (who disliked Mendeleev intensely), gave dispensation for laboratory space at St. Petersburg University to be turned over for Mendeleev's use. Architect I. I. Gornostaev redesigned and modified the University laboratory accordingly.[41] At 10,000 rubles and free lab space, this was already turning into a major endeavor. So major, in fact, that in May 1874 the Grand Prince himself came to visit the laboratory for over two hours.[42]

Mendeleev began his gas research in December 1871, but it was not until late March 1872 that he began to receive financial backing from the Technical Society. This came with strings attached: Mendeleev became a member of a larger committee, in which each member would be assigned a different aspect of the problems of gases, to be solved in his own laboratory, complementing Mendeleev's research (although in practice it appears that Mendeleev was the only member to undertake his allotted research seriously).[43] Only Mendeleev would be subsidized directly by the Society. The "Commission for Research in the Elasticity of Gases under Different Pressures, Proposed by Professor D. I. Mendeleev," was established on 21 March 1872. Its first report promised more money "to expand the experiments and to use the results achieved to research practical questions, primarily in naval and artillery technology," if the research proved successful.[44] Mendeleev was constrained to order all of his instruments via the commission and to update them regularly on his progress.[45] The Russian Technical Society opened a joint line of credit so Mendeleev could sign checks. Upon completion of his research, the instruments were to revert back to the Society. Mendeleev had full discretion to publish in any language, and a separate volume of his results was to be published at the Society's expense.

Almost immediately, Mendeleev began stratifying his laboratory in terms of material (equipment) and labor (personnel). Buying equipment was the original justification for the grant money, since precision instruments required time and funds to construct, and proper calibration—necessary in any attempt to redo Regnault's measurements—demanded a trip abroad. The funds were originally placed in the joint account in 5,000 ruble sums in April and October 1872, with an additional 2,000 added in January 1875. Mendeleev spent it almost as quickly as it came. At the time of writing *On the Elasticity of Gases,* he had disposed of 11,219.14 rubles (by August 1874), and another 4,063.10 (March 1875), exceeding his budget by over 280 rubles.[46] (The Technical Society made up the shortfall.)

His greatest expenses were also his earliest: precision instrumentation. The ether was rather elusive prey, and ordinary test tubes would not do. First, he needed an air pump that would maintain stable measures, especially under low pressures. His first publication on the gas project chronicled the construction of a "pulsating pump" that could work for months without constant need of recalibration.[47] Almost every further innovation in experimental machinery was reported dutifully to both the Russian Physical Society and the Russian Chemical Society, as Mendeleev made efforts to bridge the two groups. Building this equipment was beyond Mendeleev's abilities, and probably beyond his inclination. He contracted G. K. Brauer, former mechanic of Pulkovo Observatory, to construct precision thermometers, barometers, manometers, glass containers that could withstand extremes of pressure, and a standard kilogram and meter. All of this was supposed to cost between 2,720 and 4,440 rubles, the price depending on quality and a series of baroque clauses about malfeasance and promptness.[48]

In 1872, there was no way Mendeleev could determine from within St. Petersburg, or even from within Russia, whether Brauer's pieces of metal were accurate, and particularly if his new "meter" matched the abstract "meter" that was just then being recodified—as a metal bar rather than as a fraction of a meridian—in Paris (the actual platinum-iridium ingot was only installed at the International Bureau of Weights and Measures in 1889). Russian weights and measures had not been calibrated with their prototypes, the English system of measurement, since 1845, and there had never been an adequate comparison with French metric standards.[49] Given the tendency of exemplars of measures to fall away from the ideal standard with time and with use, the official Russian measures stored at the Depot of Exemplary Weights and Measures in the Peter-Paul Fortress in Petersburg could not provide the necessary precision. The Russian Technical Society had their own standard meter and kilogram, but those differed nontrivially from Mendeleev's recently constructed ones, so Mendeleev petitioned the Society for a trip to Paris so he could compare his measures with those at the Conservatoire des Arts et Métiers. The Technical Society readily acceded.

In June 1872 Mendeleev went to Paris and checked his kilogram and meter with those held by Gustave Tresca, and found his new ones to be better than those of the Technical Society.[50] Notice how "conventional" this process was. Mendeleev resolved, in accordance with a *choice* by several governments and scientists, that the meters at the Conservatoire were to be considered "standard" until real "standards" were made. As long as a critical mass of scientists continued to calibrate at this location, then the Conservatoire's meter would indeed be standard—the more so as more individuals chose to visit there. Now Mendeleev's local project of mediating the gas project between the military, the Technical Society, and the University began to refer to something outside of St. Petersburg: Paris. The credibility of the entire set of experiments, and of the social experiment of block grants for science funding, hinged on a fixed point outside the Empire.

That anchor worked for the instruments, since the meter and the kilogram served to calibrate all the rest. But in order to use them, Mendeleev had to construct a precision machine writ large: a laboratory where the division of labor was highly articulated and the channels of communication and authority were clearly delimited. He needed another anchor for the human capital involved—his assistants—and that fixed point had to be, as far as Mendeleev was concerned, the principal investigator: himself. His laboratory was not some archaic tyranny where the scientist-master exercised regimented and inflexible control. The actual situation was almost opposite; Mendeleev relied heavily on his assistants' special skills. As one of these reminisced after Mendeleev's death: "The boss personally did no experiments, but only monitored keenly the work of his assistants. . . . We all revered the Master and worked with enthusiasm and devotion."[51] Mendeleev had nine assistants (not including Brauer, the mechanic), each in charge of a different task. Many of these assistants considered this period of collective work one of the crowning points of their careers. The same man just quoted, for example, wrote to Mendeleev in 1881 from his new teaching post with deep nostalgia for his time in the laboratory:

> You can imagine how low the level of exact knowledge is here if I tell you that I have for two years already been reading public lectures on physics and chemistry in Warsaw with total success, and meanwhile since I left you, Dmitrii Ivanovich, I have achieved nothing and forgotten much and now I am doing what you foretold four years ago: I regret precisely that I left your laboratory, not having now the possibility to do that which I would most like. . . .[52]

The names of these assistants do not often register in histories of Russian science: Mikhail L'vovich Kirpichev, a lecturer at Petersburg University and member of the Russian Technical Society committee on gases; his brother Viktor; Nikolai Nikolaevich Kaiander, a chemist who finished Petersburg University in 1875 and worked there until 1884; Gustav Avgustovich Shmidt; Nikolai Feoktistovich Iordanskii, later translator of foreign scientific texts for Mendeleev; Fedor Iakovlevich Kapustin, Mendeleev's nephew; Iuzef Genrikovich Bogusskii, author of the above quotations; V. A. Gemilian; and Ekaterina Karlovna Gutkovskaia, a Jew and a woman, both rarities in laboratories of the Imperial capital.[53]

A comparison of the assistants' individual laboratory notebooks shows that most (with the exception of M. L. Kirpichev) were assigned no more than two tasks. After 1875, Bogusskii continued Kaiander's preliminary work on high-pressure compression. Gutkovskaia worked on the meniscus in order to account precisely for the quantity of mercury in each vial. Gemilian's 1875 notebook concentrated on various calibrations, especially at low pressures.[54] As each developed refined tables of results, they would transcribe just those results into Mendeleev's six gas notebooks. His notebooks were thus a kind of clearinghouse for the entire team's efforts. Results copied, pasted, or transcribed there were ready for further elaboration and sometimes publication. Mendeleev also scribbled plans for lectures, experiments, and calculations in these notebooks, but he only once took the step of writing in an assistant's notebook. Traffic flowed only one way.[55] Mendeleev's authority as central arbiter was not exhibited by his presence in the notes of his assistants, but by his absence. By having their results come to him, Mendeleev created a space for himself as the organizing point of the various efforts of his assistants.[56]

This impression is slightly misleading, however, since one of these assistants performed at least as central a function as Mendeleev. Mikhail L. Kirpichev served as the conceptual workhorse of the project during the first four years. He managed all the mathematical manipulation required for instrument design and data analysis. (Mendeleev had an incredible facility with qualitative reasoning, but in mathematics he needed the aid of a skilled calculator.) Throughout Kirpichev's notebooks one finds elaborate computations of differential gas laws, rates of change of pressure, and statistical distributions.[57] He was also assiduous in calibrating the differential barometer and the manometer and neutralizing the effects of temperature fluctuations and mercury vibrations.[58]

Kirpichev was born in Pskov Province, southwest of Petersburg, in 1847, son of a mathematician at the Engineering Academy. He began teaching chemistry and physics after a research trip to the Ural Mountains in 1871–1872. Unfortunately, he contracted what would turn out to be a fatal illness during that winter journey. Nevertheless, Kirpichev came to work in the laboratory even two weeks before his death, a death that Mendeleev later called "a sensible [i.e., tangible] loss to Russian science."[59] Given the central place Kirpichev had in the project, his death led to a standstill in the work for two years. The remaining assistants went off to other academic positions away from the capital.

All things considered, the project actually went quite far. In principle, Mendeleev and his collaborators had to perform simple measurements: measuring the temperature, pressure, and volume of various gases (air, hydrogen, carbon dioxide, sulfur dioxide, nitrous oxide) to determine the extent of deviations from Boyle-Mariotte's or Gay-Lussac's laws. In this way, it was hoped, the ether would be inferred as a common denominator of deviations across various gases.[60] At a meeting of the Russian Physical Society on 5 March 1874, Kirpichev and Mendeleev reported their cardinal results on low pressures from 650 to 0.5 millimeters of mercury: (a) Boyle-Mariotte was not valid at low pressures; (b) the deviations from the law became larger the lower the pressure; (c) Regnault's findings on air's expansion were inaccurate; and (d) the deviations were clearly greater than the experimental error. Their explicit justification for focusing so much attention on low pressures was "the unexpectedness of the achieved result, which goes at odds with that dominating proposition, that upon a reduction of pressure, gases approach an ideal state."[61] Hope for the ether was never far below the surface of these claims.

In 1874, Mendeleev and his technicians conducted another experiment on air to show that at very low pressures Boyle-Mariotte broke down. They took an initial gas measurement, so that pressure (P_1) and volume (V_1) multiplied together reached 10,000 milliliters-torr, and then measured the product for each drop in pressure. If Boyle-Mariotte were exact, the result should be constant, as the gas expanded to compensate for the lower pressure. (See Figure 3.2.)

P_1	$P_1 V_1$
45.394	10000
11.934	9291
2.829	8317
1.556	8997
0.663	7689
0.514	7581
0.353	6063

What Mendeleev saw here struck him. As he further lowered the pressure, the volume failed to increase as much as Boyle-Mariotte predicted it should—something constrained the gas's expansion. He wrote in his notebook: "Under zero pressure air still has a certain density—this is the ether!!"[62] His enthusiasm soon dissipated, however, and these ether speculations appear nowhere

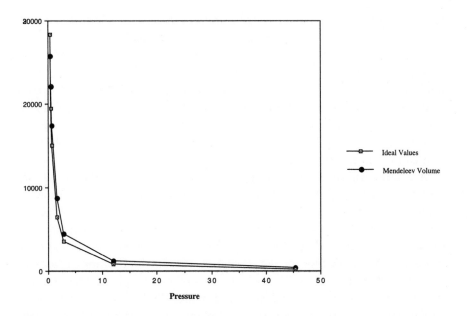

FIGURE 3.2   Boyle-Mariotte discrepancies. Mendeleev's gas data from the 1874 crucial run are shown with ideal volume values predicted from the Boyle-Mariotte law for the same pressure readings. Notice the quite substantial difference between the curves, with the ideal values being consistently lower than the actual data.

else. This is, in fact, the closest Mendeleev ever got in his entire career to actually believing he had found the ether.

A year after this glimmer of hope, Kirpichev died and the work halted. There is a clear gap in Mendeleev's notebooks from October 1875 to November 1876, during which time he engaged with local cultural disputes (addressed in the next chapter) and traveled to the World's Fair in Philadelphia, meanwhile asking Petersburg colleagues to recruit new assistants. None with Kirpichev's abilities was found. The pace of the work continued to falter as the years progressed, and never again did such furious ether-hunting excitement creep into his scrawls. Instead, Mendeleev shifted to a different way of experimenting with the various pressures upon him: publication.

CLEARING THE ATMOSPHERE: STRATEGIES OF PUBLICATION

By the middle of the 1870s, the Russian Technical Society was pushing Mendeleev to publish, and he was already losing interest in his project now that the low-pressure work was complete—and a failure. So he chose to publish in a manner that might redirect the project into areas in which the Technical Society was not currently interested. These publications on the gas work had a novel twist for him: They were for raising research money. The Russian Technical Society had agreed to foot the bill for a monograph on the gas work published in Russian, consistent with their mandate to further technical education and recruit expertise for the state. But Mendeleev's interests in publication were elsewhere. He was crafting a new social position for the scientist with respect to laboratory work and the Imperial bureaucracy, and he now wanted to project that image to the public.

Mendeleev had become increasingly interested in gases outside the laboratory, particularly their behavior in the upper levels of the earth's atmosphere. Failing to isolate the ether in his laboratory, he turned to a "natural laboratory."[63] This brought him that much closer to the atmosphere's boundary with the vaunted celestial ether. Once again, Mendeleev was pursuing unification, but instead of unification of *subject matter,* as with the transitional zone of gases, he was proposing a unification in *place*—the atmosphere. The effects in the upper layers of the atmosphere would potentially lead to discovery of the laws of refraction, weather, astronomical observation, gases—and, implicitly, the ether.[64] He was in effect delocalizing his laboratory, and in the process he made the phenomena of interest both more and less accessible: more accessible because the air was available to all; less because one could not just observe the upper layers of the atmosphere from the ground, or even, Mendeleev insisted, from mountaintops, owing to interference by the mountain itself.[65]

So how to make such observations? Mendeleev needed an aerostat, a weather balloon operated by a scientist (presumably himself) together with an aeronaut, to take observations of thermal gas expansion at various altitudes. Unfortunately, no one in the Russian government was willing to fund such balloon experiments—yet. And so Mendeleev arranged to use the publications he *had* to do for the Russian Technical Society to get the equipment he *wanted.*[66] The Ministry of Finances was already paying for the publication of *On the Elasticity of Gases,* so Mendeleev could take all profits from publication to fund an aerostat. Thanks to the expansion of a scientifically literate public, trained by individuals such as Mendeleev, there were readers who would buy accounts of technical experiments on gases, and a scientist could contribute suggestions about how proceeds ought to be spent. If the Technical Society would not give him an aerostat, Mendeleev would procure one through other means.

There were no objections to implementing his idea with the five books he published in the mid-1870s. He began the first volume of *On the Elasticity of Gases* with an account of the deviations from gas laws, and how those laws were as central as the laws of motion and gravity; the deviation problem "has a first-order significance for all of the natural sciences."[67] But before readers even reached this preface, they would have been greeted by the frontispiece, which suggested the need for an aerostat and informed the reader that purchase of this or related books would lead to the earmarking of money to build the necessary equipment. By mid-May 1876, there had been slight but encouraging sales. Mendeleev's projection of 30,000 rubles for the cost of fifteen to twenty flights, however, was a gross underestimate.[68] Money for the aerostat was not raised by this book alone (or by the sum total of the books, as it turned out), but Mendeleev did not lose heart, and continued throughout the decade to suggest repeatedly the necessity of an aerostat for the advancement of science. (In Chapter 7, we will see how Mendeleev finally managed to take his flight in a balloon, albeit not under the circumstances he foresaw here.)

Despite the positive reception of *On the Elasticity of Gases*, volume 1, Mendeleev never published the second volume, which was supposed to cover his high-pressure work. As it stood, the first volume was a monograph on calibration, with most of the gas results themselves deferred to the proposed second volume. There is no trace in his archive of a draft for this second book, although plenty of data and observations were available for such a tome. The Russian Technical Society, even after Mendeleev's resignation from the project in 1881, offered 2,000 rubles for the publication of the text and an additional 2,000 for Mendeleev personally.[69] Apparently, he did not want the money that badly. Instead, he advertised the existing volume whenever he could, such as in the English journal *Nature*:

> I have published this work only in Russian, not having means sufficient to publish a translation of a work so voluminous, and desiring to conform to the custom existing among *savants* of all countries of describing their labours in their mother-tongue, in order to present to the scientific literature of the country where they live and work a gift in proportion to their powers.[70]

Mendeleev would pursue this trope of the patriotic Russian appealing to Europe for validation with increasing frequency as he grew older.

Aerostat or no, Mendeleev was not beyond trying to pocket some money from spin-offs of his gas work. The most obvious of these efforts involved his differential barometer. In spring 1873, he announced to both the Russian Chemical and Physical Societies that he had developed a barometer that could accurately gauge differences in elevation. One could take the differential barometer, which had two tubes connected by a valve, to a certain elevation (say, sea level), and open the valve. The instrument would register the local air pressure, and the valve would be closed again. Then, after ascending (say, the spire of a cathedral), one could open another stopcock, and the difference in the former pressure and the current pressure would be displayed. Accuracies in differential elevation of as little as a meter could be determined solely through air pressure.[71] He described this ingenious instrument in several Russian journals and in Chapter 5 of *On the Elasticity of Gases*, but Mendeleev hoped to spread news of it further. In 1875 he published a pamphlet advertising the differential barometer (or altimeter), which was widely distributed to mariners, navigators, surveyors, architects, military officers, and other individuals of practical inclination.[72]

The differential barometer sold very well among tradespeople, but he was targeting an even broader audience. Each individual instrument came autographed by Mendeleev, who at this time was becoming quite a Petersburg celebrity as a result of his contemporary public campaign against seances. For 25 to 40 rubles, one could purchase an altimeter; combined with a mercury thermometer, it cost 60 rubles. The pocket altimeter and thermometer in a leather case and shoulder strap for 85 rubles or a complete altimetric set for 250 rubles were intended for popular audiences as coffee-table ornaments or amateur amusements.[73] Though this may sound bizarre, it is no more unusual than the celestial globes or telescopes that one finds in the homes of today's educated public. As Jan Golinski has shown, during the Enlightenment barometers came to represent metaphorically knowledge of the natural world by expressing variability, public access to knowledge, and easy commercialism.[74] On his trip to the United States in 1876, Mendeleev specifically advertised the differential barometer at the Philadelphia World's Fair to show the combination of technical ingenuity and popular improvement that characterized the ideology of World's Fairs in general.[75] Finally, in 1876 Mendeleev published a book on barometric surveying as a replacement for standard altimeters, the profits again to be donated for the aerostat project. This was only marketable because surveying had become more widespread during the Great Reforms, which also raised the possibility of Empire-wide barometric weather forecasting.[76]

This dream of establishing a rational and highly organized meteorology with utility for both sailors and farmers provides a crucial connection between Mendeleev's scientific speculations on the nature of rarefied gases (and the ether) and his visions of how scientific organizations should function. Some of his organizational notions hoped to decentralize the project so as to avoid the types of hardships that the Russian Technical Society had imposed on him.

THE WEATHER OVERGROUND: MENDELEEV'S METEOROLOGY

Mendeleev's efforts in meteorology highlight many of the themes of his gas project—deviations, collection of data, aerostats, large funding, public science. In the 1870s Mendeleev was concerned with establishing an appropriate public face for science, and he did this in two ways. The first was by working *as a scientist* in a properly constituted laboratory in cooperation with the government. This would show the government and the public how science secured its knowledge claims. Second, science would carve out (with generous help from Mendeleev) a *cultural* role by fairly and impartially adjudicating disputes in various arenas of public life.

Meteorology, the science of the weather, was both a very ancient and a very recent science by 1875. Its transformation to a full-fledged physical science had been the consequence of several metamorphoses from the study of climactic patterns to a science of weather forecasting, incorporating thermodynamics, the telegraph (to allow for immediate correlation of disparate data), and statistics.[77] Meteorology also underwent a profound transformation in its public status. In 1848, James Glaisher, better known for his repeated weather balloon ascents, organized a meteorological station network for the *Daily News* in London to provide reports on the weather. Because of its obvious immediate benefits, weather prediction became a very public form of science, from balloon voyages for gathering data to the libel trials of astro-meteorologists.[78]

Besides its new technical capacities and public visibility, meteorology also dispersed *institutionally* in the middle decades of the nineteenth century, in at least two senses. First, most major Western states created public meteorological institutes. Such institutions appeared almost simultaneously all over the world: France (1863), the United States (1871), Russia (1873), Denmark and Sweden (1874), Belgium and Germany (1876), Austria and Australia (1877), India (1878), Italy and Switzerland (1880), Portugal (1882), Japan (1883), and Spain (1886), for example.[79] Weather-observing networks actually dated back to the Enlightenment. In the 1770s a French cleric named Louis Cotte built on physicians' attempts to collect accurate observations on the weather, and at one point over seventy stations on three continents were reporting to him, but the inability to correlate the information quickly led to the failure of the project. A more substantial effort was the Palatine Meteorological Society, which from 1780 to 1795 had thirty-seven recording stations spanning from the Urals to North America—with two Russian observers in Petersburg and Moscow.[80]

The second dispersal was from the metropoles of Paris and London to the "provinces" of Western science. These were "outposts" where the economic barriers of modern science kept individuals from entering the experimental sciences directly, but provided an outlet for observational sciences that bordered on older traditions of natural history.[81] Imperial Russia was in the unique marginal position, especially in science, of being *both* metropole and province. The history of Russian meteorology, indeed, recapitulates the characteristics of both European and provincial meteorology: As in Europe, it was the domain of elite theoreticians and experimentalists; but in the provincial style it served as a cultural model for how to coordinate individuals—one of the central questions of the epoch of the Great Reforms.

Mendeleev drew from a long tradition of natural-historical meteorology in Russia. The first real meteorological observations in Russia date to the early eighteenth century, when Vitus Behring traveled to Siberia with meteorological observation instructions penned by Daniel Bernoulli, a Petersburg academician (that is, a member of the Academy of Sciences). These supplemented a train of twelve observation stations from Kazan to Iakutsk.[82] Meteorology fell into decline after that, and observations only began again in the 1820s and 1830s when academician Theodore Kupfer organized magnetic measurements as part of Alexander von Humboldt's plan for a global system of magnetic observatories. Kupfer's main observatory soon became the Chief Physical Observatory (Glavnaia Fizicheskaia Observatoriia, GFO), which Kupfer headed until his death in 1865. It then passed briefly to Ludwig Kämtz, inventor of the isobar, until his death in 1867. Next in line was Heinrich Wild (1833–1902), a Swiss meteorologist who, with his assistant Mikhail Aleksandrovich Rykachev, transformed the GFO into a major observational center, with a total of seventy-three calibrated observation stations across Russia by 1872. Before moving to St. Petersburg, Wild had been professor and director of the observatory in Bern, and was also the administrator of the Meteorologische Centralanstalt. Under his tenure, the GFO became one of the central observing networks in Europe.[83]

But Mendeleev, for his part, wanted to embed meteorology more deeply in the physical sciences. He began with the question of the temperature of the upper levels of the atmosphere. His several articles on this subject were not much more than a proposed aerostatic research program. His theoretical assumption, that the upper levels of the atmosphere were heated by the sun's rays as well as by reflected light, was meant to motivate an experimental program. Mendeleev wanted aerostat ascents, since freely flowing air behaved differently than air near tall bodies like mountains and observations were needed before theorizing: "Thus, if we want to arrive at an empirical law of thermometric changes in the layers of the atmosphere, we will have to rely almost exclusively on phenomena observed during aerostatic ascents."[84]

Mendeleev knew that previous ascents had been seminal to the development of the physical science of meteorology. Gay-Lussac had ascended twice in the summer of 1804, first with Jean-Baptiste Biot and again alone, in what has been labeled the first scientific expedition in a balloon. More recently, James Glaisher's repeated ascents in balloons were a highly visible form of meteorological popularization. But Mendeleev focused on the ascent by fellow Petersburger Ia. D. Zakharov two months before Gay-Lussac, in what Mendeleev called "the first purely scholarly journey with the goal of studying the upper layers of the atmosphere."[85] He declared that "France and England have already done much for the resolution of this question; now it is the turn of other states to gather the necessary data in the greatest quantity. . . . [Russia], thanks to its continental climate, is very appropriate for experiments of this sort."[86] Mendeleev had had to go to France in order to stabilize his metrological calibration; France would have to come to Russia to stabilize its meteorology.

Mendeleev's hypotheses were scathingly attacked by M. A. Rykachev, Wild's assistant at the GFO. Rykachev was irritated that Mendeleev would criticize professionals like Glaisher without himself ever having ascended in a balloon. Rykachev also said the notion of "upper layers" to the atmosphere lacked "physical meaning." This criticism struck close to home, as this was precisely where Mendeleev hoped to isolate the ether. His response was quite revealing of his philosophy of scientific concepts:

> I will put forth an explanatory example. The physical meaning of the coefficient of expansion (i.e., the coefficient of proportionality of expansive lengthenings under weight) often, even in almost all physics and mechanics courses, is determined by saying that there is a weight capable of doubling the length of a bar (whose cross section = 1, e.g. 1 cub. millimeter). But everyone knows that such an expansive lengthening can be sustained by only a few known substances. This concept has no reality, but it still carries physical understanding, so to speak, for clarity. . . . The physical understanding of absolute zero is beneficial, advantageous, but it also has no reality. Such also is the concept of electrical fluids.[87]

The point was that one used these concepts among an audience of experts who would understand when they were meant as an actual physical reality.

Mendeleev hoped to help create this expertise indirectly through a new genre of publication. Textbooks like *Principles of Chemistry* were meant to produce actual experts from the students who would study it; his edition of Henrik Mohn's *Meteorology* (1872, 1876 in Russian translation), on the other hand, was meant to create *demand* for expertise and an amateur audience of scientifically literate citizens. This book, translated from the German by Kapustin and Iordanskii, two gas assistants, was yet another in his series of aerostatical advertisements.[88] Mohn's original was a standard general text on the theory of meteorology, understood as the science of weather prediction and statistical deviations (as opposed to climatology, the study of regional statistical averages). The author, Henrik Mohn, had started out as an astronomical observer at the University of Christiania (now University of Oslo) in Norway in 1861. He was central in establishing the Norske Meteorologiske Institut, and in 1866 became the director of that institute and professor of meteorology at the university, where he remained until his retirement in 1913. His focus on Norway's crucial geographic position led to the emphasis, preserved in the Mendeleev edition, on cyclonic theory, which Mohn and Cato Guldburg had introduced in 1876 in *Études sur les Mouvements de l'Atmosphère,* a foundational text of dynamic meteorology.[89]

Mendeleev's preface concentrated on the need for an institutional organization for Russian meteorology, a parallel structure to the GFO's observation network. He was concerned with weather as the "deviations from climate"—a deliberate reference to deviations from gas laws; here, as before, he was interested in amassing data on irregularities in order to determine laws.[90] To accomplish this over the broad expanse of continental Russia, one had to train many observers. Mendeleev was delighted that the rural governing councils (*zemstva*) set up by the Great Reforms were taking an interest in weather prediction. Using the principles in Mohn's book, they could gather the appropriate data, which would then be correlated to facilitate forecasting for agriculture. "Publishing Mohn's book, I want to put the necessary guidance in the hands of our practitioners," Mendeleev declared. The next task was to organize it:

> In order to work for a science, one must have not only knowledge of the goals of science, but also a certain willpower. It is another matter when with a known means of action demanded by science, one connects immediate well-known results obvious to all. Now meteorological data, gathered at meteorological stations dispersed about Russia, are concentrated and examined in our central meteorological establishment. Such centers in essence should be many; then and only then will the scientific development of meteorology emerge into life.[91]

One needed not only more data, but a better system of coordination. In marked contrast to his later approach at the Chief Bureau of Weights and Measures, Mendeleev defended multiple centers of data calculation. These centers would collect information, and the opaque and diffuse machinery of science would order them. From his ensconced view in Petersburg, he viewed science as a decentralized solution to the problem of organizing information, people, and expertise. (The exception was the aerostat, which was too expensive for individual *zemstva* to purchase, and thus was a necessarily centralized tool.)

The dream of a decentralized, participatory meteorology collapsed for Mendeleev for two primary reasons. Most importantly, in 1880 Mendeleev lost his faith in the ability of local scientific groupings to transcend particularized interests, as addressed in Chapter 6. In addition, Mendeleev's popular *zemstva* weather program hinged on a science that was fundamentally *empirical,* and therefore accessible to a wide array of practitioners. His gas project, however, although intended to organize experimental expertise, devolved into a few theoretical results that necessarily foreclosed popular involvement.

PLAGUED BY THEORY: ABANDONING GASES

Mendeleev's gas project was aimed at finding empirical, almost curve-fitting, laws to explain the deviations in gas behavior, particularly the various irregularities in the Boyle-Mariotte and Gay-Lussac laws. He drew an analogy with the deviations from Newton's law of universal gravitation that had occupied scientists for 150 years:

> To prove that gases under very small pressures, as well as under very considerable pressures, vary from the Boyle-Mariotte law is by no means the same as to deny the truth of that law; I feel that I ought to state this most explicitly. For a long time the law of gravitation could not be made to accord with the perturbations [of the planets]; latterly these perturbations have proved the best confirmation of the laws of gravitation. In the present case it may be the same.[92]

Newton's laws were reconciled with the observed paths of planets by interpreting those paths as perturbations generated by theoretically predicted planets beyond Saturn. (The analogy with his eka-elemental predictions was clearly intentional.) However, Mendeleev did not feel that theory held the answer for gases. He thought that only through patient experimentation could the exact nature of the deviations be discovered and then reconciled with known laws. The course of events did not work in his favor. The only lasting result of his research was a formulation of the ideal gas law. For Mendeleev this was supposed to be a foundation, a formula that directed how future experiments should be conducted, and not the central achievement. The fact that this theoretical result is all that remains of Mendeleev's gas work is the bitterest irony.

Mendeleev's views on the relation between theory and experiment are rather subtle. Consider his position on the revision of the Boyle-Mariotte law. This law, remember, states that as each gas is subjected to a greater pressure, it will compress to an equal degree. Regnault had shown *experimentally* that all gases condensed more than that. That is, as pressure increased, a gas tended to occupy an even smaller volume than necessary to maintain the product of pressure and volume as a constant; likewise, at low pressures, they expanded less. It seemed that experiment had overthrown theory. Mendeleev interjected the caveat, however, that the nature of the liquid states of atmospheric gases indicated that at some pressures the compression should be less than predicted, not greater.[93] Regnault's experiments may have proven theory inadequate, but theory, especially knowledge of the critical point of liquids, showed that Regnault's experiments also had their limit. In Mendeleev's ideal, this leapfrogging of theory and experiment eventually generated correct results.

At this time, experiment was primary for Mendeleev, as can be seen in his active interest in aviation coincident with the tail end of his gas research. On 9 July 1878 the Navy Ministry authorized 12,450 rubles for Mendeleev to go to Western Europe to investigate aerostats: big science again.[94] His research did not lead to the mass development of military aerostats. Rather, most of his results focused on how to model air resistance. His central findings were published in a book entitled *On the Resistance of Liquids and on Aeronautics,* volume 1 (1880). (Volume 2, in what was clearly becoming a pattern, never appeared.)[95] This was in many ways a capstone to his gas project. The general problem of airflight, as Mendeleev saw it, was the study of resistance. Most models of flight naturally based themselves on shipbuilding, but this was inappropriate: First, in water, unlike air, the resistive material itself supported your weight; and second, one was not totally submerged in water on a ship, whereas a flying device would be surrounded by air. Mendeleev partially developed a theory of aeronautics. He felt the best theoretical study remained Book II of Newton's *Principia* (1687), designed to refute Cartesian vortices. As Mendeleev stated: "Setting myself to the study of resistance I admit that I did not expect to find the inadequacies of theory and experiments concerning it, as there turned out to be in actuality."[96] It was not until Charles Bossut's data in 1787 that Newton's theory was understood to be inadequate. Mendeleev vigorously defended further observation:

> But when the matter concerns not simply and directly only pure and free knowledge, but living and significant practical activity (such as, for example, naval and shipbuilding affairs), found in the hands of people who should possess practical and theoretical reserves of knowledge, then the premature positing of theories often brings harm and those who understand contemporary conditions should be on guard against it.[97]

In this instance, however, Mendeleev had to couch his revision of Newtonian theory as a manifesto demanding more experimentation. The various mathematical sections expounding and criticizing the extant theories—calculated by N. F. Iordanskii, one of Mendeleev's gas assistants—were shunted to appendices.

Experiment was so crucial in Mendeleev's primary focus of gas physics because one theory in particular was corrupting the entire field: the belief in an "ideal gas." While today one associates Mendeleev with an ideal law to whose specifications we marshal the results of experiment—periodicity—he consistently opposed ideal notions. Ideal gases, like the British mathematical ether, were abstract constructions of hyperactive intellects, *not* science:

> Although there is no doubt that the Boyle-Mariotte Law is not rigorously applicable even under moderate pressures, yet the prevailing doctrine, so rich in

instruction on the nature of gases, which deduces all their properties from the *vis viva* which animates their molecules, admits the supposition that in the rarefaction of gases the distances between these molecules increase to such an extent that their mutual attraction is destroyed; and in this case gases comply exactly with the Boyle-Mariotte Law. On this hypothesis the law becomes a limit towards which every gas tends in proportion as the distance between its molecules increases, and in proportion as their *vis viva* and the rapidity of their motion increases. That idea finds no support in facts.

He continued that "[t]he idea of an absolute gas belongs, then, to the number of fictions which find no confirmation in facts."[98] This notion, Mendeleev said, was invented after Regnault's deviations and defined to be a gas that *actually* conformed to the Boyle-Mariotte and Gay-Lussac laws; it "has to the present not been based on any kind of exact experimental data."[99]

That said, it appears surprising that Mendeleev is credited as one of the formulators of the ideal gas law—expressed today as $PV = nRT$ (pressure × volume = molar quantity × constant × temperature). This law is nothing if not theoretical: It is an amalgamation of the two gas laws with Avogadro's law (equal volumes of gases under identical conditions contain equal numbers of molecules). Although Soviet historians exerted themselves arguing for Mendeleev's priority here, the equation actually had a long history (of which Mendeleev was mostly unaware) by the time he came to it in 1874–1875.[100] Gay-Lussac had worked on an early version combining his law with the Boyle-Mariotte relation in his notebooks, and Sadi Carnot, one of the founders of thermodynamics, used it in 1824. It was codified in 1834 by Émile Clapeyron as $pv = R(c+t)$, where v is specific volume, p is specific pressure, t is temperature in degrees Celsius, c is a universal constant of roughly 273, and R is a specific constant. Unfortunately, R needed to be redetermined for each gas measured, since volume was specific to each gas.[101] In Mendeleev's eyes, the "law" obscured what was general behind the specifics of each gas: "The universally-known Clapeyron formula is a particular instance of this more general formula, because, having posited the nature of a gas and its mass as unchanging, the latter equation turns into the former."[102]

Mendeleev's relation is not quite the general equation known to students today, where the nature of the gas involved is essentially irrelevant for the problems to which the equation is put. Instead of removing the specific volume and other markers that the gas involved is a particular substance, Mendeleev added in *more* specific qualities to balance them out. His relation of September 1874 stood as $APV = KM(C+T)$.[103] There are two constants in this relation, K and C. C was set at roughly 273°C (what we today would understand as a calibration to absolute zero), and Mendeleev *measured* K to be roughly 62. What made this gas law specific was A, the particulate weight of a gas with hydrogen set equal to 1, and *M*, the mass of the gas in kilograms. Instead of factoring these two out into the neutral term of molar quantity, a relation that is invariant (and would be an obvious application of the Avogadro relation), Mendeleev incorporated more specific qualities of individual gases into each equation.[104]

To understand this unusual formulation we must stop thinking of this relation as either pertaining to *ideal* gases or as being a *law*. Mendeleev hoped here to unify the two central gas laws without making assumptions about gas behavior. Instead of looking at this gas law as a *result* of his research, it was similar to the periodic system in 1871: It was intended to help guide a research program to prosecute experiments under a unifying aegis. He wanted to find a law that connected all gases *experimentally*, so he set an assistant (N. N. Kaiander) to find the true coefficients of this equation for each gas, and how they themselves fluctuated based on temperature and pressure. These were not constants, but experimental quantities to be given definition in the laboratory.[105]

By the late 1870s Mendeleev's gas work occupied only a fraction of his time and still less of his interest. When offered more money in 1878 to conduct his high-pressure experiments, he declined.[106] As pressure continued for him to finish his project, he decided to abandon it, resigning publicly before the Technical Society on 21 January 1881. Mendeleev's stated reason for leaving was that he had found no one adequate to replace the late M. L. Kirpichev. He declared that he himself had given all of his time to the gas project from 1872 to 1874, and Kirpichev gave even *more* time, and still they had only completed a fraction of the work. As Mendeleev had more obligations and could no longer satisfy the demands of the work, he had to find multiple assistants to replicate the earnestness of both himself and Kirpichev—hard to do even with the 2,000 rubles a year Prince Kochubei had granted him to hire assistants. Mendeleev also stressed other commitments on his time: his trip to the United States in 1876 to investigate the American oil industry at the behest of the Technical Society; an attack of pleurisy that forced him to take sick leave from the University; and his journeys abroad to investigate military aviation for the Navy, whose interest in the topic had increased after the Russo-Turkish War of 1877–1878. All prevented him from completing his work.[107]

Mendeleev couched his resignation in terms of "freedom," a trope that should remind us that Mendeleev's efforts were a form of "big science." While lavish funding gave Mendeleev new freedom to conduct large-scale research, it also inhibited his autonomy; the tension between the two wore on him as the decade ended. As he put it: "I have no debts to the Technical Society. Freely I came, freely I go."[108] Mendeleev recommended academician A. V. Gadolin, chair of the gas committee, to take over the gas project. But he wished to spare Gadolin from the pressures he had experienced: "No, leave him [my successor] alone—without [freedom] scientific success is impossible, because science is a willful and delicate affair."[109] Mendeleev did not part acrimoniously—he returned all the equipment and they soon made him an honorary member—but Gadolin never took up the research. Mendeleev performed only one more gas experiment, and then never returned to the subject again.[110] Stymied by the aporias of organization, he threw in the towel.

Freedom, control, ether, experiment, theory, organization, money—and failure. These were the elements that increasingly structured and confounded Mendeleev's scientific research (both in the laboratory and among the scientific community) in the 1870s. The gas project gave Mendeleev opportunities he had only dreamed of earlier: access to state funds to explore the borderland between physics and chemistry. Much as the periodic system and the cultural world it was embedded in represented the foundation of a social and scientific philosophy, the ambitious gas project represents both the strengths and the weaknesses of Mendeleev's scientific and political practice. He was bold and innovative in establishing a project, but poor in its execution; he inspired loyalty from his subordinates, but could not appease his superiors; he was widely known through his publications as the model of a scientist, but was derided by his peers. Friedrich Beilstein's sneering dismissal of Mendeleev in 1872 as the "odd chap" who "was never a chemist from conviction, but rather from speculation" remained no less true at the decade's end. Two other events of this decade, Spiritualism and the Academy affair, occurring against the background of the gas project, only deepened this impression.

CHASING GHOSTS

A group of professors and teachers from the physico-mathematical faculty of St. Petersburg University, 1875. Mendeleev is in the center, leaning to the side. From Smirnov, *Mendeleev,* insert 1.

CHAPTER 4

CHASING GHOSTS:
SPIRITUALISM AND THE STRUGGLE
FOR PUBLIC KNOWLEDGE

Glendower. I can call spirits from the vasty deep.
Hotspur. Why, so can I, or so can any man;
But will they come when you do call for them?
—WILLIAM SHAKESPEARE, *HENRY IV, PART I* (III.I)

ALL OF PETERSBURG WAS ABUZZ in the mid-1870s with a raging controversy that was, quite literally, out of this world. In March 1876, a weary commentator complained:

> Whatever epithet you could think up for the passing Petersburg season, you would have to call it "Spiritualist." Nothing else has agitated society so much as to move the knocks of mediumism to the background. . . . Serious and unserious circles only livened up when the talk came to Spiritualism. . . . Many are scandalized by it and shout that a new faith has begun, that secret propaganda is being undertaken, which threatens to seize not just Petersburg, but all of Russia. . . .[1]

Spiritualism had become the rage, and this season Mendeleev was in the center of the action.

Modern Spiritualism had been transformed from a marginalized and ridiculed amusement of simpletons to the most fashionable pastime of elite

circles. The movement focused on seances, and the set of phenomena that occurred therein. Seances were gatherings in typically darkened rooms, led by a
sensitive "medium" who could elicit certain effects, ranging from physical phenomena like table rapping and levitation to the more "spirit" related—auto-
matic writing, spirit photography, and spirit materialization. Considered by
many a modernized religion more suited to the day's empirical advances, there
was substantial disagreement within the movement as to whether actual "spirits" of the departed were responsible for the phenomena. (Some believers in
the physical effects preferred the term "mediumism"—the set of effects generated in the presence of a medium.) While there are still Spiritualists today, the
heyday of the movement was in the latter half of the nineteenth century, when
it crystallized a panoply of concerns about both science and faith.[2] In Russia in
particular, where the Great Reforms had vigorously initiated debates over the
place of scientific expertise in a modernizing state, Spiritualism became the
center of a controversy about the status of religion, science, and superstition.[3]

A central episode in the history of Russian Spiritualism served as a micro-
cosm of the concerns about science's relation to the disjointed society of the
Great Reforms: the Commission for the Investigation of Mediumistic Phenomena. The Commission was set up in May 1875 at Mendeleev's instigation
by the Russian Physical Society—a newly created sibling to the Russian
Chemical Society.[4] So far, Mendeleev had negotiated his worldview of natural
laws and their rightful interpreters (university-trained scientists) with a community of like-minded peers at St. Petersburg University, the Russian Chemical Society, and the Technical Society. Spiritualism challenged this network.
Supporters of mediumism did not contest the processes of science as the
proper means to inquire into the external world; they disputed, rather, the exclusive right of Mendeleev and his peers to determine the laws of nature.
While the question of whether Spiritualism was *scientifically* defensible would
be a central focus, this investigation was simultaneously part of a broader *sociological* dispute. In response, Mendeleev established his Commission to move
this debate to the more congenial (to him) court of scientific societies. These
scientific societies, however, failed to safeguard his position, which foundered
on procedural, methodological, and social disputes.

The standard story is as follows: Mendeleev was horrified that even some
scientists were swayed by Spiritualism and so mobilized a Commission to
prove that mediums were frauds; his Commission exposed them and successfully eradicated Spiritualism.[5] This account rests on two incorrect assumptions. First, contemporary writings make it clear that the Commission did not
focus on *scientists* as deluded victims of hoaxes, but on members of the *nobility*,
which will bring us to central questions about their place in Great Reforms
Russia. Second, Mendeleev's motivation had little to do with *convincing* Spiritualists to change their minds; he did not care about their beliefs as long as he

and his colleagues at the Russian Physical Society remained the *legitimate* interpreters of the natural world for the Petersburg public. Keeping these two points in mind leads to an engrossing story of how an ambitious scientist could cut a dashing figure on the public stage of Petersburg. Throughout this story, one finds a culture vibrantly engaged with issues of elite and popular discourse in science. Conservatives and liberals, clerics and scientists, nobles and literary hacks all felt that the fate of the Russian Empire might be resolved not at the highest levels of government, but around a seance table, listening to obscure tappings in the twilight.

MADE IN AMERICA, REMADE IN RUSSIA: THE TRANSFER OF SPIRITUALISM

Spiritualism was born of humble origins in upstate New York in March 1848, when the younger Fox sisters, Katie (aged 12) and Margaret (15), evoked raps that they claimed were communications from the beyond. In 1851 a Buffalo University team concluded that the girls were frauds who produced the noises by snapping their toe or knee joints, and a relative admitted that the sisters had confessed their duplicity. Nevertheless, seances with members of the rapidly burgeoning population of "mediums"—individuals whose excess of nervous psychic energy enabled them to straddle two worlds—quickly exploded into a popular religious movement. Organized religion responded vigorously, shifting the new creed toward both mysticism and heavy involvement with progressive reform movements.[6]

A less overtly religious variant, which cribbed its metaphysical framework from the empiricist materialist tracts abundant at midcentury, emerged in America but made its greatest strides in England. Spiritualism proper was brought to England in 1852 by Mrs. Hayden, an American medium, and very quickly cross-fertilized with the Mesmerist movement, itself adapted from late eighteenth-century France.[7] Spiritualism in England took root much more slowly than the wildfire pace of the American scene, but the English strain eventually dominated the rest of Europe, as British mediums developed a reputation for authenticity divorced from the commercialism of America. English Spiritualism's success was also aided by the perceived contemporary decline of religious values. As science and modernity were thought to be eroding the traditional moral framework of the British Empire, to some Spiritualism offered a healthy compromise balancing the yearning for faith with Victorian naturalism.[8]

Its reputation as an empirical basis for faith was accentuated by several prominent Victorian scientists' vocal advocacy of Spiritualism. William Crookes, one of the most well-known chemists of his time, began studying in 1870 psychical phenomena attributed to the medium Daniel Dunglas Home.

Home (pronounced "Hume"), a Scot, was easily the most visible figure on the global Spiritualist scene. Home's return from his first visit to Russia in 1859 helped reestablish Spiritualism as a popular movement in England.[9] Crookes employed his usual experimental practices and a host of controls hoping to prove that Home was a fraud, yet ended up declaring him genuine. Since it was fairly difficult to dismiss this Fellow of the Royal Society as a crank, Spiritualists gained a respected authority to cite. The number of mediums grew, and although in 1874 Crookes withdrew his public endorsement of Spiritualist phenomena (he continued to believe and practice Spiritualism privately), supporters and opponents of the movement repeatedly invoked or dismissed his otherwise untarnished reputation.[10]

According to contemporary empiricist tracts, science was supposed to be about believing one's senses, and since Spiritualism produced verifiable effects, it should be considered genuine. Likewise, because the scientific process demanded that people trust scientific authorities who based their work on proper method, Crookes's investigations gave Spiritualists further support. Those who believed Spiritualism was merely a superstition focused on this shaky ground of authority versus the scientific method, and attempted to invoke science while ignoring these Spiritualist scientists. This specific issue became particularly sensitive in Russia.

While on a tour of Europe, Home proposed to a Russian noblewoman. He came to Russia a second time in the early 1870s to perform the ceremony, but this was blocked by Russian authorities on the grounds of religious differences, and Home was so wracked with anxiety that he was "not in power" and could perform no successful seances. He elicited support from England. Crookes wrote a reference letter to St. Petersburg University chemist Aleksandr Mikhailovich Butlerov, a relative by marriage of Home's fiancée, dated 13 April 1871: "As far as Mr. Home's character is concerned, I thoroughly believe in his uprightness and honour; I consider him incapable of practising deception or meanness."[11] Soon afterward, Home received notice from his dead mother in a dream that he would soon be "in power" again. He was then granted (through Butlerov and another future kinsman, Aleksandr Nikolaevich Aksakov) an audience with Tsar Alexander II at his suburban residence at Peterhof. The seance was a success, satisfying the Tsar's interest in the occult, and Home continued experiments in Spiritualism during his extended stay in Butlerov's apartment in Petersburg.[12] Just as Crookes proved useful to establishing Home's credibility in Russia, Spiritualists continued to invoke the names of other scientist-believers—engineer Cromwell Varley, evolutionary theorist Alfred Russel Wallace, and physicist Oliver Lodge, to name the most prominent—and the reputations of those scientists similarly became battlegrounds over the issue of authority.[13] (Even Friedrich Engels found these scientific Spiritualists important enough to lambaste.)[14]

Although the endorsements of these English scientists were important in Russia, Spiritualism owed its rise in popularity there to a nobleman: Aleksandr Aksakov almost single-handedly made Spiritualism a salient feature of cultural life in the northern capital. As American Spiritualist Hudson Tuttle declared in 1881: "We have noble and devoted Spiritualists in America, but none who can exceed [Aksakov]. He has counted rank and position as nothing, and without a thought, has sacrificed his health, feeling more than repaid, if the cause he loved prospered, and bestowed on others the happiness he had found."[15] Aksakov was a member of a very significant Russian cultural family—kin to several leading thinkers in the contemporary Slavophile and Pan-Slavist movements. He first became interested in Spiritualism by reading the works of Emanuel Swedenborg and Andrew Jackson Davis, both of whom he translated into Russian (although he was denied permission to publish the latter). Having become deeply interested in psychical phenomena, he enrolled as a free student in the Faculty of Medicine at the University of Moscow in 1855 and studied physiology, physics, chemistry, and anatomy. In 1874 he published the first issue of a Leipzig journal, *Psychische Studien,* and he later wrote a magnum opus for Spiritualists, *Animism and Spiritualism,* an elaborate response to a scathing critique of the movement.[16] Even Mendeleev owned a copy of an early Aksakov text, *Spiritualism and Science,* on the cover page of which he crossed out the word "science" and underlined "spiritualism" twice with blue and orange crayons. This book was a defense of Spiritualism against what Aksakov perceived as the "despotism of public opinion and the evidence of ignorance," by which the Russian press refused to discuss mediumistic phenomena. It contained translations of Crookes's work, among others'.[17]

Aksakov's greatest coup in increasing the popularity of Spiritualism was to persuade his cousin by marriage, A. M. Butlerov, Mendeleev's University colleague and perhaps the country's most respected chemist, of the reality of mediumistic phenomena in 1870–1871, during Home's second trip to Russia. It was not an easy task, as Butlerov was highly skeptical of all occult phenomena. He had first encountered (and dismissed) Spiritualism at the Aksakovs' house in the Abramtsevo suburb of Moscow in 1854. He returned to the phenomena at public Spiritualist demonstrations at Nice in 1868, and would eventually write quite openly about his experiments with mediums and their results.[18] Butlerov even tried and failed in 1871 to establish a commission to study Home at St. Petersburg University. Local chemists, including Butlerov's friend Friedrich Beilstein of the Technological Institute, reported to Emil Erlenmeyer in Heidelberg on 14 April 1871 that "Butlerov has become the target of mockery of even hackney coachmen. The devil-summoner Hume [*sic*] has converted him and B. believes not only everything that Hume gabs to him, but he has even publicly defended Hume!!!! Then Hume experimented before a gathering of professors at the University (including Butlerov), and of course it fell through pitifully."[19]

Butlerov is often perceived as a Russian analog to Crookes: a prominent chemist who was "deluded" into supporting Spiritualism.[20] He was born into a noble family outside the provincial city of Kazan, and eventually enrolled at the local university. There he quickly displayed his talents in the physical sciences and was appointed to a teaching position in chemistry upon graduation. After his first trip abroad in the late 1850s, Butlerov served as rector at the University of Kazan, lobbying for the development of graduate studies in science, and eventually forming a core of the first professionalized chemists in Russia. He moved to St. Petersburg in 1868 and was soon elected to the Academy of Sciences. As professor of organic chemistry in the capital, he continued his laboratory work, founding a vibrant research school.[21] This is the Butlerov most historians and scientists were comfortable with, then and now.

On the few occasions when biographers address Butlerov's Spiritualism, they treat it as a religious mania that must be separated from his scientific credentials. As usual with scientists with qualities deemed unsavory—like Isaac Newton's alchemy—Butlerov is portrayed as having a split personality: His "rational" self was the great chemist, his "irrational" self was absorbed by the occult, and never the twain did meet.[22] On the contrary, Butlerov repeatedly attempted to find scientific explanations for the events that he observed at seances, and he understood his adherence to Spiritualism to be manifestly consistent with his science: He had followed empirical methods at seances, and thus could not deny his observations.

It was precisely the *scientific* in Aksakov and Butlerov's Spiritualism that bothered Mendeleev; the occult he could otherwise ignore. Butlerov's students also saw their mentor's Spiritualism as an example of his integrity. For example, shocked by the erasure of Spiritualism already occurring upon his teacher's death in 1886, Moscow chemist V. V. Markovnikov singled it out in his funeral oration:

> We cannot silently pass by yet another side of his convictions and activity which had a paramount significance in his life. To be silent about this would mean to waive an essential portion of our respect for him. We can fail to share various convictions of a man, but we do not have the right to ignore them during a characterization of his personality. I wish to speak of the relation of Al. M. [Butlerov] to mediumism. . . .
>
> Who does not remember how in the seventies almost the entire Russian press leapt on him, and he became the laughingstock of all the petty press, who did not mince words in expressing their disdain for a person whose name stood so high in science? In order to explain this—in their opinion—contradiction of the mixture in one person of a famous naturalist and a Spir-

itualist, they were ready to recognize in him a certain moral derangement. Can one imagine how the one upon whom all these attacks were falling must have felt? Butlerov with complete reserve listened to all these mockeries and insinuations; he had expected them in advance and so calmly answered the bold attacks of his more serious opponents with his characteristic nobility of tone of the true gentleman. Against their suppositions he established facts, citing as justification names of great scientific authorities.[23]

Privately, however, Markovnikov had written to Butlerov: "Despite my toleration of Spiritualist beliefs, I look on each who exchanges his usefulness not only in science but also in any other activity for an occupation with Spiritualism with a regret close to disdain."[24]

Butlerov is significant here for two reasons: He is the dominant foil in the standard story of Mendeleev's attacks on Spiritualism; and he also figures centrally in Mendeleev's life later. However, too much emphasis on Butlerov's role in Russian Spiritualism is misleading, for it was Aksakov, not Butlerov, who served as Russia's primary advocate of Spiritualism, and thus was the primary target of Mendeleev's Commission. In addition, Butlerov was not the only Spiritualist scientist in Petersburg. Most prominently, he was joined by a University colleague and friend, zoologist Nikolai Petrovich Vagner.

Vagner often gets short shrift. He is most widely known to scholars of Russia by his pseudonym, the famous children's book writer "Kitty Cat" (*Kot Murlyka*), although he also developed a substantial reputation in entomology. He moved to St. Petersburg in 1870, taking a chair in biology in the same faculty as both Mendeleev and Butlerov. He soon became close to Mendeleev, and he and Butlerov had been dear friends since their graduate student days in Kazan. Vagner was initially just as hostile to Spiritualism as Butlerov, and he considered Butlerov's adherence to it one of the greatest shocks of his own move to the capital. Butlerov, however, persuaded him to attend some seances with Home, and by 1874 Vagner was convinced. With his literary flair, Vagner, and not Butlerov, began the "Spiritualist season" of 1875–1876.

SPIRITUALISM IN 1875: TENUOUS COOPERATION

The April 1875 issue of the *Messenger of Europe* (*Vestnik Evropy*), a popular literary and political journal, greeted its readers with an article by Vagner. The article, rather innocuously entitled "A Letter to the Editor: Concerning Spiritualism," was not a particularly virulent manifesto, but it sparked controversy because of the identity of its author. Anticipating the sensation it would cause, the journal's editor, M. M. Stasiulevich, appended an exculpatory footnote: "The name of the author and his chosen subject, which has drawn to itself recently, in all events, society's attention, and not just in our society, compels us to satisfy the desire expressed by distinguished professor of our local university N. P. Vagner to communicate to our readers his involvement with it."[25] Stasiulevich unwittingly touched upon the key elements of the ensuing year of tumultuous debate: the University, the author's status, society's attention, and the right to communicate to readers.

Vagner's article started where most such works do—with the *Hamlet* quotation that "there are more things in heaven and earth, Horatio,/ Than art dreamt of in your philosophy"—and then told the reader that occasionally in science one found evidence for phenomena that were previously believed impossible, as in his own discoveries with insects. Vagner had first encountered Daniel Home at Butlerov's apartment in 1871; he characterized the medium as a nervous somnambulist with an eager-to-please temperament. Vagner engaged Home with elaborate experiments, described them at length, and concluded:

> From all the observed phenomena I drew one clear, incontrovertible conviction: the table's movement and the knocks actually exist. They are phenomena which are purely real, objective, probably belonging on one side to physics and on the other to psychic phenomena. But there follows further, it seemed to me, another side of these phenomena. With a particular mood of all present and especially of the medium, who offers something on the order of a tuning fork in the meeting circle, these phenomena transition imperceptibly into the subjective, into the area of hallucinations, panic, psychiatry. This is why the mystical element and those strange knocks attributed to spirits play so strongly in these phenomena.[26]

Vagner's methodological position here was not strikingly different from Mendeleev's in his gas work. First the scientist would try to distinguish what was specific to the phenomena at hand, classify it, and then try to use the instrument (in this case, Home) to discern natural laws.

Vagner's mention of Butlerov in his April piece meant that the chemist was fair game for attacks by public intellectuals. Not that Butlerov had been hiding his Spiritualism before, but now that Spiritualism had been launched into the gladiator's arena, he acquired new appeal as a target. Butlerov's first piece on Spiritualism appeared six months later in a different journal, the *Russian Messenger* (*Russkii Vestnik*).[27] Here he emphasized the hostility he had encountered in trying to publish his work, particularly from most journal and newspaper editors, who had blackballed his Spiritualism articles. Nevertheless, Butlerov contended that the phenomena were real and should be treated seriously, as they were in Western Europe:

I will leave entirely aside the question of whether mediumistic experiences are harmful or not; I do not invite to them people who are unreliable, superstitious, or inclined to mysticism; I eagerly admit that they can be forged under the hands of charlatans and used as a tool for the exploitation of the gullible. For me it is enough that what I saw and described presents, in my extremely sincere conviction, real, unforged phenomena of nature. . . .[28]

This emphasis on the primacy of "fact" would form the cornerstone of Butlerov's writings on the occult. The rest of his article was typical: He stressed how many competent (especially British) scientists attested to the phenomena, and then gave a brief primer on scientific method. Butlerov cautioned that he would only respond to criticisms that were conducted in a scientific manner.[29] Vagner, unlike Butlerov, harbored no scruples about fighting in the dirt. He published a longer piece in October 1875, slamming satirists who opposed the "array of serious scientists" who held to Spiritualism.[30]

The opposition was fierce. As a general rule, all participants in the debates over Spiritualism claimed that they were underdogs who had to fight against dominant public and elite opinion. Just as Vagner argued that he was a lone voice for Spiritualism against a public unwilling to listen, so his interlocutors claimed they were only trying to halt the popular tide of Spiritualism unleashed by the zoologist. As a result, it is hard to determine whether Spiritualism was in fact on the rise or being brutally repressed by the popular press and the Russian Orthodox Church. The responses to Vagner spilled over into the *Russian Messenger,* which soon became the primary vehicle for debate on Spiritualism.[31] Some critics expanded their criticism into social commentary, blaming Spiritualism on the chaos following the Great Reforms. According to this line, the partial opening of freedom of opinion allowed intellectuals' minds to wander, but since they were not given the freedom to fulfill their dreams, they displaced their aspirations onto ethereal spirits.[32] Yet the scientific community essentially remained silent. One author prayed for intervention, saying that silence would cause "irreparable damage," and concluded that "I know that my protest is not authoritative enough: its only goal is to turn attention to this subject of people more competent than I."[33] A call for competence? As if in response, the *Messenger of Europe* announced the formation of a Commission to investigate the phenomena at the Russian Physical Society.[34]

This was Mendeleev's idea. He was convinced that the scientific society was the proper forum for resolving the problems of science. Repeatedly at the beginning of the debates over Spiritualism, and less frequently as the year wore on, Mendeleev would efface himself and defer everything to the impartiality of a *conglomeration* of scientists. The Commission's public image hinged on two scholarly institutions: the Russian Physical Society (founded 11 March 1872), which would fuse with the Russian Chemical Society in 1878; and St. Petersburg University, which employed most of the Commission's members and served as home to both the Physical and Chemical Societies.

Mendeleev proposed the Commission at a meeting of the Physical Society—of which he was a founding member—on 5 May 1875. His initial comments displayed a predisposition against Vagner's claims: "It seems that the time has come to turn attention to the proliferation of practices of so-called Spiritualist or mediumistic phenomena, both in family circles and among certain scientists." Spiritualism "threatens the proliferation of mysticism, which can tear many away from a healthy view of the subjects [of science] and can increase superstition. . . ." Mendeleev wanted the Physical Society to differentiate between any real phenomena and the residue of "daydreams and hallucinations." "This means of investigation is only attainable by a scientific society," he insisted. "Then, at least, the argument of the Spiritualists, which attracts many adepts, that these phenomena frighten scientists by their novelty, will be removed."[35] The panel of members, mostly lower-level academics or laboratory assistants, consisted of: I. I. Borgman, N. P. Bulygin, N. A. Gezekhus, N. G. Egorov, A. S. Elenev, S. I. Kovalevskii, K. D. Kraevich, F. F. Petrushevskii, P. P. Fan-der-Flit, A. I. Khmolovskii, F. F. Eval'd, and Mendeleev himself.[36] In practice, only a small subset of members conducted most of the Commission's activities.

None of the members were Spiritualists or had more than passing parlorroom experience with the phenomena. If the Commission wanted to investigate the phenomena credibly, it needed to gain familiarity quickly. At its first meeting, the Commission decided to invite Aksakov, Butlerov, and Vagner as consultants, so that they could recommend literature and introduce members to the most typical phenomena. At the second meeting, on 9 May, all three attended, but Aksakov did most of the talking. (In general, when the "witnesses from the side of the mediums," as the Spiritualist participants were called, spoke to the Commission, it was through Aksakov.) The term of action was to be September 1875 to May 1876, with no less than one session a week in the presence of a medium for a total of at least forty sessions; minutes were to be written immediately after seances.[37] Both the final deadline and control of the minutes passed here without comment. Aksakov advertised for mediums locally, but soon had to turn to English professional mediums. Apparently, no local mediums were deemed "effective" enough or wanted to subject themselves to scrutiny. No sessions took place until late in October, a full five months later.

Mendeleev spent the intervening summer dabbling in Spiritualism. On 19 May, he penned in his personal notebook—chock-full of gas experiments—that "N. P. Vagner told me that a new medium has been found—a noble girl 'from whom the table runs like a dog.' I was unable to accept the invitation—

I was traveling to the country." On 27 May, however, he sat from 9 to 12:30 in the evening in Vagner's apartment, with no results. Vagner "heard a knock which no one else heard," Mendeleev noted laconically. On 7 June Mendeleev hosted a seance with Gezekhus from the Commission, Gutkovskaia and Kapustin from his gas lab, and a friend, Liubov' Andreevna Kuritskaia, who supposedly possessed mediumistic abilities. Nothing happened. On 1 July he attended another seance at Kuritskaia's summer home, distractedly peppering his notes with meteorological scribblings.[38] Vagner himself wrote Mendeleev that Kuritskaia was not a good enough candidate for the Commission's investigations. (She fidgeted too much.)[39] And so Mendeleev had to wait for the Commission to reconvene in October.

At the third meeting, on 27 October, Aksakov explained that he had been unable to find any local mediums, and was forced import two at his own expense: William and Joseph Petty of Newcastle, England, aged 17 and 13, respectively, who arrived chaperoned by their father.[40] While he was certain that the boys were mediums, he warned that they were not the most powerful. Biweekly sessions would be held on Tuesday and Thursday from 7 to 10 in the evening in Mendeleev's apartment. The Commission also agreed to two seances done entirely under the mediums' conditions and without controls, in order to increase the likelihood that the capricious phenomena would emerge. No minutes were taken.[41]

At the sixth meeting on 11 November, F. F. Petrushevskii, professor of physics at St. Petersburg University, presided. Mendeleev did not officiate at these sessions, either as secretary or president, and he did not even attend the early seances with the Pettys, since the mediums restricted the number of attendees. At this session, the first potentially paranormal phenomenon appeared: Two hours into the seance, liquid materialized on Borgman's arms, the table, and a piece of paper. The Pettys claimed the drops were mediumistic. Chemical tests appeared to match the drops to the saliva of the younger medium. They appeared only on the side of a piece of paper facing the medium and never the opposite side, and when his mouth was covered no liquid appeared. The Commission concluded that a phenomenon that only occurred absent obvious countermeasures against fraud must be fraudulent. Here we see the first signs that the ethos of cooperation between Spiritualists and members of the Commission was breaking down. Butlerov refused to sign the minutes and instead appended a rider stating that the identification of the drops with saliva was premature.[42] Always more sensational, Vagner went to the press, writing a letter to the editor of the *St. Petersburg News:*

> The liquid whose drops appeared on the paper actually displays the chemical reaction of saliva; but under a microscope this liquid appears somewhat different from ordinary saliva taken directly from the mouth of the medium or

of saliva drops sprayed onto paper. . . . From [my additional] experiments I consider myself in my rights in concluding that this phenomenon which is being researched by the Commission is not a trick but an authentic mediumistic phenomenon, and I sincerely regret that the members of the Commission were not able to convince themselves of this.[43]

A subsequent seance clinched the Pettys' emergent status as hoaxsters. This seance, the ninth of the Commission, on 20 November, was the first step toward making Mendeleev a household name in Petersburg. Romanticizations of it are central components for constructing Mendeleev as the cantankerous individualist who would stand for no pussyfooting. Between the saliva session and the famous session of 20 November, there were two additional seances, in which the Pettys introduced what they called a "face seance." Here, the two mediums sat facing seated participants. There would be a curtain behind the mediums, with objects behind the curtain that the mediums would manipulate using mediumistic forces. During the face seance on 18 November, the members of the Commission tried to institute various precautions against fraud. In a footnote to the minutes, Mendeleev wrote of the mediums' resistance: "in my opinion it is most likely that it was expected that the participants in the darkness would forget to observe, would make up their minds according to the strangeness, scariness, and mysteriousness [of the event] and thus would achieve a satisfactory degree of that state of mind which is necessary in order to fool people."[44] When Mendeleev attended his first Commission seance on 20 November, he would make no such mistake.

Aksakov began the evening by announcing that if the Pettys failed to perform in the next two or three seances, he would return them to England. At 7:50 P.M. a table seance began with a bell in a cage on a table, and nothing happened except for more trances and convulsions on the part of the mediums. At 8:20 the seance was closed. At 9:45 they began another face seance, with attendees (seated from left to right) Aksakov, Borgman, Elenev, Khmolovskii, Kraevich, Lachinov, Bulygin, Gezekhus, and Fan-der-Flit a bit farther away. Mendeleev sat in the back "at a distance of about four meters from the curtain—and controlled the music box and also the light of a small lamp, set on the floor behind the furniture."[45] The mediums covered their heads with white handkerchiefs so they would be visible in the darkness. The light was lowered, and during a trance one of the brothers requested a few things, including that Mendeleev join the circle, where he sat between Lachinov and Bulygin. The sound of tearing paper was heard, and the older medium said this was a mediumistic phenomenon. About fifty minutes into the seance Mendeleev did something unexpected.

He lit a match, which burned for about two seconds.[46] The mediums were irate, and demanded an explanation. Aksakov laconically added: "This is not good." Mendeleev explained that he thought the elder medium was stooping, and he wished to see what was going on. The mediums insisted that if no one else lit a match, the bell behind the curtain would ring, and Mendeleev promised not to do it again. Five minutes later the members heard a moving chair and the fall of a body, and the white cloth fell from the elder medium's head as he went into a series of convulsions. Mendeleev offered to call a doctor, and the medium returned to his chair. The seance ended at 10:50 P.M. It was generally agreed that nothing mediumistic happened, but ascertaining exactly what *did* happen was somewhat more complicated.

Upon inspection of the room after the lights were raised, the commissioners found a large tear in the curtain, and they asked whether either of the mediums had a knife, but they both declined to answer.[47] After Aksakov and the mediums left, those present at the seance congregated and discussed what they had seen. Given that the match was only lit for two seconds and caught all of the members except Mendeleev by surprise, there was understandable disagreement. All of the accounts were reproduced in the minutes, and their incoherence is striking, especially given that this event was soon considered conclusive proof of the Pettys' fraudulent behavior. Five members claimed they saw the elder medium fiddling with the curtain. None of the accounts offered an evidenced explanation for the tear. Mendeleev would later claim in a footnote to these minutes that what he *thought* happened was that the elder medium had hoped to make a hole just large enough to grab the bell, but the match startled him into making a bigger tear, and when he could not fix it he fell to the floor to try to blame the tear on his convulsions or a mediumistic force. In her memoirs, Mendeleev's daughter Ol'ga made much of this incident that her father supposedly told her about the next day over breakfast.[48]

Aksakov's view of this crucial moment, written immediately upon publication of the minutes in 1876, was revealing: "The tear of the curtain occurred as a consequence of the fact that Mr. Mendeleev destroyed the conditions of the experiment and lit a match at the same time that the mediums were in a trance. The elder of them was taken with spasms and fell from his chair to the floor."[49] He had a point. Mendeleev's act was, after all, an *uncontrolled* act in *explicit violation* of the agreed-upon experimental conditions. If an analogous transgression had taken place in his gas laboratory, Mendeleev most certainly would have been enraged. The match incident was flagrantly at odds with the stated goal of the Commission, which was to investigate the phenomena in a scientific manner. Butlerov and Aksakov began to doubt Mendeleev's protestations of objectivity.

The tenth meeting of the Commission took place the next day and did not assuage the fears of the Spiritualists. Aksakov and Butlerov were not present, and Kraevich, as chair, issued the following statement on the Pettys:

Keeping in mind that in all cases when precautions were observed, none of
the so-called mediumistic phenomena occurred in the presence of the Petty
mediums at meetings of the Commission and that, on the contrary, when the
mediums were left on their own without any control, such phenomena were
observed, the Commission comes to the conclusion that the Petty mediums
constantly tried to deceive it, and thus the Commission considers them
frauds.[50]

The Pettys were subsequently shipped back to England; Aksakov was due to
bring another medium over in January. His services were needed to continue
the work of the Commission, and this unilateral declaration was bound to give
him pause. Mendeleev's next transgression did not happen in the privacy of a
seance room.

On the contrary, it happened in an auditorium packed with representatives
of all strata of Petersburg society. On 15 December, during a hiatus in meetings, Mendeleev lectured the public on the Commission's activities and read
from the minutes, including the "damning" seance of 20 November. The lecture was Mendeleev's attempt to compromise between, on the one hand, demands from the public and Spiritualists for transparency in the Commission's
actions and, on the other, the desire of the investigators (and himself) to preserve its autonomy. The public's curiosity was already intense, and Mendeleev
felt that if he did not speak for the Commission, then the Spiritualists would
do it for him—as Vagner's and Aksakov's eagerness to rush into print over details of the seances had proven. As if to justify Mendeleev's concerns, Aksakov
wrote to the *St. Petersburg News* on 7 December 1875 that "in view of the rumors and criticisms about the results of its activities, I regret that Commission
does not find it necessary to put an end to them by a short statement." Aksakov then gave a brief synopsis of the Commission's doings.[51] Mendeleev, for
his part, wrote Aksakov that he had already tried to grant limited public access
at the Physical Society meeting of 2 December. "You, it seems to me," he continued, "do not approve very much of my early effort, but I did it because my
honored colleagues, Messrs. Vagner and Butlerov, do not wait for the end of
the Commission's affairs, but print and print in journals and newspapers, in
which effort you help them." If the Spiritualists would not be quiet, he wanted
to have his say. He assured Aksakov his "intentions were and are most peaceful
(what they will be further I don't guarantee)."[52]

Aksakov let the issue of the lecture pass but was dismayed when he heard
that the minutes would be read aloud to the public. As he wrote to Mendeleev
on 11 December, there were particular aspects of the crucial evening that
needed to be explained to an audience, or they might draw hasty conclusions.
He pointed out that "the lighting of a match during an observing seance, especially when the mediums were in a trance, was a violation of the conditions of

experiment," and he demanded that Mendeleev read a rider that presented alternative theories to the conclusion of fraud: "A public reading of one set of minutes without my present statement would be, as far as I understand, a one-sided presentation of the affair, and as I don't at all doubt that the Commission . . . wants to observe the strictest impartiality . . . then I am completely sure that you will not refuse to fulfill my modest request."[53] The request was denied; Mendeleev was not about to give Aksakov authority to speak on a par with a University professor. The public would hear Mendeleev's version alone.

Mendeleev packaged the event with an extraordinary talent for public relations. He charged admission and then donated the proceeds to the cause of Balkan Slavs who were suffering under Ottoman rule in the 1870s.[54] On 15 December, he lectured to the audience at Solianoi Gorodok, on the banks of the Fontanka River in Petersburg, earning 1,548.50 rubles for charity. The talk appears, in retrospect, remarkably even-handed. Spiritualists, in fact, criticized his remarks little (although they *were* upset about the minutes). Mendeleev began with a fairly general definition of Spiritualism, which he felt would avoid some of the reckless hypothesizing and frame the issue in a more objective manner. Making an analogy with weather, whereby "atmospheric phenomena" were not phenomena generated by a specific force, but merely phenomena that took place in the atmosphere, he defined spiritualist phenomena as "*those which occur at seances, happening usually in the evening, in darkness or twilight, in the presence of special persons who are called mediums; these phenomena have in their general characteristics an affinity with so-called hoaxes and thus present a character of mystery, unusualness, [and] impossibility under usual conditions*" (italics in original).[55]

The Commission, Mendeleev argued, had believers and skeptics working together to get to the truth: "The first group will teach the second what to do. They both should trust each other." An improper forum would be to let "the leaders of public opinion"—even when they had such distinguished names and reputations as Aksakov, Butlerov, and Vagner—direct the public as "apostles of Spiritualism" outside the venue of scientific societies.[56] In order to demonstrate the importance of proper forum, Mendeleev read from the minutes, effacing himself and letting the abstract "Commission" tell the public what had been found so far. He concluded by stressing that the Commission's work was incomplete and that no final conclusions could be drawn.

The talk received average reviews, which scoffed in particular at Mendeleev's notoriously idiosyncratic speaking style: "On Spiritualism or on organic compounds, one must speak, or at least read, publicly with a certain minimum of adroitness and skill. Both are notably absent in the manner of this respected scientist."[57] The reading of the minutes, however, was a resounding success. The Pettys were often invoked by the press to prove that mediumism was akin to charlatanism, directly citing Mendeleev's lecture as the capstone of scientific debunkings of Spiritualism since the days of Mesmerism in France.[58] But, however vigorous the statements became, Aksakov, Butlerov, and Vagner were not yet attacked ad hominem; the Pettys were the villains, not Spiritualism. That would change with the end of the year.

SPIRITUALISM IN 1876: A MELTDOWN OF METHOD

On New Year's Day, Vagner wished Mendeleev the best for his gas and aeronautics research. His note also touched on Spiritualism, and his tone—earlier one of relentless optimism even in the matter of saliva drops—now became fatalistic: "Ahead will be just the same endless and indefatigable battle, leading to no agreement and reconciliation. You will stand firmly on your point of view—we [Butlerov and I] will not abandon our point of view, and both sides will be equally right, since this battle may be the oldest in the world." He admitted his own bias, but he wanted Mendeleev to do the same: "If you had really wanted to convince yourself that mediumistic phenomena existed, then your form of action would have been entirely different. It wasn't necessary for that to gather a commission of scholarly physicists and mechanics. You are the authority and judge for yourself." Where Mendeleev stressed the collective reasoning of the scientific society, Vagner emphasized *individual* judgments. He continued with the Commission only to have a hand in editing the minutes.[59]

This change of tone was not initiated by Vagner; if any single cause was responsible, it was Mendeleev's match-lighting in late November, or the popular reaction to his public lecture. The tension eventually reached such a high pitch that the subtle interplay of Spiritualists and skeptics broke down in a chaos of mutual recrimination. Not only would the Commission be destroyed, but Mendeleev's relations with Butlerov soured and Vagner refused to forgive his colleague and former friend. Discussions with Aksakov, never cordial, cooled to arctic levels.

The collapse began in earnest with the arrival of a new medium in January 1876. Unable to get the famous Mr. Monck from Bristol, Aksakov obtained the services of Madame Claire, a medium who had worked with William Crookes. Claire, like the Pettys, stayed at Aksakov's expense as his personal guest. At the eleventh meeting of the Commission on 11 January, a seance was begun at 8 P.M. Sitting around a three-legged table, Mendeleev and the other participants saw the table tilt and produce knocks. It then levitated 10 cm into the air and suddenly crashed to the ground. Mendeleev tried and failed to introduce his newly designed manometric table—which could measure both the magnitude and direction of pressure exerted on its surface.[60] A new ordinary table was brought in, and the seance resumed.

After such an auspicious beginning, things were looking up for the Spiritualists. At the next session of 15 January, however, Claire was not feeling well (apparently from drinking the local water), and Aksakov canceled the meeting. By the time Petrushevskii received the news, the members had already gathered, and this session was counted as one of the scheduled forty.[61] A *procedural* dispute now emerged, one of the three causes of the Commission's collapse. The Commission had, upon founding, established a deadline that was simultaneously specific and ambiguous. According to the reading endorsed by Mendeleev and the rest of the Commission, the binding deadline was the May closing date, since that accorded with the academic year and would mark a full year of investigations. If forty seances could be concluded in that time, all the better, but that was *Aksakov's* responsibility as procurer of mediums. Aksakov, in contrast, felt that forty seances would be sufficient to investigate the phenomena. At the current rate, however, they could not complete this many by May, and as early as November he had argued for an extension. Including the canceled seance as one of the forty brought the issue to a head.[62] At the meeting on 25 January, attended by all three Spiritualists, there was also a dispute over who had a right to take notes: The Commission absolutely refused to let a Spiritualist control such a crucial role.[63]

A second major conflict was present from almost the outset of the investigations, but became particularly acute beginning with the fourteenth session on 27 January—a disagreement that cut to the heart of Mendeleev's cherished notion of scientists as the arbiters of natural laws. In question was the ability of the Commission to impose experimental instruments on the mediums, highlighting the *methodological* concern at the core of scientific investigations of Spiritualism. According to Spiritualists, mediumistic phenomena were inherently capricious, and one had to observe them repeatedly with naked senses in order to become convinced of their existence before exposing them to experimental tests that might destroy the phenomena. Imagine, for example, an experiment on an electric field that used a measuring probe that shielded such a field. Given everyone's lack of knowledge of the precise nature of Spiritualist phenomena, argued Aksakov and Butlerov, there was no way to know what would shield them until a less interventionist natural-historical account was developed first. This was not a rejection of scientific method, but rather an *alternative* method to Mendeleev's, derived as it was from the physical sciences.

Aksakov implored Mendeleev about this issue as late as 19 February in an attempt to salvage the Commission, writing that "the use of instruments to study phenomena comprises, as far as I understand it, the capstone of the affair, and not its beginning. If we had such devices, there would have been no need for the Commission."[64] Mendeleev was vehement about the need for experimental devices, and he contended that by leaving out such instruments, the Spiritualists made it virtually impossible to persuade anyone with scientific training. He used the example of the earth revolving around the sun—an established fact, but one that contradicted the manifest evidence of our senses. At the fifteenth session on 29 January, Aksakov and Vagner allowed some controls, but no instruments. The seance was unsuccessful, and the minutes became more judgmental.[65] This was the last active session, as Madame Claire was busy for the meeting of 2 February, was ill on the fifth, and was then withdrawn by Aksakov until she left the country on 4 March.[66]

The third major cause of the demise was *social*. Claire's arrival in St. Petersburg caused some anxiety for the Commission. Precisely at the moment when the Commission began to intensify the intrusiveness of their controls on the medium, the introduction of a female medium highlighted the issue of gentlemanly behavior.[67] The Commission certainly did not ease up when Claire came under their scrutiny: In a series of appendices to the minutes, several members related their vigor in imposing controls, easily crossing the bounds of civility. For example, Mendeleev kept a close watch on her hands and feet during seances, occasionally stepping on her dress to prevent her legs from moving and feeling her leg with his own to see if it was exerting pressure on the table. He also glanced under the table and saw what looked like a spring emerging from Claire's dress.[68] He wanted to search her, but his request was denied on the grounds that this would be an insufferable insult—whether to be searched by a man, or to be searched at all, Aksakov did not specify. Mendeleev repeatedly tapped the table himself or rapped his feet, and moved the table on his own to observe the medium's reactions, all actions that Aksakov would later cite as destructive of the conditions of impartial investigation. Other members of the Commission behaved similarly.[69]

It would have been extremely hard for the gentlemen of the Commission to conduct their investigations without exposure to female mediums. According to historian Alex Owen, Spiritualism provided women with an opportunity to quietly subvert the gender roles imposed by society. Passivity, the signature of the Victorian woman, was purportedly the character that made women excellent mediums; they were passively able to receive spirits from the other world. Spiritualists also became impassioned defenders of women's rights and a slew of other progressive causes, while women's mediumship was a liberating, if somewhat unorthodox, professional role for the Victorian woman.[70] This dual status of the female medium as "active" while in a trance, and hence able to act vigorously like a man, while still "passive" in essence, and hence subject to norms of courtesy, placed Mendeleev and his peers in a bind: They were suspicious of her as they were of any other medium, yet they were constrained in the face of public opposition to "indelicate" behavior toward the "fairer sex."[71]

Mendeleev was called out by several observers of these seances for "ungentlemanly" behavior. He received a chastising English letter on 19 January from a woman in Claire's entourage, and Mendeleev replied through Ak-sakov.[72] He insisted that he had not kept his hostility to Spiritualism a secret, as the entire public knew since at least December; and Claire knew why he was being so suspicious. In the end, "in scientific research the whole point is in achieving truth, and not in trivialities." He then reinterpreted the meaning of a "gentleman":

You have carelessly expressed yourself, calling my turn to such steps [experi-ments and controls] ungentlemanly, [which is] rather the direct road to truth, which I hold is in reality a gentlemanly affair. As it suits you to introduce to the question of Spiritualism and mediums a notion of honor, and not of truth, so I throw your accusation back at you, proposing that trust and honor are indivisible from truth.[73]

Later, Mendeleev would return to this issue of gentlemanly behavior, noting that his suspiciousness was "construed by many trustworthy Spiritualists re-peatedly as indelicacy to a lady." He published extracts from the above letter and added: "True honor is indistinguishable from truth; a gentleman acts to discover it, not put obstacles before it."[74]

Mendeleev here questioned the connection between being a "gentleman" and being trustworthy, a connection considered by historian of science Steven Shapin as essential for the grounding of experimental science in early modern England.[75] According to Shapin's analysis, all knowledge claims eventually come up against a problem of trust: One cannot possibly inspect for oneself everything one hears, so one has to establish a protocol that lets one trust oth-ers' judgments. In early modern England, this was the crux of the problem of the epistemological validity of experimental science. The solution, crafted pri-marily by the Royal Society and its symbolic center, Robert Boyle, was "virtual witnessing," a set of literary techniques that simulated actual presence at the experiment, bolstered by the *cultural* credibility of the gentleman, who was al-most by definition incapable of lying. (Gentlemen, to be sure, did lie, but the cultural *ideal* of the gentleman precluded such perfidy.) In other words, the production of knowledge depended on social and cultural standards of credi-bility. Toward the end of the Enlightenment, supposedly, the epistemological validity of the experimental method was widespread enough that it no longer needed to rely on "gentlemen" reporting the results.

But just like its modern-day descendent, science in 1876 could not escape the issue of *trust*. The problem for doubting scientists, and part of the reason why Spiritualism spread so quickly among the educated, was that a great number of *trustworthy* individuals, both scientists like Crookes and noblemen like Aksakov, swore that the phenomena were real.[76] This problem became particularly acute in Russia as the nobility, traditionally the bastion of both wealth and education, lost much of its standing following the Great Reforms. Emancipation, conjoined with urbanization, began a process of gradual impoverishment of the nobility, which tried to shift its strengths from *social* supremacy to *cultural* supremacy.[77] In other words, Aksakov's Spiritualist activities were, in part, an attempt to create a newly reinvigorated public role for noblemen. He (and to a lesser extent fellow nobleman Butlerov) attempted to import the English cultural category of "gentleman" onto Russian soil to address many of the same issues about the Great Reforms that troubled Mendeleev. Aksakov, however, was trying to carve out a zone in which the nobility could still serve as the *Kulturträger* ("culture bearers") of Petersburg— and "culture" included knowledge of the natural world. He was trustworthy, and he believed Spiritualism existed, and why should anyone deny his word without a *very* good reason?

Mendeleev worked hard to defuse this aspect of Spiritualism. Instead of arguing that Aksakov was untrustworthy—a tack taken by many journalists— Mendeleev offered a vision of a culture ordered by meritocratic science, a culture in which *trust* was no longer a necessary basis for making decisions about the world. During a dispute at the fourteenth session of the Commission, Mendeleev had told Aksakov and Butlerov that "it is impossible to base research into phenomena on trust in people; doubt is mandated here, because the conditions of the experiment allow the possibility of deception."[78] Mendeleev combined the worry about gentlemanliness with the concern over method, arguing that this linkage solved the Spiritualists' concerns about the Committee: Use of a manometric table obviated the need to trust in social status.

The Spiritualist consultants remained unpersuaded, and since they felt betrayed by the dispute over procedural deadlines, deadlocked over the methodological controversy, and insulted by the lack of "gentlemanliness," they withdrew their cooperation. Aksakov wrote a letter to the Commission on 4 March detailing his objections to further cooperation, and simultaneously sent it to the *St. Petersburg News*. He listed a series of violations by the Commission of its own rules, including delays in writing the minutes and the appendage of individual statements of the minutes. In the end, this nobleman claimed that the Commission's very *elitism* doomed it:

> To confirm a phenomenon is easy, to study it very difficult. Thousands of people confirm that mediumistic phenomena exist; the duty of the Commission, if it had grasped the public question, was to lower itself to the crowd and first see what the crowd sees and see how the crowd sees, in order that then, with knowledge of the external side of the matter, [it could] put forth corresponding tests.[79]

In this letter Aksakov began to paint Mendeleev as the Commission's *éminence grise,* an image that Mendeleev would perpetuate. Vagner and Butlerov sent in similar resignations.[80]

Aksakov's withdrawal precluded the possibility of procuring further mediums, so the Commission began to wrap up its investigations.[81] On 16 March they revised the final conclusion of the seances in a meeting in Kraevich's apartment, and on 21 March the Commission finalized its concluding statement.[82] Mendeleev and his peers believed they had defended the boundaries of legitimate science against incursion by reasserting law and method. Communicating their victory came next.

PUBLIC SPIRITED: SPINNING THE COMMISSION

The Commission published its official statement first in the widely circulating Petersburg daily *The Voice,* and then later in Mendeleev's publication on the Commission, *Materials for a Judgment about Spiritualism.* The statement was designed to encapsulate the Commission's history in easily digestible nuggets for the casual newspaper reader, and was *also* supposed to function as a scientific document. As a result, it possessed a hybrid nature that drew on both the sensationalism of the daily press and the dry tone of scientific journals. In a series of numbered points, the basic conclusions about the Pettys and then Madame Claire—that neither exhibited authentic mediumistic phenomena—were outlined, and various contradictions in Aksakov's positions were implied. The Spiritualists were painted as opponents of proper method:

> The majority of followers [of Spiritualism] do not have either tolerance for the opinions of people who do not see in Spiritualism anything scientific or new, or a critical attitude to the subject of their beliefs, or a desire to study mediumistic phenomena with the help of research tools common in science. And meanwhile, Spiritualists with especial persistence disseminate their mystical views, passing them off as new scientific truths.[83]

The conclusion of the Commission, later widely cited, was simply worded: "On the basis of the totality that was discovered and seen, the members of the Commission unanimously arrived at the following conclusion: *Spiritualist phenomena occur from unconscious movements or conscious deception, and the Spiritualist doctrine is superstition.*"[84] Signatures of all participants with their qualifications followed.

Mendeleev granted *The Voice* and any other periodical carte blanche to reprint this piece as long as they included the entire document.[85] The responses to it by Butlerov and Aksakov, on the one hand, and Vagner, on the other, appeared in the same periodical shortly afterward. Aksakov and Butlerov's joint response is notable for its calm. They claimed they noticed the absence of a "normal attitude" to Spiritualism long ago, and that the printed statement of the Commission "fully justified what we had expected and what was pointed to in advance" in their resignations of 4 March. Instead of arguing with the "dogmatism" of the Commission, Aksakov and Butlerov chose to print the minutes of a successful *private* seance with Claire on 29 February, which used Mendeleev's manometric table (on loan from the chemist). They also urged the Commission to publish the minutes "*most quickly . . . with all* appendices," so that everyone could see that Claire actually produced mediumistic phenomena.[86] Aksakov and Butlerov opposed the Commission's improper exploration of Spiritualism by juxtaposing their own scientific seance.

Vagner's response took the opposite approach. Instead of trying to out-Commission the Commission in scientific rectitude, he painted Mendeleev as a manipulating despot who had orchestrated a theater of mirrors to elevate his own importance. In other words, Vagner tried to overturn Mendeleev's solution to the problem of trust by rejecting him as untrustworthy. He emphasized anything that could be understood as ungentlemanly, such as "destroying even a woman's good name," lying, or manipulation of evidence. After acknowledging that he and Butlerov did indeed try to convince the "court of the public" about Spiritualism with their initial publications, he asked rhetorically: "As if it were not to this same court to which the minutes of the meetings of the learned Commission turned?" Throughout, everyone had been courting public opinion; Vagner argued that Mendeleev was unworthy of it.[87]

It was precisely anxieties about Spiritualists' appeal to the "court of the public" that spurred Mendeleev to form the Commission and engage with Spiritualists in the first place. He appeared to be almost entirely indifferent as to what Butlerov and Vagner *believed* about Spiritualism. He was upset that they were *publishing* in journals, marring the public image of scientists. He was not worried that more Petersburgers would become Spiritualists, but that the two had subverted the appropriate forum for investigation: the scientific society. Mendeleev explained in reference to Vagner's *Messenger of Europe* piece:

It displeased me above all because it appeared in a literary journal, putting forward the name of my colleague into the newspaper arena and a matter worthy, in the terms of a scientist, of scientific investigation was exposed immediately not among scientists, but where exact scientific concepts are extracted or already formed, or where they are applied to public life. . . . In a word, I thought that N. P. Vagner did not address himself where he should have and forgot that we already have the possibility to turn with new scientific questions to scientific societies, the development of which in recent years in Russia characterizes our age.[88]

Mendeleev was upset because Spiritualists were not treating Spiritualism scientifically *enough* by making it a matter for public, and not expert, investigation and discussion.

Mendeleev continued his attack in a second public lecture, given in two parts on the evenings of 24–25 April 1876. Just as the tone of the Commission shifted after the new year, Mendeleev's tone now became much more unforgiving. The public lecture had been reinvigorated in post-Emancipation Russia, and Mendeleev took full advantage of it.[89] Advertisements were placed in the major local dailies, selling tickets for the 8 P.M. talks at prices that ranged from 50 kopecks to 5 rubles, fairly modest sums.[90] Aksakov tried to disrupt the first lecture when Mendeleev began to "expose" Madame Claire's fraud, but was prevented from speaking by both the crowd and the authorities, who considered his rebuttal to be unauthorized public discourse. Afterward, Mendeleev's stenographer reported that Aksakov held an impromptu seminar outside accusing Mendeleev of faking noises during Claire's seances (which Mendeleev admitted), and claiming that Mendeleev did not read the minutes of those seances because he knew they would undermine his cause. Mendeleev's next lecture proceeded without demonstrations from Spiritualists.[91]

The focus of Mendeleev's second talk was the particular ways in which Spiritualist mysticism operated through simple-minded manipulation of the rhetoric of "facts," a concept that he insisted Spiritualists understood incorrectly. In an effort to persuade the public, he argued, Spiritualists confused the evidence of common sense with "facts," appealing to a transparency in public knowledge belied by facts that contradicted superficial sensory data, such as the earth's revolution around the sun. Perhaps the most central comments concerned the ability of the scientist to speak *qua* scientist. He explicitly analogized the authority of the scientist with other public discourse:

> Science exists separately from scientists, it lives autonomously, it is the sum of knowledge worked out by the whole mass of scientists, similar to how the acknowledged political order of a country is worked out by the mass of persons who live in it. Science is authoritative, separate scientists are not. A scientist can only and should only use this authority when he is following science, just as in a well-ordered state the authority of power is used only by the person who observes the laws.[92]

The relation of science to free discourse here is interesting. Although science is reported through free discourse, it is *not* pursued through open discussions;

privacy was an essential element in the formation of public knowledge. Mendeleev dismissed Spiritualism as an "error of judgment" on Vagner and Butlerov's parts, but one that did not invalidate their legitimate scientific pursuits.[93] Overall, Mendeleev thought that "there will definitely be a benefit from our discussion of Spiritualism, because both sides speak and write about it freely: they will see the relation between science and scientists. . . ."[94]

For Mendeleev, then, the Spiritualism debate demonstrated to the public not just the active and vibrant role that science *properly pursued by professional scientists* could have for Russian culture, but also the success of the Great Reforms in opening the Russian landscape for free discourse. Free discourse meant that Mendeleev could speak his mind, citing his *professional* authority to battle the *cultural* authority of Aksakov's elevated noble status. Whatever Mendeleev's views on this matter, there was certainly not a level playing field for the Spiritualists. Despite demonstrable public interest in Spiritualism, it took the scientific reputations of Vagner and Butlerov to convince journal editors to publish their articles, the likes of which had been earlier excluded from the mainstream press. After Vagner's article was published in the *Messenger of Europe,* the editor, M. M. Stasiulevich, balked on publishing any more pro-Spiritualist material. The attacks on Vagner in that journal specifically targeted Butlerov as well, and the chemist wanted a chance to defend himself. Stasiulevich received Butlerov's letters but never answered them, refusing him space in the journal—an action Butlerov decried as censorship.[95] Only the *Russian Messenger* was willing to provide Spiritualists a hearing, an idiosyncrasy of its editor, Mikhail Katkov. Butlerov in an undated letter forcefully condemned Stasiulevich as working contrary to the Great Reforms ideal of "openness" (*glasnost*), and reiterated his belief that "the best path to truth is free investigation done in good conscience, upon which one lets both *pro* and *contra* speak equally loudly."[96] For Mendeleev, on the other hand, this freedom of discourse could only take place within a certain restricted elite who engaged with the public only *after* achieving consensus.

The Spiritualists received even more forceful rebukes from the Russian Orthodox Church, institutionalized in the ministry-level Holy Synod and controlled by the influential Konstantin Pobedonostsev.[97] In 1883, years after the Commission had closed shop, Butlerov proposed reading lectures of his own on Spiritualism. Pobedonostsev opposed Butlerov's plans long after the program was already printed in the papers, declaring:

Among our people and even in so-called educated society even without this there is a strong inclination towards the emergence of sects of various types and the dissemination of pseudo-religious doctrines and prejudices. Mr. Butlerov's public lectures could give this inclination new fuel. . . . Mr. Butlerov's lectures will awaken, doubtless, a newspaper polemic of a very seductive character. Devout people will come from these lectures and polemics into great temptation and will begin to grumble for the government to allow superstition in public display. The Orthodox Church considers it its duty and condemns in print from the church pulpit both these doctrines and society's attraction to them, and the mass of the public's curiosity to research mysterious phenomena. . . . [F]rom the Church's point of view the phenomena of so-called mediumism enter into the area of sorcery, strictly condemned by the Church under the threat of Church expulsion and excommunication, and there can be no doubt that our Church has precisely this, and no other, attitude to Mr. Butlerov's subject. For all these reasons I can in no way consider it suitable to allow Mr. Butlerov's public lectures.[98]

The only public lectures on Spiritualism presented by our main characters remained Mendeleev's.

The media reaction to his second lecture was overwhelmingly positive. After the first lecture, the press was frustrated by Mendeleev's restraint and his balanced comments on Spiritualism; this time they cast his partisan words as exemplars of objectivity: "Mr. Mendeleev's entire lecture, from beginning to end, was shot through with the firm intention to expose to the listeners a dispassionate analysis of Spiritualist theory, to show the abyss into which people might be attracted who don't know how or don't want to differentiate black and white, the existing and tangible from the mythical and supernatural."[99] P. Boborykin explained that the Spiritualists were being "unscientific":

> [N]ot one of them will agree anew to subject to strict scientific analysis those sensations which led them to the belief in the reality of mediumistic phenomena. This last point seems to me to be the most essential characteristic which differentiates mediumism from any other conviction based on positive data. . . . No Newtons, Lavoisiers, Keplers, Liebigs, Claude Bernards have ever stated that their discoveries must first be *empirically* confirmed by everyone, and then be subjected to strict scientific verification.[100]

The absence of protesters from the second April lecture rendered his speech a victory from start to finish.

Mendeleev's next move was his *Materials for a Judgment about Spiritualism.* The text begins with a preface, stating Mendeleev's reasons for publishing, and then reproduces all the minutes, unedited, followed by the final official statement and the appendices for the various meetings. Then he added a second part, with a second preface, which included an array of scholarly and historical articles on Spiritualism, as well as his own two public lectures.[101] The book's structure itself expresses Mendeleev's views on regulating public discourse very clearly.

His most salient method for controlling discourse was through a series of carefully crafted, looped juxtapositions in the text. They formed part of a general popular style that Mendeleev cultivated to present science to the Russian public as authoritative knowledge, a style that he fashioned in contrast to the impersonal style of his scientific publications, although nowhere else did he use this method so extensively or with as much force. He spent significant time developing his footnotes to the minutes and collecting the appended articles, as well as editing his own public lectures and giving *even them* a set of footnotes to explicate their meanings.[102] The title was no joke: One clearly could deploy the materials in the *Materials* for a "judgment about Spiritualism." But it was equally true that Mendeleev held the reader's hand while that judgment was formulated. The minutes continue for over 100 pages and are followed by itemized appendices, thus providing the "materials for judgment" advertised by the title. But Mendeleev did not let the reader's judgment run too freely. Almost every entry of minutes was extensively footnoted (each one dutifully signed by the chemist), telling the reader how to interpret it. He also sparred with public comments, newspaper articles, and other interventions that might "distort" the reader's *proper* interpretation of the text:

> I wanted to illuminate with my notes certain short and fragmentary segments of the minutes, to supplement certain places and to compare and present at places my form of thoughts, because in this way I think [I will] assist the distinctness of the impression, which can be taken from a familiarity with what was done by the Commission[.] In other places I wanted above all to present my views against the results brought forth by the Spiritualists in defense of their doctrine.[103]

The echoing of the "objective" voice of the minutes with Mendeleev's authoritative voice was meant to convince readers while letting them believe that they were deciding on their own. Mendeleev included the Commission's official statement only at the very end of the minutes, so that the reader would approach it *after* having been guided along the way by his juxtaposed notes. The appendices in the second half of the materials on topics "related" to Spiritualism also worked through this rhetoric of juxtapositions.[104]

Consider a juxtaposition from the book's preface. The book declares on its title page that all proceeds are earmarked for the construction of an aerostat, just as in Mendeleev's publication ventures discussed in the previous chapter. The *Materials*, as the facing page illustrated, was meant as another in the series of gas/weather books.[105] The foreword explains:

However far apart these two subjects, Spiritualism and meteorology, appear, there exists between them a certain connection, a remote truth. "The Spiritualist doctrine is superstition," thus concluded the commission which examined mediumistic phenomena, and meteorology has also battled and will still battle with the superstitions which dominate with respect to the weather. In this battle, as in any other, material means are necessary. Thus let one superstition help against another however it can.[106]

Throughout the footnotes and in the April lectures, Mendeleev made meteorological analogies to suggest how Spiritualism *should* be studied, just as he had made analogies to the Great Reforms to argue for how meteorology should be studied. Mendeleev not only wanted to use the debunking of one superstition to end another, he also wanted to boost science's status by juxtaposing pseudoscience (Spiritualism) with a real science (meteorology).[107]

There was another implicit connection between Mendeleev's interest in meteorology and his opposition to Spiritualism that helps explain his hostility to Spiritualism as a movement: the ether. A great many Spiritualist texts invoked the luminiferous ether as the domain in which spirits lived, so seances were suggested as a way for scientists to investigate ether–matter interactions.[108] For Mendeleev, who at this very moment was heavily invested (both financially and personally) in an experimental attempt to investigate ether as a form of matter, these claims had the potential to discredit his whole program. His attack on the Spiritualists assumed the rhetoric of defending scientific method and proper forum partially to prevent noble dilettantes like Aksakov from meddling with a research program that was already showing signs of disintegrating. His references to meteorology were merely the overt expression of this ether concern. One can see direct evidence of this in Mendeleev's library. He classified the books on "Spiritualism" (25 works) together with those on "Chemico-Physical Cosmogony" (97 works), and "Luminiferous or interplan. ether" (15 works). Maxwell was next to Aksakov, who in turn was next to Helmholtz.

But two could play this game. Such juxtapositions were used in other publication ventures to discredit Mendeleev, most notably by Aksakov himself. Furious at the way Mendeleev treated his mediums and himself during the seances and upset at the tone of the *Materials,* Aksakov republished the minutes of the Commission with his *own* juxtapositions. He began his volume with an essay entitled "Science and Charlatanism," criticizing Mendeleev's Commission in detail. Claiming that Mendeleev's ordering of the minutes distorted the results of the Madame Claire seances, he placed the Commission's special appendices immediately following the seances to which they referred. Following each appendix, he included, in larger type, his own comments about what was *actually* happening. On both the cover page and as an epigraph to his essay, he quoted Mendeleev, ironically commenting on the latter's own perversions of the scientific method. Aksakov claimed that the *Materials* themselves were "documentary proof of that passionate and personal, but consequently *not scientific* character, which was inherent in our Commission; therefore, possibly, Mr. Mendeleev considered it useful to assure himself repeatedly of its dispassionateness." He also offered a mini-riposte to the Commission's own official statement:

> My respected colleagues, also witnesses from the side of the mediums—professors Butlerov and Vagner—are not inclined, and also have neither time nor desire to busy themselves with this affair. The essence of *my* "examination" is contained in the following statements: 1) Mediumistic phenomena occurred before the Commission, but they were as far as possible hidden, kept secret, left unnoted, or attributed to deception. 2) Deception of any sort from the side of the mediums did not exist; in any event it was nothing like what was attested to *by the Commission:* singular displays are not proofs. 3) There was deception and there was slander, but from the side of the Commission and its members; that which the Commission blamed against the mediums is *unsubstantiated,* and the same accusations appear now against themselves, but now *documented.* 4) The action of the Commission was, therefore, entirely biased, systematically pursuing one predetermined goal of battling mediumistic phenomena no matter what.[109]

This alternative logic of scientific investigation agreed with Mendeleev's views about the value of method, but questioned Mendeleev's use of that method. On the other hand, critics such as novelist Feodor Dostoevsky, among others, agreed with Mendeleev's conclusions, but disapproved of his method.[110] For his part, Aksakov did not feel that the results of the Commission necessitated any change in his beliefs on Spiritualism, or in his efforts to publicize his creed.[111]

AT WIT'S END: SPIRITUALISM AFTER THE COMMISSION

All of the Spiritualists involved with the Commission remained committed to the movement; Vagner even began exploring new areas of the occult such as spirit photography and performed experiments on the star medium of the turn of the century, Eusapia Palladino.[112] Butlerov, for his part, engaged in a series of highly public newspaper polemics with conservative journalist and general grouch Nikolai Strakhov.[113] After the 1905 Revolution, interest in Spiritualism skyrocketed. The Spiritualist periodical *Rebus* sponsored a congress in

Moscow for the Society of Russian Spiritualists on 20–27 October 1906, and more than 400 attended. It is possible that by the turn of the century there were more than 1600 spirit circles in Moscow and Petersburg alone. The movement had broken out of the confines of the capital to become a general Russian phenomenon. By the eve of World War I, there were more than thirty-five officially registered occult groups in the capital alone, with Theosophy and Spiritualism the two most important. To this day, the occult maintains a strong attraction for the Russian public.[114]

Given that Spiritualism had such an astonishing success in Russia despite the efforts of Mendeleev's Commission, one must address the issue of how "the public" responded to the efforts of Mendeleev and his friends. I put the term "public" in quotes because it is imprecise in Imperial Russia to speak of any in-dependently acting public in the modern Western sense. The few traces of re-sponses to the Commission were confined to the small stratum of the literate cognoscenti of Petersburg society.[115] As expected, most of the editorialists found the Commission's work satisfactory, and a series of scientific societies—largely medical ones—undertook their own scientific investigations of mediu-mistic phenomena.[116] A collection of the more respected citizens of Petersburg, however, found the Commission's behavior to be biased and dis-graceful, and argued that, "[r]estricted to 8 seances, the Commission did not have a respectable basis to declare the research concluded." (They demanded that the Commission's work be resumed.)[117] It is hard to draw conclusions from such an eclectic selection of letters and articles, but one could easily ar-gue that Mendeleev's Commission failed. Not only did he not convert his sci-entific fellows from Spiritualism (if that was ever his intention), he also did not prevent Aksakov from publishing or slow the spread of Spiritualism in Russia.

How did Mendeleev himself respond to Spiritualism after his Commission? Perhaps surprisingly, given his volatile nature, he seemed to ignore all refer-ences to Spiritualism after this year-long foray. Shortly after his April lectures, he was sent to the United States to attend the Philadelphia World's Fair and investigate the success of the Pennsylvania oil industry, and on the long boat passage across the Atlantic he passed the time with cards, chess, and "spiritual-ist experiences and hoaxes." While in America he noted the addresses of vari-ous mediums, but he did not investigate further.[118] Even as Butlerov and Vagner became bolder in their defense of Spiritualism, Mendeleev seemed to recede into the background.

Yet the story does not end here. Nearly twenty years later, in 1894, Viktor Pribytkov, editor of *Rebus,* and Aksakov claimed that Mendeleev had finally re-canted and admitted the existence of mediumistic phenomena. Apparently, Mendeleev had approached Pribytkov at a party, brought up Spiritualism, and discussed some of the various hoaxes he had seen among professional mediums in America. Despite his mocking tone, Pribytkov asked Mendeleev if he now believed in the phenomena, and Mendeleev was reported to have said: "They exist . . . I saw . . . But they are rare. . . . It isn't worth it to pay attention to them and not a single serious, busy man would get involved with them." When Pribytkov expressed surprise, Mendeleev responded: "What? You don't understand? All this is garbage, nonsense!" Exultantly, Pribytkov claimed that this was a recantation.[119] Had Mendeleev changed his mind? Unlikely. The beginning of the conversation was almost certainly Mendeleev pulling Pribytkov's leg; a joke entirely consistent with the chemist's character. Second, Mendeleev would hardly have made such an admission to a Spiritualist journalist. And, finally, in 1904 Mendeleev published a final article about Spiritualism in which he reaffirmed the Commission's conclusions in opposition to a reinvigorated Spiritualism. Even after almost thirty years, he framed the issue in terms of scientific method.[120]

Although Mendeleev's faith in the power of natural laws to provide some sense of stability and resolution to local debates in Petersburg remained unshaken, the failure of the Commission to stop Spiritualism from spreading—or even to delegitimize claims by the nobility and amateurs concerning his coveted ether—eroded some of his hopes for the panacea of decentralized scientific societies. His popular image, however, was bolstered by his escapades against the Spiritualists and left him primed to become a media celebrity again.

Yet the whole affair ended up as much ado about (next to) nothing. In this sense, the history recounted in this chapter is an allegory to the Great Reforms as we see them in 1875: While equipped with all the trappings of modernity, progress was stalled in sterile discussions. Perhaps this is not surprising, given how trivial—by today's standards—the controversy over Spiritualism in the winter of 1875–1876 was. Soon Mendeleev was to become the center of a whirlwind of controversy around the Academy of Sciences, where the tools generated by the Great Reforms formed the kernel of a powerful critique of the status quo. But at mid-decade, while waiting for a breakthrough in his gas research, Mendeleev—just like the Spiritualists—was sitting in the dark hoping for something miraculous to happen.

THE GREAT REACTION

This caricature of Mendeleev appeared on 7 December 1880. It features the chemist dreaming underneath a tableau of scheming academicians, none of whom appears to represent an actual member of the Academy of Sciences. The cartoon is entitled "Daydream" and the caption reads: "D. I. Mendeleev. Will they elect him? . . . Won't they? . . . They didn't! . . . What is this: a dream that is similar to reality, or reality that doesn't differ from a dream?. . . " From "Son na iavu," *Strekoza,* 7 December 1880, #49: 1. A copy may be found in A. M. Butlerov's personal files in PFARAN f. 22, op. 1, d. 38, l. 10.

THE GREAT REACTION: EVERYONE AGAINST THE ACADEMY OF SCIENCES

Idols should not be touched; their gilt comes off on the hands.
—GUSTAVE FLAUBERT[1]

IF IT HAD NOT BEEN SO PERSONALLY INSULTING, he might have appreciated the irony. In November 1880, Mendeleev was subjected to a personal humiliation that became a national scandal, with hundreds of Russia's most vocal intellectuals entering the fray. Four years earlier, after placing himself at the center of a commission to debunk Spiritualism, he had been for a time the darling of the liberal media. Now, he would become their darling once again—but not under circumstances of his choosing. The cause of his fame and also his embarrassment was the very institution whose recognition he had coveted for so long: the St. Petersburg Academy of Sciences.

SOCIAL CLIMBING: THE ACADEMY AND THE PHYSICO-CHEMICAL SOCIETY

For two decades now, Mendeleev had believed scientific societies were the best way to organize expertise in the rapidly developing Russia of the Great Reforms. In his somewhat tempestuous involvement with the Russian Technical Society, Mendeleev encountered the dangers of getting too close to the power and money such a quasi-official organization could provide. He experimented with a semiprivate commission within the Russian Physical Society as the appropriate way to combat Spiritualism. Yet during this period Mendeleev constantly eyed the Academy of Sciences, the long-recognized apex of Russian natural sciences and humanities.

The Academy of Sciences was founded in 1725 as one of Tsar Peter the Great's final acts. In its first years, the Academy was intended to serve as an exemplar to the newly Westernized Russian nobility for how to behave in a gentlemanly fashion.[2] After some turbulent growing pains, the Academy eventually settled quite comfortably into its quarters on Vasil'evskii Island in St. Petersburg, sitting on the banks of the Neva river adjacent to St. Petersburg University (across a narrow street that is today called Mendeleev Line).

Given the absence of native scholars with the requisite credentials, the original academicians were appointed by Peter from a list of central European academics recommended by Gottfried Leibniz's disciple Christoph Wolff. Later in the century the Academy was rocked with debates on the continuing dominance of foreigners (mostly German-speaking) in the Academy hierarchy despite the increasing Russification of the institution, confrontations orchestrated and dominated by M. V. Lomonosov, the first Russian academician and often-lauded father of Russian chemistry.[3] The Academy nevertheless maintained its status as the Olympus of science in Russia into the nineteenth century, which it began with a new charter proclaimed by Alexander I in 1803. The Academy was granted greater authority over scholarship with substantially reduced government supervision.[4] Its membership was still largely foreign-born, although supplemented by a growing population of native Russians.

As the number of individuals interested in science expanded in the nineteenth century, and newly minted Russian scientists (often trained abroad) began to staff Russia's major universities (chiefly Moscow, St. Petersburg, and Kazan), the situation started to change. By midcentury, the Academy began to lose precedence over the universities. As the population of Russia expanded, so did the universities, while the Academy's membership remained defined by its charter, revised in 1836. The rise of student enrollment in the natural sciences meant hiring active young scientists to university posts, much as the young Mendeleev was in the 1860s. St. Petersburg University now began to dominate its next-door neighbor in scientific research as the majority of academicians languished in what some decried as comfortable sinecures.[5] Multiple attempts to reinvigorate the Academy through a new statute were staged from 1841 to 1866, all intended to make the Academy a more practical institution. These efforts, initiated by the Imperial bureaucracy, were widely perceived as a potential Great Reform. The Ministry of Popular Enlightenment halted these efforts after the attempted assassination of Alexander II in 1866, leaving the Academy mired in obvious faults of the old charter.[6] By the late 1860s the Academy had already acquired the reputation of a dinosaur clinging to the privileges of a vanished social order. The pent-up frustration would find its release on 11 November 1880.

Mendeleev always saw the Academy as the ultimate destination of his scientific career, and so it is somewhat ironic that he would become an unwitting martyr to this reformist cause. He certainly would never have let himself undergo the humiliation of rejection merely for a salubrious effect on Russian national consciousness. The ideal of what the Academy should have been, a naïve ideal that he had appropriated from the Karlsruhe Congress of 1860, was Mendeleev's perpetual motivation for attempting to organize expertise in the Empire. So far, he had only one point of comparison: the Russian Physico-Chemical Society.

Mendeleev was among the most active members of the Physico-Chemical Society. He had presided over the first meeting of the (then) Russian Chemical Society on 6 November 1868, when N. N. Zinin was elected president, and served himself as president in 1883–1887 and 1891–1894. Up to and including 1908—a year after his death—there were only five years in which the Society's *Journal* did not include at least one publication by Mendeleev.[7] The early Chemical Society was almost exclusively a Petersburg affair, and was wrapped in the symbolism of the capital, as was noted on its twenty-fifth anniversary in 1893:

> The Society was opened at Petersburg University on the grounds that Petersburg, as the capital, would always have a large number of scientific forces. And Petersburg University then had two chief chemical forces: Butlerov and Mendeleev. Founding their Society, Russian chemists had in mind not just the satisfaction of their personal demands. The Society was supposed to serve at the same time as a propaganda weapon for chemistry in Russia.[8]

Not all scientists were pleased with the Society's progress in the early years. The author of the above quotation, V. V. Markovnikov, a fiery Moscow chemist who later suffered rejection at the Academy's hands, was frustrated that some Russian chemists still published their original results abroad first, and threatened to withdraw if the Society could not somehow force its members to support Russian scholarship by publishing in their native tongue.[9]

Before November 1880, Mendeleev had enjoyed some success at the Academy. Its Demidov prize, awarded for his *Organic Chemistry* textbook in 1861 by academicians J. Fritzsche and Zinin, put him in the black in those desperate early months after his return from Heidelberg. And before 1880

Mendeleev regularly had his new research read into the minutes of the Academy. On 8 October 1874, Mendeleev's Petersburg University colleague and academician (and aforementioned Spiritualist) A. M. Butlerov, together with academicians Zinin, A. N. Savich, and O. I. Somov, proposed Mendeleev as an adjunct academician in physics and chemistry for the first division of the Academy, the physico-mathematical division. (The second division concentrated on Russian history and philology, and the third on philology and the humanities more generally.) The citation pointed to Mendeleev as an expert in the crucial unifying area of physical chemistry. Of course in 1874 gallium had not yet been discovered, and so Mendeleev's work on the periodic system did not have the same status it gained only a few years later. Given the importance of the periodic law for the construction of Mendeleev's reputation, it is interesting to see how periodicity was discussed in the 1874 citation proffering Mendeleev's candidacy:

> We further point to Mr. Mendeleev's articles relating to the atomic weights of elements. These works laid a foundation for a new rational system of elements, allowing the prediction of many relations and summoning new views; they have given the foundation for working out one of the basic subjects of chemistry. After Professor Mendeleev this question has been subjected to study by many foreign scientists.[10]

Two points are of note here: First, the periodic system was expressed in modest terms; second, foreign recognition was singularly important in endorsing a Russian candidate, an insecurity present in almost all Academy elections of this period. But Mendeleev was not fated to be an academician. Adjunct chairs, unlike regular chairs, were not allotted in the charter, and so it was possible to hold a preliminary vote to see whether any specific field actually needed the chair. For reasons discussed below, the Permanent Secretary of the Academy, K. S. Veselovskii, held such a vote for dismissal on 29 October 1874, and 11 voted for, 8 against, out of the 18 people at the meeting (the president of the Academy, F. P. Litke, wielded two votes).[11]

Mendeleev was again proposed for the Academy on 16 November 1876 (after gallium's discovery), this time as a corresponding member, backed by the recommendations of academicians G. P. Helmersen, N. I. Koksharov, F. B. Schmidt, A. V. Gadolin, and—again—Butlerov. With Mendeleev on the ballot were chemists G. V. Struve, Marcellin Berthelot, and Edward Frankland. Mendeleev was elected with an impressive 17 for and 2 against.[12] So, in 1880, having combated the Spiritualists, riding a crest of public acclaim, and with two successful predictions to his credit, Mendeleev was confident of nomination for the Academy's full chair in technology.

THE BALLOT BOOTH: VOTING ON MENDELEEV

One might consider the chair in technology not the most appropriate for a chemist, but Zinin's death in early 1880 left it vacant and, besides, technology was then generally considered to be applied chemistry. Zinin himself did practically no work in technology, focusing instead on pure organic chemistry (although his research did have direct implications for the development of synthetic aniline dyes). Zinin had moved to the St. Petersburg Medical-Surgical Academy from Kazan in February 1848, and was elected an adjunct academician in 1855 with a momentous 23 votes for, 1 against. He was elevated to extraordinary academician in 1858, and became an ordinary academician (the highest level) in 1865 by another striking vote of 28 to 3.[13] As former president of the Physico-Chemical Society, his death was a blow to Russian chemistry, and Mendeleev seemed a logical candidate for his successor.

On 11 March 1880, Butlerov headed a committee composed of Koksharov, Gadolin, and meteorologist Heinrich Wild. On 22 March, Butlerov proposed two potential candidates, Mendeleev and N. N. Beketov, professor of chemistry at Khar'kov University.[14] Less than two months later, on 9 April, Russian-born German Friedrich Beilstein, Mendeleev's successor at the Technological Institute, was added to the list. Butlerov—somewhat surprisingly given the recent Spiritualist battles—pushed for Mendeleev.[15]

Butlerov and Mendeleev had long had a complicated relationship. Although Mendeleev had brought Butlerov to St. Petersburg University in 1869 from Kazan, he had not done so entirely willingly, stalling for over a year before finally succumbing to pressure. Upon Butlerov's arrival, Zinin led the drive to get him admitted as an adjunct academician in chemistry (achieved on 16 December 1869 by a vote of 16 to 3). He then became an extraordinary academician in October 1871, and on 18 January 1874 was elevated to the level of ordinary academician.[16] When Mendeleev finally brought himself to endorse Butlerov for the post of professor of organic chemistry at the University, he gushed. In this often-quoted letter to the faculty council, Mendeleev spoke of a "*Butlerov school*" in organic chemistry, a school that was purely Russian in its origins. (Mendeleev omitted draft statements about the seminal importance of Frenchman Charles Gerhardt as the origin of Butlerov's innovations and the following back-handed compliment: "I consider myself impartial in this judgment since I do not adhere to the Butlerov school.")[17] Butlerov became a very popular professor, and was Mendeleev's ally in the defense of professorial autonomy.[18] Yet Mendeleev still harbored his resentments, and his public ridiculing of Spiritualists often had the quality of a personal vendetta. A former student recorded the local saw that there were two chemists at St. Petersburg University: one smart and one talented.[19] Mendeleev was the latter, and it must have rankled.

Butlerov was more willing to let bygones be bygones. He was displeased by Mendeleev's rejection in 1874 (or, more correctly, the circuitous rejection of the adjunct post), and as the decade wore on he became more vocal in his opposition to the politics of academic elections.[20] From early 1880, Butlerov began a behind-the-scenes negotiation by urging Mendeleev to write a letter, as per the statute of the Academy, that he would accept the post if offered. Hearing that Beketov had already sent one, Mendeleev, in an uncharacteristic display of cold feet, wrote that he was sure that Butlerov preferred Beketov and he himself would step aside. Mendeleev's tortured doubts were eventually put to rest by October 1880, when Butlerov pushed Mendeleev's candidacy over Beketov's.[21] On 28 October, Mendeleev was promoted as the candidate for the post of extraordinary academician of technology by sponsors Butlerov, P. L. Chebyshev, F. V. Ovsiannikov, and Koksharov.

The vote did not go well. The election took place on 11 November 1880, and the final tally was 10 against and 9 for. With the double vote of the president cast against Mendeleev, this meant that Mendeleev lost by a single vote, or, to be more precise, that one more vote was cast against him than for him. One vote the other way would *not* have elected Mendeleev; a two-thirds vote was required for full approval. The "one vote" margin did possess a marked rhetorical power, however. The minutes of the election were simple: "Conclusion: *not considered elected.*"[22] In the eyes of the intellectual elite of Petersburg, however, the conclusion was not nearly so straightforward.

There is no way to be certain—even for the academicians in 1880—who voted for Mendeleev and who against. The vote was conducted by a secret ballot whereby academicians placed either a white ball (pro) or a black ball (con) in a box. This did not prevent endless contemporary speculation about the exact breakdown. The most common tally was that worked out by Butlerov himself on his copy of the election protocol:

Not elected: 10 black, 9 white.

Obviously—blacks are [President] Litke (2), [Permanent Secretary] Veselovskii, Helmersen, Maksimovich, Schrenck, Strauch, Schmidt, Wild, Gadolin.

White: Buniakovskii, Koksharov, Butlerov, Famintsyn, Ovsiannikov, Chebyshev, Alekseev, Struve (!), Savich.[23]

The reason for Butlerov's mark of surprise for Struve was that he was the only academician of non-Russian heritage to vote for Mendeleev. All the seemingly Russian names on the "black" side, with the very significant exception of Veselovskii, were Finns or Balts. In his personal papers at the archives of the Academy of Sciences, Veselovskii kept a file on the Mendeleev rejection. On the first page, he criticized tallies such as Butlerov's:

In the Imperial Academy of Sciences all elections of its members are carried out by a secret balloting with balls, and consequently, by the very nature of such a balloting, it will always remain unknown to anyone which of the voters cast which kind of ball. Therefore, the named lists of academicians which have been appearing in various newspapers, giving votes for or against the candidate recently proposed for one of the vacant seats in the Academy, do not have any foundation, and conclusions drawn from them belong to the realm of daydreams.[24]

Such caveats, even when they were made public (this one was not), were ignored.

The rejection of Mendeleev sent shock waves throughout elite Petersburg, not least within the Academy itself. The next scheduled election was of Viktor Baklund as an adjunct astronomer for late November. Before the election, Vladimir Lamanskii, University professor and long an open critic of nepotism and pro-German chauvinism in the Academy, wrote an open letter in the local newspaper *New Times* (*Novoe Vremia*) urging Baklund to refuse the chair as a respectful protest against the Academy's insult to Mendeleev. He wrote a similar open letter to Struve, Baklund's employer at the Pulkovo Observatory. Struve's status as the only "German" to vote for Mendeleev was mentioned.[25] Baklund was elected by the division, as expected.

Criticism was very severe. The *New Times* sarcastically noted that Baklund clearly had all the qualifications for being a member of the Academy: He was a foreigner and a foreign subject; he had no Russian academic degree; he could not read or write Russian; no other academy paid any attention to him; and he had not achieved anything in astronomy at his young age. They used his election to call for two fundamental reforms to the Academy statute: that all academicians must be or become Russian subjects; and that they take a university-level test on competence in the Russian language.[26]

Butlerov, for his part, took action behind the closed doors of the Academy before its General Assembly. Typically, after a division of the Academy had elected its nominee, the decision was rubber-stamped by the Assembly, only after which was the nominee truly an academician. On 5 December, Butlerov raised statutory objections to appointing a scholar as young and inexperienced as Baklund. If there were really no Russian astronomer adequate to the task, it would be more honorable for Russia, he argued, to admit its inadequacies than "to decorate ourselves, from the outside only, with that which is not organically connected—and usually remains unconnected—with national enlightenment." He urged the Assembly to reject the candidacy. It did not. Veselovskii refused to publish Butlerov's objections in the minutes.[27]

The chair of extraordinary academician in technology, however, was still vacant, and needed to be filled once the popular outrage for Mendeleev died down. Early in 1882, the first division of the Academy settled on the third name on the original roster for election to the technology chair: Friedrich Konrad Beilstein. Beilstein was born on 5 February 1838 in St. Petersburg, the child of German immigrants who ran a local grocery store. He began his chemical career early, traveling abroad to study in Heidelberg with Robert Bunsen already in September 1853, and in 1855 moving to Munich to work with Justus von Liebig, to Göttingen in 1857 to study under Friedrich Wöhler, and in 1858 to Paris and Adolphe Wurtz. Beilstein's biography thus had some of the most prestigious stops in the training of a nineteenth-century chemist. For a time he ran the *Zeitschrift für Chemie,* a chief conduit for Russians to publish chemical works in German. On 24 September 1866, he was named professor of chemistry and head of the laboratory at the Technological Institute in St. Petersburg, becoming Mendeleev's successor, and he was sworn as a Russian subject in June 1867.[28]

Long an active member of the Russian Chemical Society, he was in many ways emblematic of the internationalism that had been traditionally at the heart of Russian chemistry. Yet personal relations between Beilstein and Mendeleev were unfriendly, especially as Beilstein perceived himself as cleaning up Mendeleev's mess at the Technological Institute.[29] When Beilstein was reconsidered as a candidate, he had just published the first edition of his *Handbuch der Organischen Chemie,* a comprehensive reference work that is still (in revised form) the standard in organic chemistry. With his post at a Petersburg technological institution, and his smattering of works in technical chemistry (the bulk of his work was in pure chemistry), he seemed a perfect candidate.

Butlerov, however, by now interpreted all elections at the Academy as purely nationalist and political, and he determined to embarrass the Academy by fighting off Beilstein—once a close friend—as recompense for his humiliation by proxy in the Mendeleev debacle. At the 22 December 1881 meeting, Butlerov, arguing that he was the only academician competent to judge on chemical matters, objected to the committee's choice of Beilstein and again suggested N. N. Beketov. He was ignored and on 19 January 1882 the first division elected Beilstein by a vote of 12 to 4. Beilstein's impressive credentials prevented many of those who had supported Mendeleev from rejecting him on nationalist grounds. Butlerov persisted, telling the first division publicly that "[c]ompetent specialists are convinced that by contributions to applied chemistry, Prof. Mendeleev undoubtedly stands above all other Russian chemists," and made a point-by-point comparison in their qualifications. Unable to call for a reballoting, Butlerov again moved to the General Assembly to repeat the maneuver he had tried with Baklund. This time he was successful. On 12 February 1882, the General Assembly rejected Beilstein as an academician in technology.[30] Permanent Secretary Veselovskii made an informal poll of the academicians at the General Assembly, and found that seven of the dissenters were professors at the University. Beilstein had 17 white ballots, just one short of the necessary two-thirds for election.[31] Only after Butlerov's death was Beilstein finally admitted to the Academy in the technology chair (while Beketov filled Butlerov's chair in chemistry).[32]

Butlerov's actions in these two elections, Baklund's and Beilstein's, were direct responses to the Mendeleev election. He repeated some of these general themes in his later journalistic attacks on the Academy: the invocation of academic procedure, the emphasis on a nationalist interpretation of the Academy, and continual reference to the competence in chemistry he had by virtue of his status as a *University* professor. His reactions reflected broader responses in two circles. Butlerov was, after all, a Russian chemist, and his beliefs were in accord with those of most members of the Russian Chemical Society. And he was also, in the end, a member of the Petersburg intelligentsia, a group that conducted perhaps the most furious of the pro-Mendeleev uproars after the election.

TEMPEST IN THE TEAPOT: RUSSIAN CHEMISTS

There was substantial agreement among Russian chemists that Mendeleev *deserved* election to the Academy, and that the failure to elect him reflected some deep bias of the Academy. The chief points of the chemists' reactions appear on three fronts: first, the response within the Russian Chemical Society, whose president at this time was Butlerov; second, Butlerov's personal rebuttal, which was published in a widely read article that remains the orthodox interpretation of the affair; and, finally, chemists' public avowals of Mendeleev in the popular daily newspapers. These three fronts—institutional, personal, and journalistic—provide a panoramic view of how one issue managed to galvanize what had truly become a Russian chemical *community*. From meager origins just over ten years previously, this community, itself a product of the Great Reforms, was able to organize a far-reaching protest that newspapers would in turn appropriate to formulate a criticism of those very Reforms.

The Russian Chemical Society traditionally met in the first week of the month, so there was no meeting directly following the 11 November Academy election to allow for discussion of ways to demonstrate support. Almost immediately after he heard the news, Society secretary Nikolai A. Menshutkin, editor of its *Journal,* drafted a statement for local newspapers:

In a meeting on 11 November 1880, the Physico-Mathematical Division of the Imperial Academy of Sciences rejected D. I. Mendeleev. . . . The indubitable merits of the candidate, whose equal Russian science cannot offer, [and] his fame abroad make his rejection completely inexplicable. In view of

the repeated failure to elect the best Russian scientists to the Physico-Mathematical Division of the Academy of Sciences, we consider it necessary to
draw public attention to this.[33]

Most chemists responded to the statement positively, merely signing their
names. Two responses, however, are noteworthy for criticizing Menshutkin's
draft from opposite extremes. Both of these reactions were premised on the
idea of a "Russian chemical community." V. V. Markovnikov stood firm on the
Russian position:

> The composition of the statement you sent is not entirely appropriate. In *this*
> affair reference to the recognition of the achievements of D. I. [Mendeleev]
> abroad is not appropriate. *Until we ourselves claim enough authority in the res-*
> *olution of such questions, the Germans will have the complete right to scorn our*
> *scientists.* We are not deciding the question of a candidacy to the Berlin or
> Paris academies; but who is worthy of being a Russian academician—this is
> our issue.[34]

Beilstein, on the other hand, felt that a telegram to a newspaper violated the
integrity of the *chemical community:*

> We are people of science and our field of action is not any newspaper, but the
> *scientific milieu,* and therefore I allow myself to make the following proposition:
> our colleague—a famous scientist—is offended, but this is in the background.
> The question is purely personal. We all feel the need to express to our colleague
> our respect and sympathy. For this the most appropriate form is an *address,* in
> which it is *silently* stated that although there are persons who consider D. I. un-
> worthy of the highest scientific position, we *chemists,* and thus more competent
> judges than anyone, consider D. I. a premier scientist among us. We ask him
> not to be distressed by what has happened; he has done much and probably
> will still do many wonderful things. The recognition, the sympathy of all who
> love the science of D. I. is certainly to be placed higher than the rejection by
> certain persons, etc. We can deliver the address to him immediately upon sign-
> ing it at the next meeting of the Society. After this, if the newspapers feel like
> communicating to the public what happened in our milieu, we will eagerly re-
> lease all materials. Thus, it seems to me, we will act *independently,* originally,
> and with the preservation of all our merit. Our calling and position do not al-
> low us to busy ourselves with petty newspaper abuse.[35]

He was the only Society member who did not sign the letter, a silent move that
spoke volumes to both Russian chemists and to Beilstein's rival, Mendeleev
himself.

The Society under Butlerov took additional steps. Butlerov released to the newspapers the original written endorsement of Mendeleev's candidacy submitted to the Academy. This document chronicled Mendeleev's achievements in the sciences, stressing the periodic law and littered with endorsements by foreign scientists. In addition, given that Mendeleev was nominated for a chair in technology, Butlerov and his colleagues emphasized Mendeleev's contributions to the Baku oil industry and the applications of his dissertation work on alcohol solutions to liquor production.[36] At the 5 December meeting of the Russian Chemical Society, the first after the events at the Academy, Butlerov delivered a speech of support for Mendeleev and, out of a moral obligation to honor Mendeleev because of "the well-known event of 11 November," moved to make him a distinguished permanent member of the Society.[37] Mendeleev accepted, and became the third such member—along with Zinin and his mentor Voskresenskii—and the only one living.

Butlerov later publicly criticized the Academy's establishment in detail. He was by no means hostile to the Academy per se, and would not have been so vehement had he not revered its status. He had won the Academy's distinguished Lomonosov prize in 1868, after which he became a member, fulfilling a lifelong dream. In Markovnikov's congratulatory note to Butlerov, he wrote: "Do you remember that you once told me that you would want nothing more than to be an academician, i.e., to have the possibility to work to your heart's content and not to be cowed by any obligations?"[38] Butlerov did not abandon teaching at the University, but he did convert his chair into a pulpit for an expanded research program. For a variety of motivations—nationalism, gratitude for the services of other chemists, a desire to develop chemistry in the capital—Butlerov agitated almost immediately for the election of more of his Russian peers.[39]

Butlerov had for years been keeping a list entitled "Academic transgressions, as witnessed by me personally (since I have been a member of the Academy)."[40] In bullet-point form, he outlined instances of bureaucratic interference in the administration of the Academy, micromanagement by Secretary Konstantin Veselovskii of the work of individual divisions, academicians' lack of productivity, and nationalist gambits whereby qualified individuals (in Butlerov's opinion) were denied academic chairs by virtue of their Russian nationality (again in Butlerov's opinion). After about a year of public silence on the Academy affair, the Beilstein election finally prompted Butlerov to publish the manuscript that had emerged from his notes, originally entitled "Contemporary Materials Toward a History of the Academy of Sciences."[41] With few changes, the article appeared in the nationalist journal *Rus'* under the more inflammatory title "Is There a Russian or Only an Imperial Academy of Sciences in St. Petersburg?" Ever since, this document has structured the Russian understanding of the events.

The piece's venue helps account for its prominent reception. *Rus'* was the mouthpiece of Ivan S. Aksakov, a prominent intellectual with Slavophile tendencies, and a relative of Butlerov's wife. (He should not be confused with his cousin *Aleksandr* Aksakov, the leading Spiritualist discussed in the previous chapter.) Ivan Aksakov approached Butlerov in early 1882, telling Butlerov that his voice would be crucial in finally eliminating "Germans" from the Academy.[42] Butlerov let Aksakov title the piece, which the latter adapted from an outburst by an academician of Prussian extraction who once yelled at Butlerov: "The Academy is after all not Russian, but rather an Imperial Academy!" He also added an editor's preface. After an extended criticism of Peter the Great's destruction of native Russian traditions—a Slavophile trope—he deliberately assimilated Butlerov's criticism of the Academy into an indictment of Westernization in general:

> . . . Mr. Butlerov's article has an interest of the highest importance not only for specialists, but for Russians generally. The Academy in its current form is the legitimate child of that nonnational or, more accurately, anti-national direction in all spheres of administration of not only our foreign policy, but our domestic as well, which so distinguishes the Petersburg period in Russian history.[43]

For Aksakov, the Academy represented everything that was wrong with Russia since Peter. Butlerov would certainly not have agreed, since he was just as much a part of the Westernizing project as the Academy was.[44]

Simply put, Butlerov's piece clearly distills the chemists' point of view through an insider perspective on the Academy. Butlerov began directly with a discussion of the academic "majority": "I must admit that, despite the absence of personal observations, I had reasons from the very beginning to relate with a certain caution to the activities of the academic majority."[45] He claimed that he initially resisted identifying this majority with the Germans, since he had come to the Academy with a great deal of respect for foreign scientists, but developed his anti-German stance through experience. Butlerov then began to catalog a series of incidents in which Russian scientists were denied prizes or places in the Academy because they were Russian. Why had he kept silent for a year and half? "Last year, during the journalistic noise concerning the failure to elect Prof. Mendeleev, an attack was stated in a meeting of the division, without being introduced into the minutes, on those members of the Academy who were ruining the secrecy of the deliberations."[46] He had then mentally reserved the right to voice his views eventually, and now he felt compelled to speak.

The article begins with quotations from the Academy's charter, which provides that Russians should be elected over foreigners (article 30), and the proper procedure for elections; then he detailed the ways in which these explicit rules had been repeatedly violated. Butlerov's successful efforts to get botanist A. S. Famintsyn into the Academy alienated the academic "majority," and only succeeded in 1878 after eight years of effort.[47] After that, Butlerov had little political capital to spend on Mendeleev's case. After Veselovskii's successful effort to block Mendeleev's election to adjunct status, the permanent secretary became the dominant villain in Butlerov's narrative of academic injustice. When Sanskrit scholar Leopold Schröder was rejected in 1879 due to Russian agitation that there was no need for a *second* adjunct in that language, Veselovskii spat out at Butlerov perhaps the most revealing lines in the article: "You are entirely responsible for this! You dragged Famintsyn into the Academy; you want us to ask [St. Petersburg] University's permission for our elections. This will not be. We don't want university types. Even if they are better than us, we still do not need them. As long as we live, we will fight against it."[48] So when Butlerov put Mendeleev, yet another "university type," up for election, Veselovskii, Butlerov claimed, blackballed him out of spite. The problem with the secretary, in Butlerov's mind, was that the post was permanent, nearly ensuring that "the scholar vanishes in the bureaucrat"—a point demonstrated by the fact that Veselovskii had submitted no scientific works in twenty years.[49]

After setting up the principal elements—a history of German domination, violations of the charter, and the *bête noire* of Veselovskii—Butlerov turned to the Mendeleev election. For Butlerov the Academy affair was fundamentally about the nature of arbitrary will (*proizvol*), without which the pernicious German national bias would have had no foothold in the Academy.[50] The term *proizvol* had a specific connotation by 1880 that was hardly lost on his readers. The Great Reforms were initiated in part to eliminate the state's arbitrary power in favor of the devolution of certain responsibilities onto a modern citizenry, replacing *proizvol* with lawfulness.[51] That the Reforms did not touch the Academy was in itself an indictment of the course of those Reforms, which otherwise Butlerov—like Mendeleev—supported strongly. This veiled criticism of the Reforms became a recurring theme in mainstream attacks on the Academy.

For chemists at large, however, the issue was not so much nationalism or the failure of the experiment of the Great Reforms, but the wounding of corporate pride, and they took to the broadest possible forum, the capital's daily newspapers, to express their solidarity. In each of the main dailies, telegrams appeared from chemists spread all over the Empire (and even abroad) full of respect for Mendeleev and scorn for the Academy's action. The telegrams usually originated from individual scientific societies or universities, small corporate centers that made up the Russian scientific community. Throughout the two months following the Mendeleev rejection, various societies and universities either elected Mendeleev as an honorary member or issued heartfelt statements of sympathy and honor, including: Moscow University, its physico-mathematical faculty, and its students of the natural sciences; Kiev chemists; the Petrovskii agricultural academy; the Russian Industrial Society; Kazan University; the St. Petersburg Society of Russian Doctors; the Czech Chemical Society; the South Slavic Academy in Zagreb; the Society for the Success of the Merchant Marine; and even the out-of-the-way Society of Kostroma Doctors.[52] *The Voice,* Mendeleev's favorite paper, offered a subscription for a prize in Mendeleev's name, to be administered by the Russian Chemical Society, which yielded 14,666 rubles, 83.25 kopecks by Mendeleev's death in 1907.[53] On a more personal level, on 5 December Mendeleev was honored by local professors with a dinner at the fashionable restaurant "The Hermitage," and it was lovingly covered in the major newspapers.[54]

This outpouring was not without repercussions. Although the Academy would soon begin to elect many more Russian members, it initially ostracized Butlerov in particular and Russian chemists in general. For example, when Butlerov proposed two of his former students, Markovnikov and Aleksandr Zaitsev, for corresponding memberships, neither was elected.[55] Despite these setbacks, Russian chemists looked on the Academy affair as a valuable reinforcement of feelings of solidarity that had been building since the birth of the Chemical Society.

OUTSIDE THE TEAPOT: THE GREAT NEWSPAPER WAR

It is not surprising that the Academy's rejection of Mendeleev sparked resentment and hostility from Russian chemists. He was one of their own. Truly striking, though, is the extensive coverage the daily newspapers lavished on the debates over Mendeleev—coverage that soon transformed the affair from a human-interest story to a question of national pride to a criticism of the regime's pace of reform. The message of the daily press was in many ways a product of the medium itself.

The Spiritualism debates had been conducted largely in the tradition of "thick journals," wide-circulation general-interest periodicals for popular consumption that favored lengthy intellectual analysis and endless self-references. By the time of the Academy affair, more and more Russians, especially in the metropoles, obtained their information from daily newspapers. This meant not just a change in format, it meant the style of presentation irrevocably altered as well, in a way that was new across Europe but more rapid and intense in Russia. Now, instead of detailed philosophical discussions, arguments had to be focused into slender columns. The rhetoric had to be pointed and the argumentation transparent. Obviously, the material should be somewhat sensational. The Mendeleev rejection was a fast-paced, simple controversy that met many of these needs. Editors in turn nourished the affair by expanding it into a criticism of the Great Reforms, the very acts that had created the modern Russian press.

Ironically, Russia's first periodical newspaper was created by Peter the Great under the auspices of the Academy of Sciences, although the true dawn of the daily newspaper and its ascendancy over thick journals can be dated very precisely to the Russo-Turkish War of 1877–1878. The war sparked massive popular interest, and the need for day-to-day information, impossible in the monthly cycle of thick journals, was met by the on-site correspondents of daily newspapers. In 1863 there had been fourteen Russian-language dailies in the Empire as a whole, eight more than just three years previously. After 1879, there were seventeen in Petersburg alone.[56] A. A. Kraevskii's *The Voice* and A. S. Suvorin's *New Times,* both Petersburg papers, soon dominated the market for daily news even outside the capital. A new social caste of journalists was created to meet the demands of punchy and accurate information, and a new breed of feuilletonists crafted a form of social commentary that emphasized sarcasm and brevity to create a rapport with subscribers.[57] (In a complementary move, thick journals began to break apart the link, almost unquestioned since the 1830s, between social and literary criticism. As newspapers devoted themselves to social criticism, thick journals concentrated on literature, a step in the rise of modernist Russian prose.[58]) The change in media climate had tremendous implications for the popular understanding of the Academy affair.

Although prominent Russian writers like Feodor Dostoevsky and M. E. Saltykov-Shchedrin, who still published primarily in the world of thick journals, commented on the Academy affair, the dominant traffic occurred in newsprint.[59] Of the five preeminent newspapers of the day, only one, the *Moscow News* (*Moskovskie Vedomosti*), contained no prominent references to the affair during the last two months of 1880. The Petersburg newspapers— *The Voice, New Times, St. Petersburg News* (*S.-Peterburgskie Vedomosti*), and *Rumor* (*Molva,* which had previously been the stock-exchange newspaper, *Birzhevye Vedomosti*)—were, on the other hand, full of vitriol until beyond New Year's Eve. The distinction between Moscow and Petersburg was crucial. Mendeleev typically had engaged in local disputes, like Spiritualism and student unrest, that were considered peripheral to the Empire as a whole. And while the dispute between Mendeleev and the Academy of Sciences was similarly an affair of Petersburg personalities, bearing little interest to outsiders, it would evolve dramatically. As it unfolded, the affair assumed Imperial tints that would become central for Mendeleev's changed focus in the second half of his adult life, a turn from local Petersburg to Imperial Petersburg.

While each newspaper covered the events to a different degree (*The Voice* and the *New Times* were the most prolific), all were unanimous in their support of Mendeleev's candidacy, and all interwove broader political themes with their criticisms of the Academy. Each paper displayed some individuality. The *St. Petersburg News* typically had excellent coverage of scientific societies, and focused less on Mendeleev in his own right. *Rumor*, in turn, emphasized popular scandal and stock quotes, and devoted few articles to Mendeleev, but those were pointedly aggressive. The oddity of this popular outrage and mass support did not pass unnoticed by the very intellectuals and editors who fanned the flames:

> We are so little used to appropriate displays of public opinion that we are unwittingly surprised by its persistence in the given case, moreover since the matter concerns not sympathy for an actor, singer, or dancer, whom society is used to applauding loudly, but to a scientist, and one who studies chemistry at that, a science which speaks some kind of double Dutch, boring and known to few.[60]

Of course, support was not unanimous, but it was not a coincidence that *every* major daily in the capital constructed a political pro-Mendeleev position. This position can be summarized in two contemporary strands: anti-bureaucratic and nationalist, the latter more emphatic than the former.[61] Each position reflected the tensions underlying public discourse at this late stage of the Great Reforms; the first emphasized that the Reforms had not gone far enough, and the second that the Reforms had ignored nationality.

The anti-bureaucratic slant of various newspapers painted the Academy as a calcified institution, cloaked in secrecy and hamstrung by decrepit regulations and bureaucratic *proizvol.* Interestingly, this position was taken by the widest *variety* of newspapers in the capital, including French and German publications. For example, Permanent Secretary Veselovskii kept this cutting from the French-language *Journal de St. Pétersbourg:* "It is thus that our Academy does not have a president, but a veritable *boss,* so powerful as to paralyze the entire independence of the academicians if he wishes. He is a boss without control, striking equally well at the authority of a scientific body as that of a ministry."[62] This attack emerged directly from the vocabulary of the Great Reforms: Administration crippled independence through arbitrary power. Other critics attacked Veselovskii for being *unable* to control the institution he was supposedly in charge of.[63]

A caricature of Mendeleev made the anti-bureaucratic case even more forcefully by depicting the supposedly democratic process of voting as a secretive cabal of stylized bureaucrats (see frontispiece to this chapter). Yet others approached the matter from a legalistic view, arguing that the Academy violated its own charter by not electing Mendeleev.[64] The Academy came to stand for all that had gone wrong since the Great Reforms had slowed after 1874: "The Academy of Sciences is the best proof of what becomes of an institution outside of any control."[65] These attacks were pointed but not consistent: In the first of these objections, Veselovskii was bad because he exercised too little *proizvol*, in the second, because he wielded too much.

The nationalists couched their position in the language of Empire. The expansion of the Russian Empire in the nineteenth century had brought into its realm larger and larger populations of non–Great Russians. (This group includes not just Armenians, Tatars, and Turkic peoples, but also Ukrainians and White Russians—anyone who was not traditionally classed as "Great Russian.") The state had traditionally obtained the loyalty of these different ethnicities by incorporating them into the bureaucracy with rights and privileges equal to other residents of the Empire. In this fashion, Russia was a truly multinational state at the time of the Academy affair.[66]

The expansion of industrialization during the Great Reforms and tremendous social mobility, however, raised anxieties about national identity in Russia. Two specific solutions are of particular interest in this episode. The first focused outward, emphasizing the Slavic roots of all "true" Russians, and therefore urged an expansion of the Empire (or at least its influence) over the Slavic peoples of Europe—a heterogeneous and complex set of agendas known as Pan-Slavism. Mendeleev's rejection was occasionally invoked as a rallying cry of Pan-Slavism, as an editorialist for the *New Times* noted when the Czech and South Slavic Academies elected Mendeleev as an honorary member in the face of his domestic rejection: "Now one can say that the scholarly representatives of the entire Slavic world have expressed their reproof to the spirit which reigns in our Academy; only Polish scholars have yet to join the Slavic protest."[67] As was the case with much of Pan-Slavism, this expansionist view of Russian culture was encouraged by smaller Slavic nations, which in turn used it as a cudgel against perceived alien aggressors at home. For example, the Czech newspaper *Pokrok* wrote vigorously on Mendeleev's behalf, decrying German dominance from within the Austro-Hungarian Empire.[68]

The second attempt to solve the problem of nationality faced inward, attempting to make those foreigners who already lived within the confines of Russia more "Russian" through Russification. In the case of the Academy affair, newspapers blamed the Academy's behavior on the pernicious influence of "Germans" (both Baltic German subjects of the Russian Empire and foreign nationals) on the institutions of Russian culture. The rejection, according to one anonymous editorialist, was symptomatic of the inability of Russian Germans to relate to Russian society, especially after German unification in 1871.[69] *The Voice's* subscription for the Mendeleev prize was framed in deliberately nationalistic language, asking Russians to show "respect" for Mendeleev while the "Academician-Teutons" ignored him.[70] Thus honoring Mendeleev became a litmus test for patriotism. Rarely was the Academy affair hooked on the basis of the supposed *inter*-nationalism of science.[71]

To be sure, there was in Petersburg a long-standing resentment of foreign academicians. Foreign dominance at the higher levels of the Academy of Sciences had just begun to fade by 1880. The scars, however, ran deep, and talk of the "German party" in the Academy was ubiquitous in newspaper critiques of the institution. This was a troubled topic for the Academy. Count Sergei Uvarov, Minister of Popular Enlightenment under Alexander I (1801–1825) and then president of the Academy under Nicholas I (1825–1855), rejected nativist policies in the Academy of Sciences and deliberately brought Baltic Germans and other non–Great Russians into the Academy. Arguing that science was international and that academicians should be outside Russian politics, he led the Academy to its last peak of non-Russian representation.[72] (One should note, however, that many of these "foreign" scholars were Russian in the sense of being subjects of the Tsar, albeit not "Great Russian" by ethnicity or Orthodox religious confession.) The situation shifted somewhat back in favor of ethnic Russians when the *Russian* Academy of Sciences, founded by Catherine II in 1783, was abolished in 1841 and its membership fused with the *Imperial* Academy of Sciences. These members were largely packed into the second division of the Academy on Russian language and literature, next to the largely non-Russian third division for history and philology, and the almost evenly split first division on the physical and mathematical sciences. This signaled the beginning of the end for foreign dominance in the Academy. In the 1880s just over half the Academy members were native Russian.[73]

By 1880, however, newspapers had repeatedly raised criticisms concerning cases of Russians who were supposedly kept out of the Academy by Germans, or of those Russians who relinquished the study of Russia and its environment to these foreign carpetbaggers.[74] The most sustained expression of resentment was an article by F. I. Bulgakov in February 1881. According to Bulgakov, the Germans rejected Mendeleev in order to defend their privileges in the Academy, as they had since they hijacked the "original intention" of Peter the Great for the Academy to enlighten Russia.[75] This action, however, had mobilized Russian scientists:

> The knowledge of scientists, their contributions to science, their influence on society, now will not go to waste. We are convinced of this by the attitude of Russian society to the Academy Germans' escapade in connection with the evaluation of Professor D. I. Mendeleev's merits. . . . By [our scientists'] unanimous protest against the mentioned escapade they have clearly shown that they recognize their power, and the more they do, the stronger their con-

nection amongst themselves. And no parties will destroy this connection, which is strong in faith in the sympathy and cooperation of society in the aspirations of our scientists to liberate the development of Russian science from favoritism, nepotism, and the sinecures of uninvited *Kulturträger*.[76]

Bulgakov was rather vague about just who those Germans were.[77] However, one specific group, and one emblematic of the failures of the Great Reforms, was the Baltic Germans.

Baltic Germans hailed from what are now Latvia, Lithuania, and Estonia, the three former Soviet republics that border the Baltic Sea, but they should not be confused with the ethnic residents of those areas. They descended from German settlers who served as able local administrators of the Russian Empire's commercial interests, and were suitably rewarded. As a result, a small but significant portion of upper-level bureaucrats in Petersburg were of Baltic German descent, a point often used by novelists like Dostoevsky and Tolstoy to criticize the bureaucracy.[78] These Germans were quietly tolerated until the 1860s, when the Great Reforms spurred a public campaign against Baltic German culture and institutions. The group came to be seen as subversive and thus Russification of Baltic German educational and political institutions was promoted by Slavophile social commentators. Alexander II was opposed to forced Russification of the Baltics, but after his assassination in 1881, Alexander III began a wholesale transformation of the traditional administration of this region, decreasing Baltic German influence in the state.[79] Baltic German presence in the upper bureaucracy became an easy target for Academy-centered attacks, even though very few academicians in the first division were of Baltic German origin.

Yet Baltic German voices were prominent in the capital. By the mid-1870s, the main German paper in Russia was the *St. Petersburger Zeitung*, founded in 1727 as the newspaper of the Academy of Sciences. In 1858 it was leased to a moderate liberal, but in 1874 Baltic publishing figure John Baerens took over the lease, merged the paper with the failing *Nordische Presse*, and assembled a conservative editorial board. In 1877 he retired and turned the reins over to Paul von Kügelgen, who controlled the paper as a bastion of right-wing elitism and Baltic nationalism until October 1904.[80] The *Zeitung* was the only newspaper in the capital that excused the Academy by labeling Mendeleev a charlatan, thus earning the undying enmity of Russian papers.[81] The *Zeitung* saw in the Academy dispute the same potential for social commentary as the Russian nationalists; they defended the Academy as an isolated bulwark of tradition and privilege, similar to those they defended at home. While the Academy stood for Russian nationalists and liberals as an instance of the Great Reforms not having gone far enough, the Baltic Germans took the affair as a sign that reforms had gone too far.

Another chink in the pro-Mendeleev armor came from the left. While liberal and radical intellectuals now tended to support Mendeleev as an icon of rational scientism fighting off the anarchy of the Spiritualists (they would object later to his conservative economic policies), and while they certainly were not pleased with the Academy of Sciences, they found the media blitz surrounding the chemist a distraction from important political issues:

> "On guard! The German academicians voted against the Russian chemist Mendeleev!" And suddenly there arises in all the Russian land and even in the whole Slavic world such a rumpus, such wails of indignation and protests, that one rightly might think that the Germans knocked the earth off its orbit and that Bismarck took Moscow, trampled and insulted its holy objects and sacred gates. "In the person of Mendeleev a deep insult has been perpetrated on the entire Russian people, a mortal offense to our national feeling." And a tireless agitation began to proclaim under this banner and with these militant cries! Ovations and protests from all sides: protests from newspapers, journals, professors, scientists, scientific corporations and societies, even various protective, cooperative, and encouragement societies, etc. How much fire and energy, money and dinners, dedications, shouts, and agreements do you think have been tossed away—and for what?
>
> Tell me kindly, is the rejection of Mendeleev really more important and does it have more essential significance than the rejection, for example, of an entire press or this or that system of training and education, i.e., the fate of our children, is it really more important than the rejection of popular well-being and instruction, state finances, etc., etc.[?] However, not one of these questions has worried us as much as the German rejection of a Russian chemist.[82]

The human-interest component of Mendeleev's case thus managed to sideline other issues even while it flexibly absorbed a wide variety of criticisms of the regime. Again the medium drove coverage: The desire for higher circulation trumped the political agenda that often molded the contents of thick journals.[83]

There remains one issue that clamors for investigation. After having examined the dominant hypotheses of why Mendeleev was rejected by the Academy, the *real* reason is still unclear. Neither the simple "Mendeleev vs. the bureaucracy" story nor the jingoistic "Mendeleev vs. the Germans" attitude is adequate, if only because Germans could be found on both sides of the affair, and because Mendeleev's rejection was orchestrated by the Permanent Secretary, a Russian. So why was he rejected? Although there is no definitive answer to this question, its exploration requires us to examine Mendeleev's position in Petersburg culture. As Mendeleev became increasingly aware of the figure he cut on the public stage, he was in turn shoehorned into a constructed persona. The election debacle was partially a consequence of a particular image, the image of the *individualist Russian scientist,* and its backfire when a crisis in Mendeleev's personal life coincided with the Academy election.

BACK ROOMS: WHY WAS MENDELEEV REJECTED?

Why Mendeleev was *actually* not accepted into the Academy is substantially less important than the reasons why people were upset about it. Mendeleev failed to obtain his Academy seat for a variety of reasons, some institutional and some personal. These form a different narrative than the newspaper reaction—and even the chemists' reaction—since they were important not for the public formation of science and politics in Russia, but for the private formation of Mendeleev's views. When four factors stymied his efforts to enshrine his expertise in the Academy, Mendeleev reformulated both his persona as Russian scientist and his reform program for Imperial Russia so that those factors would be less relevant.

The first reason was scientific: The Academy may simply not have considered Mendeleev to be qualified, or at least not more qualified than other candidates, such as Beilstein.[84] To readers today, with knowledge of the periodic law (and who have probably never heard of Beilstein), this seems absurd. On Beilstein's behalf there was a long series of important experimental studies on isomerism of aromatic organic compounds, as well as the first edition of his seminal *Handbuch der Organischen Chemie.* Mendeleev, on the other hand, had only one of his eka-elements to his credit (scandium, discovered in 1879, was still inadequately verified). Further, the priority dispute with Lothar Meyer over the discovery of the periodic law had just flared up again in Berlin in early 1880, which made Mendeleev's claims for recognition temporarily suspect. And many chemists still found the pedagogical utility of the system debatable. In one Munich discussion, reported by a visiting Russian, only one chemist strongly advocated the system.[85] Meanwhile, Mendeleev's gas project was ignominiously collapsing, with Gadolin, an academician, one of the central monitors of the project. Many foreign chemists also considered Beilstein the better candidate. As chemist Albert Ladenburg recalled in 1907:

> In my first years in Kiel I usually used the Easter vacation to visit my parents in Mannheim, and from there went for one or several days to Heidelberg to visit with my former teachers and friends. On one of these visits I met also with Hermann Kopp, who told me: "Just now we ('we' meant Bunsen and Kopp) got the question of whom we consider more worthy and appropriate

to be a member of the Imperial Russian Academy—Beilstein or Mendeleev. We, it stands to reason, proposed Beilstein; surely you agree?" I immediately responded that I entirely disagree, and although Beilstein is my dear friend, without a doubt I would have proposed Mendeleev.[86]

Yet Ladenburg did not consider Kopp's suggestion to be ludicrous. It might not, therefore, be too surprising that several members of the Academy agreed.

Even if some of the academicians believed that Mendeleev was not qualified, this was dwarfed by the second, and most important, basis for the rejection: the strained relations between the Academy of Sciences and St. Petersburg University. Both Butlerov and the newspapers hinted at this. Consider one of the responses by German journalists in Russia to attacks on their loyalty to the Russian state, as argued in the *Neue Dörptische Zeitung*:

> [T]he Academy must be free of the yoke of university terrorism. Well-known foreign scientists, under current financial and other conditions, even without this are hard to tempt with an invitation to the Petersburg Academy; young scientists not yet having a position, and one which allows them to devote themselves unfettered to science, and who for a beginning, of course, took an adjunct position in the Academy, are not allowed into the Academy by their enemies and if the Academy is not to be closed, if it will still drag out a similar existence, then all that will remain is to fill all of the vacancies with Russian professors.[87]

This defense was symptomatic of a larger anxiety: that the Academy's position vis-à-vis the University was slipping.

There is much to endorse this account of Mendeleev's fate in the Academy. First, given the dominance of Great Russians in the University and non–Great Russians in the Academy, interpreting the event as an opposition between the two neighboring institutions captures some of the affair's nationalist resonance. This interpretation also provides a satisfying sense of *locality*. Given Mendeleev's public visibility, his election was a fortuitous case for the Academy to oppose him as a representative of the University.

The best evidence for the institutional aspect of Mendeleev's rejection comes from its architect, Permanent Secretary K. S. Veselovskii. As quoted earlier, when Veselovskii became exasperated with Butlerov's efforts to elect more professors into Academy chairs, he exploded: "[Y]ou want us to ask [St. Petersburg] University's permission for our elections. This will not be. We don't want university types. Even if they are better than us, we still do not need them." A look at the academicians themselves reveals why this feeling persisted. The members of the first division sort into three categories: six pedagogues, eight administrators, and four indeterminate. The first group tended to be pro-Mendeleev, were active researchers, and often University professors. The second group was generally anti-Mendeleev, scientifically inactive, and shared a fairly common biography. Five of the eight had graduated from Dorpat (now Tartu) University in the Baltic provinces, and usually began their career with a research expedition—commonly geographic or natural-historical—in the 1850s and 1860s, and then spent the rest of their careers heading academic institutes. They did not want Mendeleev in the Academy because they had no desire to be aggravated by another Butlerov-style activist.[88]

Veselovskii's files reveal traces of this anti-professorial attitude. Worried about the negative repercussions of the affair on the Academy's status, he sent a letter to the Academy's president on 21 December 1880 showing how Kazan University students—led, he contended, by the bad example set by professors—protested for Mendeleev and he feared that the 29 December public assembly of the Academy might be interrupted by protests.[89] (It was not.) Veselovskii's unpublished memoirs give even more insight into his agitation. Although from the perspective of the pro-Mendeleev press, those responsible for Mendeleev's rejection were a homogeneous block, there was persistent hostility between Veselovskii and President F. P. Litke, to mention just one example. Motivated by resentment of his superior's status, and also genuine lack of respect for the latter's intellect, Veselovskii penned this retrospective, self-serving, and at times consciously deceptive rendition of the incident:

> . . . Still more embarrassing for the Academy turned out to be the consequences of Litke's incompetence concerning the failure to elect Mendeleev as academician for technology. Academician Butlerov, being at that time a professor at the university, led a constant open war against the Academy, and to please his university colleagues several times tried to introduce Mendeleev as an academician, against the wishes of the majority of members of the physico-mathematical division. The first time he proposed him as an adjunct in physics, regardless of the fact that Mendeleev had no achievements in this science, but even directly embarrassed himself by his attempt to correct Regnault's tables on the expansion of gases. But since it was a matter of an adjunct post, and adjuncts are not allotted to sciences by the charter, and the Academy retained the right to select which science it considered necessary, then in view of the very great probability of a negative result of the voting, if it were allowed[,] the voting on Mendeleev was averted with the help of a preliminary question, i.e., the question was put: should the vacant adjunct post be allotted to physics[?] The negative resolution of this question by voting obviated voting on Butlerov's proposed candidate.
>
> Some years later, when a vacant post of an ordinary academician in technology opened up, Butlerov, [being] stubborn and malicious towards the

Academy, proposed Mendeleev for it, knowing full well that this candidate could not generate the necessary majority of votes, but with malicious glee counting on generating a scandal that would be unpleasant for the Academy. It was impossible to avoid the danger as before with the help of a "preliminary question," since the position of a technologist was ordained by the charter and was at that time vacant. The only way to avoid the scandal of a rejection was the right of *veto* granted to the President by the charter. Therefore, by the wish of a majority of the academicians, I went to Litke, pointed out to him the almost entirely indubitable negative result of the voting, to the scandal which it could produce given the malice towards the Academy of those persons who had pressed Butlerov to make the specific proposal, and explained that only by the *veto* right allotted to him could the danger be averted. As much as I pressed that uncomprehending geezer, he never agreed. . . . Nothing helped; the voting took place, Mendeleev was rejected to the great delight of those who had arranged the entire scandal in the form of a declaration of war on the Academy.[90]

Here Veselovskii presented himself much as his critics from the anti-bureaucratic strand did: a man incapable of controlling the Academy, a pawn buffeted by forces beyond his control. This was clearly not the case. Yet his account does demonstrate the prevalence of anti-University sentiment among academicians.

The third major reason why Mendeleev was not elected to the Academy was personal. He was a rather difficult man to stomach, both politically and personally. Politically, Mendeleev's activities on behalf of student protesters made him a somewhat undesirable figure for a staid institution like the Academy. He was no rabble-rousing agitator, but his sympathies occasionally leaned a little too much toward the students—and the defense of the 1863 statute, which liberalized universities and gave considerable autonomy to the professoriate—for hardliners in the Academy. Academician and professor of botany A. S. Famintsyn, the most prominent Academy supporter of Mendeleev's candidacy besides Butlerov, was secretly arrested after he gave a report at the University in December 1878 blaming the 27 November student unrest on social and political structures.[91] During this same brief uprising, Mendeleev was cited by Tsarist authorities for similar sympathies to the students, according to spymaster A. R. Drentel'n's notes to Alexander II: "General-Adjutant Gurko told Professors Mendeleev and Menshutkin, who, judging by our agents' reports, responded disrespectfully to inspections, that if any kind of demonstration took place from the side of the students, they would both be exiled from Petersburg immediately." (The Tsar noted in the margin: "And well done.")[92] Mendeleev was also a cantankerous character, certain to cause personality-clash problems in the Academy just as he had at the University in the Spiritualism battles. One of Butlerov's students overheard a conversation in the Academy library the day after Mendeleev's rejection between an academician and a librarian, in which the only reason mentioned for rejecting Mendeleev was "his difficult character."[93]

The fourth reason for objecting to Mendeleev was linked to his acerbic personality: his supposedly dubious moral character tied to his highly public divorce, an event that generated much public notoriety for Mendeleev at precisely the moment when his candidacy for the Academy was on the line. Mendeleev decided to marry his first wife, Feozva Nikitichna, shortly after the 1862 Demidov prize money from the Academy of Sciences relieved him of some of his more immediate financial pressures. He was 28 years old; she was already 36. After the death of a first daughter in childbirth, Feozva Mendeleeva bore two children, a son Vladimir and a daughter Ol'ga, before the marriage started to sour. By the end of the 1860s, just as Mendeleev was beginning to achieve social and scientific visibility at St. Petersburg University with his newly formulated periodic system, the married couple was no longer living together. While Mendeleev was in Petersburg, his wife and children lived at their country estate Boblovo, near Moscow, and when Mendeleev went there, Feozva returned to Petersburg. This lasted until 1877, when a young art student from the Don region of Russia, Anna Ivanovna Popova, came to stay in Petersburg with Mendeleev's older sister Ekaterina I. Kapustina.

Kapustina had met Popova in August 1876, when the latter took on lodging in the widow's home in order to attend courses at the Academy of Arts, down the Neva embankment from the University. In April 1877, while Kapustina was changing residences, she moved with her family—and her lodger—into Mendeleev's apartment at the University. The 43-year-old Mendeleev became infatuated with the teenage student, writing her secret letters every day (and never posting them), helping her carry her art materials, commissioning her to copy paintings for him, and in general earning the snickers of gossips on Vasil'evskii Island, who dubbed the pair "Faust and Margarita."[94] Worried about the damage this infatuation of Mendeleev's could have on his reputation (he had attempted an affair with his daughter's governess almost a decade earlier), the Kapustins moved out in November 1877. Mendeleev continued his courtship, however, initiating his "Wednesdays," salon evenings at his apartment, where he began his acquaintance with various artists. The goal of these events, admitted by Mendeleev to Popova later, was to advance her in the art world. In 1878 Mendeleev contracted pleurisy and was sent abroad to heal, "and I thought that there I would manage to get over Anna Ivanovna."[95] Apparently not. Mendeleev proposed to Popova in 1879, but she turned him down, and her father visited the capital to warn Mendeleev off. He continued his pursuit, and in December 1880 her father sent her to Italy to put distance between the two.[96] Thus, at the time of his rejection by the Academy, Mendeleev was the subject of many rumors—not the least of which was that he had been conducting an affair with the young student.

After Popova left for Italy, Mendeleev, already crushed by the collapse of his dream of entering the Academy, contemplated suicide and wrote a will naming her as a main beneficiary. Then he went to Rome in 1881 and proposed again, saying he would kill himself if she did not agree. She did, and divorce negotiations commenced.[97] Mendeleev's first wife opted for divorce on the grounds of adultery. The Church officially dissolved the marriage on 23 February 1882.[98] (The birth of Mendeleev's daughter with Anna, Liubov', was postdated to the spring of 1882, although she was born in December 1881.) By the divorce agreement, Feozva received Mendeleev's salary from the University, so Anna and Mendeleev had to live on the profits from *Principles of Chemistry* and his economic consulting.[99]

Mendeleev's problems with the Church over his divorce were not finished, however. According to Orthodox law, Mendeleev had to wait seven years before remarriage, any violations of which were considered bigamy. In January 1882, even before the official dissolution of the marriage, Mendeleev was married in an Orthodox ceremony at the Admiralty church, a privilege he secured by bribing the prelate with ten thousand rubles. (The offending cleric was later defrocked.) Thus Mendeleev lived through the early 1880s as a publicly acknowledged bigamist, officially overlooked because of his scientific reputation. Later, when a prominent bureaucrat wanted to accomplish a similar marriage transfer and asked the new Tsar, Alexander III, for a pardon, citing Mendeleev as a precedent, the Tsar refused, reportedly announcing: "Mendeleev has two wives, but I have only one Mendeleev."[100]

TO THINE OWN SELF: THE MAKING OF A NEW MENDELEEV

The story of Mendeleev's divorce and the Tsar's comment brings us to the crucial issue of reputation. If there was a positive outcome to the Mendeleev protests, it was that they finally solidified a scientific reputation for Mendeleev at home in Russia. At first glance, this sounds ridiculous. Surely Mendeleev's reputation was created by his formulation of the periodic law (1869–1871). If that was not enough, then Mendeleev's attack on the Spiritualists must have made him into the respected scientific figure we now take him for. As it happened, perhaps nothing catapulted Mendeleev to such public prominence as the fact that the Academy of Sciences did *not* consider him worthy of membership.[101]

Before the Academy affair, most of the periodical-reading public would only have known of Mendeleev because of his efforts against Spiritualism. In terms of his science, even observers at the University were occasionally skeptical. Recall that Mendeleev devoted the entire 1870s to an experimental research program to discover the laws of gas expansion. This venture in physics was damaging to Mendeleev's scientific credibility, as Veselovskii's statement above indicates. The discovery of the predicted gallium in 1876 granted him status only among chemists. As for his public ventures, they bore the stamp of the dilettante. Future Khar'kov University professor of zoology A. M. Nikol'skii, who had worked at St. Petersburg University in 1878–1879, found Mendeleev amateurish: "I am embarrassed to say that I did not like him, either as a professor or as a person." Nikol'skii found Mendeleev more eager to win a cheap laugh than to communicate scientific facts, and contrasted him unfavorably with the more established Butlerov.[102]

The Academy affair was to change all that, crafting for Mendeleev a new persona as a *public* scientific genius. In order to make the case against the bureaucracy and the Baltic Germans, the newspapers needed to argue that Mendeleev *would* have been elected if bias had not distorted the matter. As a result, editorials extolled the virtues of the periodic law and his work for the oil industry—in short, whatever they could find. The horde of local scientific societies that rushed to honor Mendeleev made the case even more solid. At his death, Mendeleev possessed 131 titles, awards, and memberships in a variety of scientific societies. Plotting the quantity of awards across his life, one finds a peak among domestic societies around the Academy incident. (Foreign societies honored him predominantly in the late 1880s and early 1890s.)[103]

Mendeleev came from a *raznochinets* family—a family of "various ranks" that populated the lower rungs of the civil bureaucracy—to become the epitome of an *individual Russian scientist:* the unofficial label that would define him thereafter. The "scientist" reputation was constructed through journalistic demand for a more sensational stick with which to beat the Academy. From that coverage, Mendeleev's "Russianness" vis-à-vis the "Germans" emerged unquestioned. His rejection by the Academy made him already "individual," appealing to a long-standing Russian popular-literature tradition of siding with the underdog against a dominating institution, an impression he intensified by dismissing his rejection before his students on 12 November with a cavalier folksy comment.[104] The fanfare made it clear to a wider public what it meant to "do science" in late Imperial Russia. All of these aspects of the public persona were created without Mendeleev lifting a finger.

In fact, Mendeleev has been noticeably absent from this story. Whether the debates were raging in the Academy, among chemists, or on the pages of Petersburg dailies, Mendeleev's own views received little attention. He was understandably quite wounded by the rejection, and he did not capitalize on the public notoriety created for him in the popular newspapers. As he wrote to his future wife on 21 January 1881: "Things with me, dear Annie, are not good, and I don't recognize myself. . . . These different addresses and honors are as repellent to me as all the attacks were."[105] The one exception to Mendeleev's withdrawal was his acceptance of *The Voice*'s efforts to subscribe a prize in his name, which he at first wanted to use to build his coveted aerostat, but ended up donating to the Russian Chemical Society.[106] He observed quite astutely that the Academy affair was really "an issue of a Russian name, and not about me," and that "what has occurred shows more clearly than anything that here in Russia a feeling of scientific autonomy has grown. . . . I am very happy that I served as the simple, external cause of the awakening of societal consciousness."[107] Despite his seeming goodwill, Mendeleev also expressed sour grapes, writing to chemist P. P. Alekseev on 23 November 1880: "I didn't want to be elected to the Academy because they do not need what I can give."[108]

His relations with the Academy continued to deteriorate. Despite Butlerov's protest against the Beilstein election and Famintsyn's later effort to give Mendeleev Butlerov's chair, Mendeleev never achieved any higher honor in the Academy than his corresponding membership of 1876. In Mendeleev's personal chronology of his life, he did not even mention the Academy vote in 1880, although his trip abroad with Anna Popova was dutifully noted.[109] Mendeleev next appeared in the minutes of the Academy in 1895, when the Ministry of Popular Enlightenment asked the Academy to delegate Mendeleev as their representative at the International Committee of Weights and Measures, since he already represented the Chief Bureau of Weights and Measures. The Academy selected F. A. Bredikhin instead. On 19 January 1898 Mendeleev returned the insult by denying permission for academician Baklund—the same man elected in Mendeleev's wake—a visit to the Bureau.[110] Minister of Finances and Mendeleev's close friend Sergei Witte wrote to the Academy in 1899 in outrage that Mendeleev was *still* not a member of the Academy, while there was a widespread impression in Western Europe that Mendeleev was.[111] How could the most famous Russian scientist *not* be?

Mendeleev's perspective on the Academy disappointment had enormous consequences, transforming how he viewed the project of reforming Russia through enlightened use of expertise. Before 1880, Mendeleev had applied his reform efforts on a more-or-less local scale. His interests and passions were focused on Petersburg as a *city*, his home and place of research. Becoming an academician, the pinnacle of his profession in Russia and thus a coveted honor for local sociability, would have allowed him to advocate his interests more effectively. That dream, however, was not to materialize, and Mendeleev—spurned by the Academy—began to look at Petersburg through new glasses: Imperial ones. For the Academy of Sciences was a double institution: It was *local*, as the official center of science among the many institutions of the capital; and it was *Imperial*, the crown jewel of the Empire's civilizing mission.

Around this fulcrum Mendeleev began to traverse the local networks of St. Petersburg differently. His concerns were no longer for decentralized local reform, but for Empire-wide reform. Spiritualists would no longer fret a Mendeleev concerned about Imperial tariff policy and the introduction of the metric system—a change in emphasis that I call his "Imperial Turn." Mendeleev's life may have entered a personal crisis at the moment of the Academy affair, but this was dwarfed by the cataclysm that transformed not just Mendeleev's faith in the ability of the Great Reforms to serve as a model for change, but the faith of Russian society at large. On 1 March 1881, Tsar Alexander II, the Great Reformer himself, was silenced by a terrorist's bomb. Under Alexander III, there was no longer need for Mendeleev's rejection to function as a criticism of the Great Reforms. The Great Reforms were over, and Russia entered its own Imperial Turn.

THE IMPERIAL TURN

Mendeleev, on left, with metrological assistants at the Eiffel Tower, 1895, from Makarenia, Filimonova, and Karpilo, *D. I. Mendeleev v vospominaniiakh sovremennikov,* 2d. ed., insert.

THE IMPERIAL TURN:
ECONOMICS, EVOLUTION, AND EMPIRE

Any system—in the good and the bad sense of the word—is not a
Russian thing.

—IVAN TURGENEV[1]

ASIDE FROM SOUR GRAPES and a stifling depression of several months,
Mendeleev had two productive reactions to his rough treatment at the hands
of the Petersburg Academy of Sciences. First came disillusionment with the
potential of local scientific societies to overcome personal prejudices. If the
Academy could reject him, perhaps such societies were not the best way to mediate cultural conflict in Imperial Russia. So Mendeleev began to reinterpret
the legacy of the Great Reforms, believing they were no longer about turning
state power over to a newly created public sphere, but about the power of the
bureaucracy to transcend personal animosity for the greater good: Mendeleev's
"Imperial Turn." His second reaction to the Academy affair was to take to
heart the image of a rugged individualist given to him by the Petersburg
dailies. On face, this persona was incompatible with the Imperial Turn: The
Byronic hero could not but chafe under the rigors of the bureaucratic control
of expertise. The consequences of these opposing "conservative" reactions to
the events of 1880 define much of the rest of Mendeleev's public life.

In February 1882, after his return to Russia from travels abroad with his
new wife, Mendeleev dictated a reform program for an institution he had previously exempted from criticism: the Imperial Academy of Sciences. Echoing

Butlerov, he entitled his manuscript "What Kind of Academy Does Russia Need?"[2] Mendeleev discreetly avoided discussing his personal rejection by the Academy; instead of trying to find out how the Academy had become so estranged from Russia, he wanted to make it relevant again.

Originally, he argued, the Academy was intended to train Russians in the individual sciences, and it had accomplished this goal admirably. But the Academy had since suffered from a rupture in its function, just as the nobility had since the Great Reforms:

> After all, everyone knows that the Russian nobility is a service court establishment, that each nobleman was obligated originally to serve and through this, so to speak, obtained his significance and state position. Then the nobility was freed from service and remained a nobility with the same estates that were given them for service.
>
> This, one could say, also happened with the Academy. Summoned to a pedagogical matter, to the fulfillment of obligations, it received the rights, so to speak, as reward for duties which it must carry out. The duties have ceased, but the privileges remain and have even increased.[3]

How should one modernize old institutions? This was the central problem of the Great Reforms. Yet Mendeleev's proposal bore a different slant from his models of the 1860s.

According to Mendeleev, Russia no longer needed the Academy as a pedagogical institution, since universities now could provide "the local academies that Peter, founder of the Russian Academy, wanted." He maintained that science had historically moved from the secrecy of the monastery into the academy, and then from the academy into the university. Each of these institutions had been progressive at first but then became conservative, since "conservatism in science is completely inevitable, because science in essence is a legacy, unthinkable except as the wisdom of centuries past, and thus cannot be passed on without conservatism." As a result, universities should now assume the mantle of Russian science aided by the proliferation of scientific societies, which existed only to increase knowledge, not to increase privilege.[4] Furthermore, the Academy inadequately provided the technical expertise desperately needed by the state. Instead, expertise was fragmented within each government ministry, which hindered wide-ranging solutions.

Mendeleev proposed unifying the scientific roles of universities, scientific societies, and ministerial technical committees into a renovated Academy. Academicians would be recruited from all over Russia, but they would not be paid. Remuneration would be covered by the local universities where the scientists worked, along the model of the Russian Chemical Society. The state could solicit (and pay for) advice from this already-assembled body of experts, which could likewise pool resources for expensive projects that no small society could afford on its own—encouraging research in areas like coal, oil, and gold exploitation. Meetings of the Academy would be voluntary and public; the small but consistent attendance of interested citizens would provide "the best control of the correctness of the conduct of affairs in the Academy."[5]

Certain of these elements will recur everywhere in Mendeleev's programmatic thinking after 1880: Empire-wide organization; evolutionary justifications; rational, compensated allocation of expertise; state-led development of the economy; pooling resources for large-scale projects; restricted public control of bureaucratic institutions. These are the hallmarks of Mendeleev's shift from local Petersburg scientist to Imperial technocrat. This simple perspectival transition from Petersburg as bustling city to Petersburg as nerve center of a sprawling Empire had tremendous repercussions. Following in the footsteps of other chemist-economists like Antoine Lavoisier and Justus von Liebig, Mendeleev in the 1880s began to develop its implications for restructuring the national economy.[6] From consulting on oil to the "Mendeleev tariff" of 1891 to demographic studies, he would propound an evolutionary economic model fundamentally opposed to emergent Marxist doctrines; instead of a model based on conflict, then popular among Russian intelligentsia circles, Mendeleev offered a theoretical political economy of constant circulation. In all areas, the systematic aspect of the post-1880 Mendeleev sought stability in the circulation of vital coordinators—labor, statisticians, metersticks, capital—throughout the Empire. These journeys produced no qualitative change in either the coordinators or the system; the metersticks would not change by being universally distributed. In fact, a purely predictable and lawlike system was *defined* by being invariant under this kind of circulation. Coupled with an autocrat's transcendent will, Mendeleev's framework purportedly allowed the Empire to modernize gradually along a universal evolutionary path. But there was a necessary complement of haphazard will that underlay such smooth structures in Mendeleev's thought. By embedding an irreplaceable, unaccountable genius at the base of his structure, Mendeleev undermined his own foundations. The edifice was liable to come crashing down in both politics and science.

THE TWO PETERSBURGS: MENDELEEV'S EARLY ECONOMICS

Mendeleev never offered a complete systematic outline of his views, but an internally consistent and relatively stable position emerged in his writings from the 1880s until his death in 1907. This consistency contrasts sharply with the heterogeneity of his views from the era of the Great Reforms (1861–1880). This new and complicated framework of the Imperial Turn emerged from a simple change of perspective.

After the series of personal crises that buffeted Mendeleev in late 1880 to early 1881, he began to look at the same city—Petersburg—in a radically different way. Earlier, he saw the city as composed of a collection of *local* institutions (St. Petersburg University, the Russian Technical Society, the Russian Physico-Chemical Society, the Commission on Mediumistic Phenomena) that would marshal the necessary expertise for the state in a decentralized fashion. Mendeleev had been worried about *Petersburg* Spiritualists and *Petersburg* educational institutions, not *Russian* ones. The Academy of Sciences changed all that; it served as the transition point from looking at things as collections of Petersburg personalities to seeing them as Imperial institutions. When his rejection showed him that the Academy's locality trumped its Imperial agenda, he moved from local perspectives to Imperial ones. Now he saw the city as a connection of Imperial institutions: the Navy, the Ministry of Finances, the Bureau of Weights and Measures, the Winter Palace. He no longer wanted to mediate expertise through the public sphere. Instead he went to the source and sought to control expertise through the bureaucracy, the institution that embodied the spirit of the Great Reforms: transformation through reasoned policy.

Mendeleev's economic thought changed markedly after November 1880, maturing from ideas in his early career about evolutionism and science.[7] By "evolution" I do not mean the Darwinian ideas that were percolating through Russian intelligentsia circles in this period; Mendeleev was oddly silent regarding biological evolution. Of chief import to Mendeleev were his views of inexorable *social* evolution, which lay at the base of many of the sociological doctrines of his contemporaries.[8] For example, during his gas research of the 1870s, Mendeleev proffered analogies between the physics of gas molecules, scientific societies, and the Russian context. As he mused in a notebook while on a trip abroad:

> The molecules of gases are revived organisms running from each other. Expansion. The savage is thus. Civilization is unity, connection. The beginning of Christianity and civilization is family and love, communality (*obshchenie*) is born and multiplies. Thus in the beginning there were Poles, Russians, Little Russians, Georgians, melding into one. Egoism vanishes and does not develop. Union.[9]

Mendeleev actively constructed such analogies between scientific work and political and social structure, to some degree derived from his voracious reading of social and economic works.

His early agricultural reform proposals are a prime example. In 1866, under the auspices of the state-sponsored Free Economic Society, Mendeleev composed a questionnaire so that farmers could gather information about soil quality. But it was not enough for farmers to have the data; it had to be organized somehow. "Only a . . . scientific society rich in both moral and capital resources can organize such experiments," he claimed.[10] Similar to what he would propose for meteorological observations in the 1870s, local farmers would gather data (after being properly instructed by an expert—himself) and then he would coordinate all the results through the Free Economic Society. Inadequate sampling, however, prevented these experiments from generating the comprehensive results Mendeleev had anticipated.

His proposal for a decentralized farmers' credit union (1870) also met with an inauspicious fate. This organization was meant to pool the producers of different crops into societies, generating both a more effective distribution network and a cushion of diversified risk for farmers during hard times. Mendeleev suggested locating the first bank, of course, in Petersburg. Petersburg's "centrality" was not geographical but economic. When A. A. Klaus criticized Mendeleev for eschewing grassroots efforts, he responded: "First of all I turn your attention to the fact that we are dealing with Russia, in which everything begins at the top. What is to be done—such is the order of things! . . . In order to begin matters from the bottom, we need a greater development of units than we now have."[11] Even at this early point, the dynamics of Imperial politics underwrote his utopian dream of local societies. Historian Beverly Almgren has suggested that the empirical failure of Mendeleev's agricultural projects led him to believe that Russia must shift to industry. And indeed, in 1899, while reflecting on these early writings, Mendeleev concurred: "These are important to me because they warranted the entirety of my later attitude to industry." While writing this work, he said, he became convinced of "one possibility—protectionism—to create a new class of people and a new perspicacity—but these sleep even now."[12] But both Mendeleev and Almgren make the transition rather too smooth.

The discontinuity in Mendeleev's thought was conditioned by the opportunity to reform the Russian state, and on 1 March 1881 that opportunity presented itself—quite literally—with a bang. On that day, the Tsar-Liberator, Alexander II, was assassinated by a terrorist's bomb. The People's Will, the group behind the killing, believed that once the Tsar was shown to be mortal, Russian peasants would unify and help create a revolutionary utopia.[13] Indeed, a sudden and almost universal cultural transformation, a massive cultural shock, followed in the wake of the killing. That shock, however, rebounded against the terrorists, who were soon captured, prosecuted, and executed. In addition, the assassination promoted a siege mentality in the government and a yearning for stability among the public. The *Moscow Times*, two days later:

Freedom and a truly human life in society are possible only if there exists a unified and unquestioned authority above all others. Its diminution inevitably breeds trouble. To the degree that the sovereign power shows weakness, primitive and beastly forces grow. Social decay sets in, violence commences, the foundations of all morality are shaken, the spirit of corruption overcomes everyone's minds. . . .[14]

Already in despair in the aftermath of the Berlin peace settlement of 1878—the pyrrhic victory of the Russo-Turkish war—Russians were now completely bereft of the sense of optimism that had followed Emancipation. Mendeleev's personal desire for stability after 1880 thus had a much broader cultural backdrop.

Politically, the state retreated from the ideals of the Great Reforms. The movement toward Counter-Reform had been paved by the agricultural crises of 1878–1880 and the attempts to control them single-handedly by Minister M. T. Loris-Melikov, but the assassination provided an additional justification for reevaluating both the condition of the countryside and internal security needs.[15] Although the new Tsar, Alexander III, was motivated by similar interests as his father—maintaining fiscal and military stability in the Empire—he governed in a substantially different style. Insecure in Petersburg high society, the new Alexander hearkened back to the reign of his grandfather, Nicholas I, who had emphasized direct monarchical control of the bureaucracy. Alexander III was so jealous of his monopoly over the state that he even refused to purge Great Reformers from his administration for fear that this would make him too dependent on the rightist lobby.[16] As a result, only *he* could broker settlements between factions.

This new style suited the Imperial Mendeleev just fine. Under his new benefactor, the chemist achieved a meteoric rise in the Imperial hierarchy, and he remained loyal to the memory and policies of Alexander III even under his successor, Nicholas II, who was much less amenable to Mendeleev's approach.[17] Before the Imperial Turn, Mendeleev had considered the administrative structure of the Great Reforms best suited to his goals. In the framework of ministerial politics under Alexander II, interest groups lobbied the relevant ministry and, through elite participation and solicitation of opinions, influenced government policy. Instead of ceding control over parts of the state, Alexander II assembled information so that he could make informed decisions. The idea of direct, specialized access to the Tsar attracted Mendeleev when he was still a relative outsider.

Under Alexander III, Mendeleev had already become an important consultant for the Ministry of Finances, his voice heard within the halls of power. By centralizing decision-making, the new Tsar resolved Mendeleev's primary criticism of his father's administrative style.[18] Rather than perceiving the increasing restrictions on public opinion as an abandonment of lawfulness and a return to arbitrary will (*proizvol*), Mendeleev interpreted them as a law-based program closest to his own. Furthermore, the Empire experienced a rare time of peace during the reign, despite the leadership's aggressive tendencies; thus, Mendeleev's advice on economic reform through the offices of Ministers of Finances Ivan Vyshnegradskii and Sergei Witte risked little interference from the Minister of War.[19] Mendeleev had finally become an insider. He would remain one until his death.

REAL ECONOMICS: MENDELEEV AND THE RUSSIAN ECONOMY

Economic history remembers Mendeleev best for his practical actions in a variety of areas that actively contributed to developing Russia's industrial capacity. He did some of this work as a consultant to private firms, some as a bureaucratic advisor. At the start of his consulting career, his suggestions were specific and not at all couched in a system of economic concepts. That would come later. Before and even after his creation of a macroeconomic model in the 1880s, Mendeleev consulted across the commercial spectrum, with an emphasis on heavy industry.[20] These personal experiences formed the empirical base for grounding specific points of his abstract evolutionary economics; they also altered the economic landscape of Imperial Russia—mostly through the so-called "Witte system"—becoming an intrinsic part of the context that Mendeleev's theoretical work meant to change. These fragmentary specific policies, however, fit only with difficulty into a coherent framework. To distinguish Mendeleev's earlier practical work therefore demonstrates the innovations of the later theory.

Mendeleev's economic policies are central to the major debate of whether late Imperial Russia possessed a developed capitalist economy. Historians agree that Russia lagged far behind the major Western European economies, even after the Revolution of 1905, but substantial disagreement remains about exactly how far. The industrial economy was highly concentrated, with one-third of corporations located in Petersburg and another fifth in Moscow. Enormous obstacles to capitalist development remained: merchants delayed in adopting new business techniques, bankers granted credit with reluctance, locals resented foreign investment, fragmented internal markets and poor transportation stunted profits, and so on.[21] Nevertheless, the net national product in the late Empire experienced 3.25% growth (1.7% per capita), making Russia one of the fastest-growing economies along with those of the United States, Japan, and Sweden.[22] Important here is the role of the *state* in promoting this growth. Did the state promote growth through "priming the pump" and wise policy-making, or did political measures hinder even greater capitalist development?

The debate has crystallized in the last several decades around the modernization theories of economic historian Alexander Gerschenkron and his view of the benefits of "backwardness." According to Gerschenkron, in 1855 Imperial Russia was economically "backward" with respect to Western Europe. Beginning with the Great Reforms, but more particularly with Alexander III's reign, state policies generated substantial growth, but could only have succeeded under such conditions of relative backwardness. As Gerschenkron put it: "Economic development in a backward country such as Russia can be viewed as a series of attempts to find—or to create—substitutes for the factors which in more advanced countries had substantially facilitated economic development, but which were lacking in conditions of Russian backwardness." Thus the government's budgetary policy substituted for a deficient internal market and such "Asiatic" measures were withdrawn once that market had emerged.[23]

Mendeleev's case has implications for this thesis. So far most of the discussion on Gerschenkron has centered on competing numbers (total growth, growth per capita, and so on), but few have stopped to examine the analytical concepts being used. At issue in Mendeleev's theoretical economics and the Witte system more generally was what "state intervention" and "private initiative" *meant* in historical circumstances when the economic function of the state and the meaning of "private" remained unclear. Mendeleev saw a tabula rasa in the Russian industrial economy—not that industry was absent, but that this industry lacked a conceptual framework. It helps to look orthogonally at the Gerschenkron thesis: Given the sizable amount of growth and state intervention, it is better to examine the ways in which contemporary economic thinkers—decades before Gerschenkron—saw their policies not as "state drives private" or "state inhibits private," but rather "to what degree does the state *make sense* of the private?"[24]

This question moves the discussion beyond the arcana of historical statistics and directly to the crux of a contemporary controversy between Russian populists (*narodniki*), who argued that Russia was undeveloped and could thus "skip over" capitalism on its own path to socialism, and Marxists like Lenin, who claimed that Russia had already gone so far into capitalism that it was ready for the dictatorship of the proletariat.[25] Mendeleev considered the Marxists to be utopian dreamers who misunderstood the role of private initiative and individual natural capacities. They had to be fought in the realm of theoretical economics. He saw the populists, however, as more dangerous, since they resisted capitalist development—Russia's only savior—altogether (and, incidentally, they were the seed of the violent terrorist movements like the People's Will that had assassinated Alexander II in the first place). They had to be confronted by the specifics of the "Witte system" Mendeleev helped build.

Named after Mendeleev's sometime friend Sergei Witte, who assumed the portfolio of Minister of Finances in 1892, the "system" was not a coherent set of principles but a loosely defined set of economic measures that were retrospectively grouped together.[26] The most important policies emphasized the primacy of industry for the economy. First, the state primed heavy industry by guaranteeing profits in railroad construction, offering incentives for oil development, and other fiscal measures.[27] Then, the state encouraged foreign investment to supplement a domestic shortage of capital through protectionist measures, which made joint ventures with Russian companies more economical than importing.[28] Finally, Witte moved Russia to the gold standard in 1897–1899, which stabilized the wildly fluctuating exchange rates and ameliorated the need to export large amounts of agricultural goods to achieve a positive balance of payments.[29] All of these were intended to maximize the predictability of industrial development: the less random the environment for capitalism in Russia became, the more the economy would grow. This was a man and a doctrine after Mendeleev's own heart. Witte himself interpreted Mendeleev, "our distinguished but unappreciated scientist," and his own "devoted associate and friend," as a tragic case of good intentions gone awry (much as Witte felt about himself).[30] In his memoirs, he spoke fondly of Mendeleev as a kindred spirit:

Mendeleev was a man of both theory and practice. He contributed greatly to the development of our oil and other industries. But, as often happens with outstanding men, he was the subject of wide and bitter criticism, partly because of his personality, partly because he was more intelligent, more talented, more learned than those around him. His writings on economic development, which favored protectionism, were the object of derisive criticism, of [unfounded] charges that he was in the pay of industrialists.[31]

Both Witte and Mendeleev offered similar concrete proposals and wielded gruff personalities and autocratic managerial styles, making the two a good match.

There has been understandable emphasis on Mendeleev's consulting for the growing Baku oil industry and its relation to the Witte system.[32] Mendeleev was affiliated with the oil industry (one of the most important sectors of the economy) from the start of his career, beginning with private work for V. A. Kokorev in Baku in 1863, and he published on oil throughout his life—including a monograph comparing the Pennsylvania and Baku oil fields as well as newspaper pieces advocating the elimination of the government lease system and the excise tax.[33] Oil became a focus of his scientific work as well: Mendeleev argued (incorrectly) that oil was not derived from microorganisms, but rather from metallorganic reactions with water.[34]

But this emphasis on oil severely distorts our understanding of Mendeleev's economic thought.[35] Although Mendeleev consulted on oil throughout his life, his efforts dropped to almost nil in the early 1880s after losing a dispute with Ludwig Nobel (brother of Alfred of Nobel Prize fame) over the economics of constructing an oil pipeline: Nobel argued for refining oil in the field and piping out kerosene and Mendeleev defended piping out crude and refining it at industrial centers. Nobel would soon become the largest oil magnate in the region, owning or controlling one-third of Russian crude, two-fifths of refined oil, and two-fifths of Russian consumption by 1916. Thus, just as Mendeleev's economic theories were coming together, oil was fading from his attention. Mendeleev's oil work does not provide a *systematic* view of his economic thought precisely because he began working on it so early. Starting as a hired hand in the 1860s, he developed a statistical approach that did not generalize from the oil sector to other areas of the economy.[36] Moreover, in oil Mendeleev generally held to a laissez-faire attitude, the opposite of the protectionist stance he advocated in almost every other sector of the economy. They key to Mendeleev's economics is not oil; it is international trade.

THEORETICAL ECONOMICS: THE EVOLUTION OF SOCIETIES

On 11 May 1901, Sergei Witte had his subordinate V. I. Kovalevskii send a "personal and secret" missive to Mendeleev, then director of the Chief Bureau of Weights and Measures, demanding that the chemist read the enclosed note, as "it is impossible to leave it without a response." Witte wanted Mendeleev to write a "*short, but bold response*" by the next Friday. The note in question, entitled "The Workers' Question in Russia," created, Kovalevskii observed, "a depressing impression and can act strongly on the consciousness of those who are unclear on the foundations, the essence of the matter, the facts, and the teaching of history."[37] Witte and Kovalevskii kept close tabs on this project, visiting Mendeleev at home the next two evenings to answer questions about the format and style of his response.[38]

What was this document that had Witte so agitated? A brief propaganda pamphlet from an anonymous populist, it blamed the Witte system for causing the disastrous famine of 1891–1892 by promoting grain exports to finance industrialization.[39] The pamphlet argued that Russia's backwardness with respect to the West was beneficial, since Western capitalist governments destroyed workers for the sake of expanding markets. It concluded, however, that Russia differed from the West by unifying the will of the people through an autocratic Tsar who transcended any particular class, and so Russia had no "historical soil" for any revolutionary movements. The author blamed contemporary Russian revolutionary activity on Emancipation, which drew peasants away from the land. The rural question thus lay at the center of the economic question, and three of Witte's and Mendeleev's proposals were responsible: the tariff of 1891, railroad guarantees, and credits to heavy industry. The author proposed to undermine the revolutionary movement by destroy-

ing the "alien" capitalist projects: "Time is a great teacher. It has shown that
the experiment undertaken to create a Western European economic structure
here has suffered a total fiasco. This experiment has ruined us and, mainly, has
seduced us from the true path of economic development bequeathed to us by
the epoch of serf emancipation: the construction of the countryside."[40]

Solicited to compose a brief response no longer than a third of the original—which itself ran for roughly two brief pages—Mendeleev wrote an essay
twice its length. He began by pointing out that rural life was not specifically
Russian, but existed everywhere and merely reflected "the youth of nations."
All nations, "like the American Indians," erroneously wanted to preserve their
old habits and ways of life. But history inexorably forced peoples through
stages: first "disordered" life close to animals; then into families and small
groups, driven by the need to *work* together; then the patriarchal early state;
then the agricultural state (represented by the Middle Ages and contemporary
China); and finally the industrial state. These stages of history could not be
held back. The Great Reforms and now Alexander III recognized the need for
Russia to develop its industry. Purely agricultural nations, like China, were
poor and could not defend themselves. As for famines, Mendeleev had four responses: Famines had been a great deal more frequent and lethal in Russia before industrialization; agricultural India had a better climate yet more severe
famines; Western Europe had no famines because it could afford to import
bread; and heavily industrialized Moscow also had no worries. Mendeleev
charged that the author of the pamphlet operated with a faulty metaphor.
Capital did not "invade" Russia, it was invited in a controlled and rational
fashion.[41]

This "Workers' Question" exchange highlights some central aspects of
Mendeleev's theoretical economics. A stray statement in his notes exposed the
main fault of the "revolutionaries": "We have ruin everywhere. This is the revolutionaries' song."[42] Mendeleev was an economic optimist; his optimism,
rather than specific data, motivated his theoretical economics. His response
emphasized historical evolution through phases, and Russia's place in that
process. And he had faith in the power of reason to control capital so as to
maximize Russia's industrial potential. Curiously juxtaposed with his procliv-
ity to rational state action was his assertion that private initiative actually drove
economies through history. Historian Francis Stackenwalt, for one, has placed
private initiative at the center of Mendeleev's economics, arguing that
Mendeleev designed state intervention to provide private individuals with a
space in which to operate.[43] This interpretation, however, minimizes the place
of the *collective* in Mendeleev's thought; for Mendeleev, political economy and
political theory emerged together. The appropriate approach to the economy
would simultaneously create the perfect state for Russia. This state would in
turn grant a little room for private participation and initiative. The two, there-

fore, must be studied as a piece. In Mendeleev's economics, the dominant language was statistics. In his politics, the dominant *agents* were statisticians.

Mendeleev's first postulate was the evolution of societies. Every society, from England to Africa, followed the same stages of development, each stage being characterized by the size of its productive units. The factory-based industrial economy, as the largest sustainable productive unit, was for Mendeleev the end point: "Thus, the beginning for human societies is everywhere uniformly patriarchal, and the end should also be uniform, and, whatever might change, factories will inevitably and obviously participate in it."[44] This process could not be stopped, but certain measures could make industrialization more "rational" and thus minimize negative aftershocks. The linchpin of rationally managed historical evolution was the autocrat. Mendeleev cast Alexander III as the successor to Peter the Great: a new enlightened ruler who modernized Russia through the force of will. Alexander III was ideal for Mendeleev because he recognized this need for change and picked the right advisors (meaning Mendeleev and people Mendeleev approved of) and followed their advice to make change happen. Without factory development soon, "Russia must become either a China or a Rome, and both are dangerous, according to the verdict of history."[45] Mendeleev's historicism was a virtuous circle: History was both the process and the justification of economic development.

He believed the mechanism of historical change was individual initiative, for the state could not develop a truly capitalist economy without producers and entrepreneurs. These entrepreneurs, however, needed the state to give them incentives to develop appropriately. An "individual" alone could not do anything without the broader society (and state): "[A man] alone (a separate 'I'), is in essence no more than a philosophical abstraction or an entirely exceptional phenomenon, like Robinson Crusoe."[46] Individuals thus resembled the chemical elements: abstractions from a collective set of phenomena that only made sense when encased in an explanatory framework. It was their diversity that made them interesting. Whereas chemical elements interacted through their chemical properties, diverse social individuals combined through labor: "Labor is the death of extreme individualism, it is life with obligations and with the rights that come only from them; it presupposes an understanding of society not as a bedlam, meant for the benefit of separate individuals, but as a milieu or a necessary space of human activity."[47] The state thus necessarily set the conditions for capitalism, a fact that Adam Smith and other liberal economists refused to recognize.[48] Echoing his views on labor, Mendeleev in a November 1898 letter to Tsar Nicholas II pointed to capital's singular importance by virtue of its circulatory powers: "[T]he problem is not with [capital], and it, no matter where it came from, will everywhere give birth to new capital; thus it moves around the entire earth, bringing nations together and then, apparently, loses its contemporary significance. In this sense it is a peacemaker, a governor, and an agent of progress."[49] When linked with recent scientific advances, growth truly had no end.

Science and technical expertise gave the crucial proper direction to private initiative. Given that people were naturally unequal (citing examples like the separate sexes), Mendeleev argued the impossibility of the dream of "communism" to equalize all, and that one could only balance out inequality through the wise use of knowledge and economic power.[50] For the present, "bourgeois and kulaks" (Mendeleev's terms, the latter referring to prosperous peasants) served as necessary intermediaries for the state until the scientific cadres created by the new technical education began to grow.[51] In time, industry would harmoniously mediate between educated classes and simple people:

> In my opinion, the expansive development of factory activity in Russia is not only the only true means for the further development of our well-being, but it is also the only path for the agreement of the interests of the mass of people with the interests of the educated classes, because in the factory the simple folk and the nobleman will be working side by side, and in the factory the real utility of education, which now can seem like a whim and luxury of the bureaucrat and landowner, will become obvious to the people.[52]

Scientific intellectuals would create the very conditions that would generate greater demand for these same intellectuals. The cyclical path of the economy was directed by a circular motion of the demand for specialists (not to mention the motion of the specialists themselves).

A public relations campaign could increase economic enthusiasm. Mendeleev's own central contribution to public economic education was his most complete work on theoretical economics and politics, *Cherished Thoughts* (published 1903–1905).[53] Mendeleev's almost confessional *profession de foi*, this book covered a wide range of economic, educational, and political topics. Here he introduced his economics by distinguishing "work," a purely physico-mechanical concept, from "labor," a social concept. "In the days of Smith and even Marx," he added, "they confused this."[54] Indicative of Mendeleev's growing abstraction, he argued that this distinction classified all societies, since while the same work might be going on everywhere, each society differed by how it organized that work. Industry was the capstone of advanced societies, and its inevitability offered solace.[55] The age of rural utopias, such as those advocated by Jean-Jacques Rousseau and Leo Tolstoy, had ended, although it was "charming, like childhood" to reminisce about them, and instead Mendeleev wanted to promote urbanization.[56] By trusting in the gradual evolution of unity, "[w]ithout revolutions, without the fulfillment of the hasty and poorly substantiated utopias of the communists, those disadvantages which are often pointed to in industry will be gradually corrected."[57] Mendeleev valorized labor maximized through *trade*, the greatest of all cycles. Economies grew because populations increased, which increased aggregate demand, which in turn caused private individuals, taking advantage of state protection, to generate more goods. Industry completed the circuit.[58]

Trade, as a process fundamentally *about* circulating goods and labor and openly controlled by state policy, offered a crucial locus for coordinating the diversity of his theoretical views, culminating in his work on extensive sections of the 1891 tariff, later dubbed the "Mendeleev tariff" by its critics.[59] This tariff, triggered by an increase in German duties a few years prior, stood then as the most protectionist in Europe. Most Russian tariffs in the nineteenth century, such as those imposed in 1822, 1824, and 1841, were designed to increase state revenues and not to effect active transformations on the domestic market. While the tariffs of 1857 and 1868 lowered duties on some products and abolished them on others, lack of consensus persisted about the goals of trade policy under Alexander II. If the West could always flood the Russian market with cheaper products created by long-established, more efficient industry, native Russian industry would never bloom, so Alexander III's 1891 tariff, the first since 1868, raised duties on many items, especially in heavy industry. Duties increased from 5 to 15–52.5 (kopecks per unit weight) on pig iron (depending on mode of transport), from 20–50 to 90–150 on regular iron, and from 75 to 300 on locomotives and other machinery. Duties as a percentage of the total value of imports reached 24.3% in 1851–1856 and 40% by 1902.[60] The industries most heavily hit were those in which Russia wanted to displace foreign, mostly German, imports with domestic products.

Mendeleev became the tariff's public mouthpiece.[61] His most visible defense of it, an article in the *New Times* in 1897 entitled "A Justification of Protectionism," argued for the policy in a straightforward, pedagogical fashion. Using accessible examples such as bread prices, Mendeleev tried to demonstrate the necessity for a protectionist tariff over a "free trade" policy. The notion of "free trade," Mendeleev insisted, served as an ideological mask for covert protectionism on behalf of whichever nation controlled the modes of transport. The English advocated free trade because they made most of the advanced goods and could ship them cheaper to underdeveloped countries. As a result, it would be impossible for the poorer nations to compete with the dominance of British industry:

> Therefore you should look at the English doctrine of free trade as a variant of protectionism, that is, as a policy designed for the protection of English industry and trade. That's how it is everywhere: some goods are allowed without duties, others with customs levies. In general a pure free-trade system,

strictly and rigorously applied, does not exist; its existence is even unthinkable under the current organization of the life of peoples into states.[62]

Even a cursory examination of Britain's tariff showed that England did impose duties on foreign imports designed to stimulate domestic production, exactly as did the new Russian tariff: "The system of peaceful protectionism, begun for us only under the last Tsar, has obviously raised Russia in its internal and external relations, is clearing a path to the East, is opening the doors to true, living enlightenment and, of course, allows the wide dispersion of Russian genius, increasing national assets, as is seen if only in the growth of deposits in the protected reserves."[63] Like capitalism—and, as we will see, like the metric system—the internal coherence of this Imperial system of protection lay in the perception of *inevitability*. Given unequal development, it was conceptually and practically impossible for the international economy to operate in a totally free market. The question was not whether to be protectionist, but how: fiscal, protective, or "reasoned" (a protective tariff focused by experts on certain sectors of the economy).[64] Despite his propaganda efforts, however, the tariff ended up pleasing no one. Russian merchants wanted even higher duties, while public opinion quickly shifted against the tariff on the grounds that it raised prices and generated inefficient production of substandard goods.[65]

Nevertheless, the tariff was an unqualified personal success for Mendeleev. His experience working on it helped to organize his economic thought into an articulated system.[66] Protectionism, variously defined, would become his favored mechanism by which the state could create propitious economic conditions, integrating his theoretical politics and theoretical economics. Even as early as 1867, he insisted that economies that had developed without protectionist measures did not exist, echoing Friedrich List.[67] After the Imperial Turn, Mendeleev could finally enact his desired program through his access to decision-makers. In a diary entry of 19 July 1905, Mendeleev fumed about recent public attacks on his protectionism:

Here I—through I. A. Vyshnegradskii, in conjunction with S. Iu. Witte—became a protectionist. My chief goal—to give work to all classes or strata, starting from the capitalists and technicians down to the crudest day-laborer and all sorts of workers. Let judge me whoever wants to, I have nothing to apologize for, because I did not in the least serve because of capital, or because of brute force, or for my own advantage, but I only tried—and will continue to try as long as I can—to give fruitful, industrial-realist affairs to my country in the conviction that politics, construction, education, and even the defense of the country now are unthinkable without the development of industry, and the crown of the transformations I desired—all the "freedom" we need—is intimately connected with this.[68]

Notice that "freedom" was tied directly to the state's willingness and capacity to guarantee it. Private initiative might be the means to move Russian industry forward, but it had to be channeled by the state. Mendeleev argued that trade policy proved that people did not recognize their proper interests until directed by the state:

> In industrial affairs an individual entrepreneur can be directed only by personal advantage, and thus, given the value of capital, [he will be moved] only by those which will soon give a return on the expenses, and therefore he cannot consider [what would yield] percentages only in the future, as is possible for an entire state. A state and a government has this opportunity and thus exerts a positive influence on the course of contemporary life, which is completely unattainable by the efforts of private activity.[69]

According to Mendeleev, the state, by pooling resources and thinking about society as a whole, functioned like an extended scientific society. Meanwhile, two threats—external belligerents and internal miscalculations—required people to become dependent on the state.[70] Historically, the Russian state had always led its people to advancement in science, economics, and culture. Trade was no different. Even the construction of the Russian Empire proceeded through consolidation of trade barriers: As the state eliminated trade restrictions between two areas, those areas entered into its political sway.[71] To defend this position, Mendeleev needed a coherent theory of statehood.

THEORETICAL POLITICS: GOVERNMENTS AND POPULATIONS

Mendeleev's career always centered on a variety of institutions dedicated to the organization of knowledge. In the 1860s, he lobbied for a decentralized organization of learned chemists, the Russian Chemical Society, as an organ of scientific progress, and in his attacks on Spiritualism in the 1870s, he worked through its sister, the Russian Physical Society. Although those organizations retained their autonomy from the state, officially sponsored Congresses of Russian Physicians and Natural Scientists created both. Those organizations may have been separate from the state, but that very state underwrote their existence and constrained their realm of action. In those two decades, Mendeleev grew more concerned about how those organizations, in a similar fashion to the British Association for the Advancement of Science ("one of the most important engines of scientific progress"), could generate the necessary expertise for a modernizing Russia.[72] Back then, he focused on the scientific establishment as a problem of personnel. After the Imperial Turn, Mendeleev directed his attention to the political environment of scientific organizations, analogous to his view of private entrepreneurs functioning in the environment created by the tariff. On the twentieth anniversary of the periodic system, Mendeleev wrote to a foreign correspondent to *deemphasize* his own personal role in Russian science: "I hasten to assure you that the great honor bestowed on me will not lessen my conviction in the possibility of the movement of science not by individual efforts by separate persons, but only by the friendly cooperation of many organized forces."[73] The organization created the scientist and guided his productivity, and not vice versa.

Similarly, when Mendeleev thought about the state after 1880, he elevated its institutions over its population. Although he became more vociferous in advocating autocracy, he had never been a democrat even before the Imperial Turn, writing in 1876, after his trip to the United States, that "[t]he fundamental thought of republics is a lie: the government and the people are one."[74] He argued that the state was not composed of the people, but that it created *them.* Democrats did not realize that the tradition embodied in state institutions and the agency granted to people *by* the state made government possible. After the Imperial Turn, Mendeleev became increasingly explicit and unapologetic about his conservatism.[75] He believed that change should happen gradually, without cataclysmic revolutions: "In nature there are no gallops, all is gradual—a continuous function. Earlier everyone thought it was better to do everything at once—by revolution, coup, [or] sudden rupture. And they see that [everything] changes, but slowly. Truth and work will wear everything down."[76]

The modern state would emerge through gradual and inexorable evolution, just as the modern capitalist economy had. States evolved from smaller units to larger ones, resulting in the late nineteenth-century European political order in which several large world empires dominated international relations. In Mendeleev's monographic defense of the 1891 tariff, he promulgated his evolutionary vision of state-formation, drawing a familiar parallel with chemistry:

> If there are differences in atoms or in separate persons, then all the more do they and must they exist in different molecules or states. It is impossible to blend them together, to destroy the distinction, or to confuse the divided—it will be chaos, a new Babel. People, through development, separated out states and nations from chaos. One should pass by this nationalism, this statehood, carefully and attentively not only when it concerns mankind, but precisely because state interests are in between individual [interests] and humanity's [interests].[77]

For Mendeleev, states evolved into greater degrees of interdependence just as economies did: They began as small units like families or tribes, and then they aggregated into larger groupings until they reached the greatest economically sustainable size.[78] Instead of seeing this as a gradual evolution toward world government, Mendeleev insisted that national character would keep these large states from melding into one super-state.[79] This was a vision of political change fully consistent with his gradualist economics.

Although he devised his economic framework first and then moved to political theory, he considered the state to be historically prior to the economy. The state appeared as a protective force long before the economy emerged under its protection, since, as his justification for the tariff made clear, economies could not develop on a large scale without a political concord of laboring classes, an inversion of Marxist theory.[80] His modern state would consist of a meritocracy of all kinds of experts, not just scientists. Commenting on the recent French elections of 1902, Mendeleev argued against a franchise that would elect only scientists or capitalists. Instead, he advocated a selection process that would place the right experts in the right posts.[81]

In Mendeleev's economics, statistics helped experts decide where to target efforts to rationalize industry. Likewise, the science of statehood was statistical: demographics, the study of populations and population growth. Russia's population growth, according to Mendeleev, had caused a massive economic boom, a pattern of growth that could be further accelerated with proper understanding.[82] Mendeleev developed his views on population in two monographs, *Toward a Knowledge of Russia* (published in an astonishing six editions between June 1906 and June 1907) and the posthumous *Supplement to Toward a Knowledge of Russia* (two editions in 1907). The first text concerned domestic demographics, and the second expanded this method to foreign nations. Mendeleev wrote these books as continuations of *Cherished Thoughts,* developing the "political" elements of his political economy.

As a primary task, his work exposed the errors in Thomas Malthus's late eighteenth-century predictions that population growth would outstrip a country's ability to feed itself. Using demographic data from the United States in particular, Mendeleev argued for population growth as one of the best ways to jump-start an economy. In this, he was merely one of many Russians who resisted the Malthusian implications of unchecked population growth.[83] In fact, population expansion possessed a cultural imperative. For Mendeleev, human civilization had no purpose if one did not think about leaving a legacy to one's descendents. Communists and anarchists went against human nature by wishing to destroy both state and family, the very *institutions* that one left to one's descendents.[84]

The study of population required accurate and precise statistics, implying that statisticians were the best experts to advise an advancing state. Mendeleev began *Cherished Thoughts* with a chapter on statistics, and he claimed it was there in order to show thematically the quantitative empirical grounding necessary to speak concretely about the national good.[85] Because Russia straddled two halves of the world as a continental power, Russia must have a developed statistical capacity: "Russia is—more than any other country—a Middle Kingdom, interested in the course and development of international relations. Therefore we Russians must know other countries and the entire world well, no less than the English, who have possessions in all parts of the world."[86]

He wanted to organize a "central statistical establishment," set up under the State Council, to be supplemented with local bureaus to provide raw information.[87] To bring this about, Mendeleev lobbied Alexander directly for a unification of industrial affairs under one ministry. At the time of writing (1888), Mendeleev pointed out that many functions necessary for industry were divided among the Ministry of Finances, the Ministry of State Properties, the Mining Department, the Ministry of Communications, the Ministry of Internal Affairs, the Ministry of Popular Enlightenment, and various military ministries. Elsewhere he argued that only a unified ministry of industrial affairs could "give the Tsar the material which no current fully enabled minister" or "local gathering of scientists or representatives of trade and industry" could. The Tsarist censorship consistently excised all of Mendeleev's attempts to publicize calls for such an organization.[88]

The scope of this theoretical vision is impressive. Starting with a fundamental faith in the power of history to give meaning to human affairs, Mendeleev formed a theoretical edifice in which population growth led to mutual human interdependence, which led to the consolidation of large states through the elimination of trade barriers, which in turn developed their economies by erecting new trade barriers. Even if Mendeleev had not implemented aspects of his theoretical economics in the 1891 tariff or other sectors of the economy, his writings would still be of substantial interest as works of political economy. But these views had real-world consequences that help us better understand the reforms of late Imperial Russia. The Counter-Reforms of Alexander III were not merely reactive, but possessed just as consistent an ideological structure—often using the same basic concepts—as his father's Great Reforms. Mendeleev's ability to move so smoothly as an advocate of one philosophy of reform to the other highlights the intellectual continuities at hand. To take one example, after Mendeleev left St. Petersburg University (1890), he served for fourteen years—until his death—as the director of the Chief Bureau of

Weights and Measures. Jumping out of chronological order here and exploring his metric reform helps show Mendeleev's political economy in practice.

MEASURE OF ALL THE RUSSIAS: MENDELEEV AND THE METRIC REFORM

Mendeleev had a long-standing interest in metrology, the science of measurement, a science that was becoming all the more crucial as the nineteenth century drew to a close. Electricity, steam power, telegraphy, heavy artillery—all required scrupulous attention to the fine details of calibration and standardization. In practice, metrology is constituted by taking the conventional and persuading people that its conventionality is irrelevant to its utility. Standards are constructed by people who agree to take them as standards, and then an elaborate process of acclimation begins that persuades those who were not part of the original decision that they should adhere to the broader agreement.

The metric system, in particular, was created by French academicians in the late Enlightenment to convert a wide variety of standards of measurement to a uniform standard supposedly rooted in nature.[89] For Mendeleev, the metric system represented almost the perfect application of his economic and political theories, and the perfect emblem for a Westernizing Russia. When he became director of the Chief Bureau of Weights and Measures in 1893, he found that Russia was no further along in introducing the metric system than it had been early in the nineteenth century, when agitation for its introduction began.[90] Although state agencies had expressed intense concern for the standardization of Russian weights and measures since the medieval period, it was not until 1797 that Paul I proclaimed the first Empire-wide law on standardization. The second such law, which standardized the Russian units of measurement—the *sazhen, arshin,* and pound *(funt)*—and related them to British units of measurement (a *sazhen* is seven British feet exactly), passed in 1835, during the period of codification under Tsar Nicholas I. The third law, planned and written largely by Mendeleev, and promulgated on 4 (16) June 1899, both allowed optional use of the metric system and standardized all Russian units with respect to it.[91]

Mendeleev had publicly advocated introducing the metric system to Russia since the late 1860s. In his first statement on the system on 28 December 1867, he argued for its use to foster international scientific unification. Easy to work with because it used the decimal system, the metric system had exact exemplars, standardized electrical and mechanical international systems (like the railroad), and would prove "best suited to universal distribution." As a result of its "uniformity in all relations," the metric system, along with printing, trade, and science, would prepare the path for "the unification of all nations . . . the dream of the world."[92] Playing the backwardness card as he did so often, Mendeleev pointed out that a host of nations, including almost all European powers, had either converted to optional use of the metric system or were about to. He urged Russian scientists and teachers to employ the system in their lectures so as to accustom the public to it gradually. In his own *Principles of Chemistry*, composed the very next year, he immediately introduced the metric system and employed it throughout.[93]

Mendeleev was active in numerous metrological projects. Since 1863, he had been heavily involved in attempts by the Ministry of Finances to reform the state's liquor monopoly through the recalibration of Russian volume and weight measures to metric ones. Extrapolating from his doctoral research on alcohol–water solutions, Mendeleev urged better alcoholometry to improve excise tax collection and preserve uniform quality in vodka production.[94] This has been widely misinterpreted as Mendeleev creating the 80-proof vodka standard (and thus supposedly "inventing vodka"), a view that has recently been compellingly dismissed with evidence that the 40%-by-volume metric had been standard long before Mendeleev.[95] Likewise, throughout his gas work, Mendeleev fiercely defended the metric system as the only way to conduct precise scientific measurement. Thus the selection of Mendeleev to head the Central Bureau signaled a determination for reform on the part of the regime.[96]

Much of the early movement for the introduction of the metric system into Russia, and the international standardization of metric units in general, came from grassroots pressure by natural scientists, not bureaucrats.[97] In 1869 the St. Petersburg Academy of Sciences sent a communication to the Paris Academy of Sciences to propose the establishment of an international metric system. On 8 August 1870, after an initially hostile reception by French scientists, who did not want other nations meddling in French units, a Parisian commission was set up to establish reliable units for the meter and kilogram, culminating in the Convention du Mètre in 1875. The Russian Technical Society, which often functioned as a coordinator of technical and business interests, also set up a commission on the introduction of the metric system by 1877, but did not press for action "in view of the political state of affairs" (meaning the Russo-Turkish War).[98] There was a risk in this early period that the Academy of Sciences in Petersburg would monopolize scientific efforts on standardization. In his typically provocative fashion, Aleksandr Butlerov in 1879 demanded that the Academy share input on weights and measures with faculty at higher educational institutions (like St. Petersburg University), but was rebuffed.[99]

In 1889, Russian delegates attending an international conference in France signed an agreement committing to eventual conversion to the metric system, and then took home a representative meter and kilogram from the collection of prototypes.[100] Mendeleev was appointed to the International Metric Commission already in the mid-1870s, and was an active member until domestic metrological work drew him away in the 1890s.[101] Despite all this effort, complete obligatory adoption of the metric system in Russia did not take place until the Soviet state instituted conversion on 14 September 1918.[102]

Mendeleev did not begin his work as director of the Chief Bureau of Weights and Measures by embarking on the introduction of the metric system. The Chief Bureau itself was newly created with Mendeleev's appointment, replacing the earlier Depot of Exemplary Weights and Measures, and the post of Director of the new institution replaced that of Learned Storekeeper of the old. As Witte recalled in his memoirs, with a whiff of self-congratulation: "With my support, [Mendeleev] put the board back on its feet."[103] Mendeleev was given a chance to fulfill his dream: creating an institution that realized his grandiose political economy on a small scale. The Depot had not been such an institution, being, as the name implies, little more than a warehouse. Originally located in the Peter-Paul Fortress beside the Mint, its second and final Storekeeper, V. S. Glukhov, moved it to specially constructed quarters on Zabalkanskii Prospect (now Moskovskii Prospect), across from the Technological Institute.[104] Mendeleev moved into his new quarters there in the early 1890s, and died there at work in 1907, facing out onto the educational establishment in which he had begun his professional Petersburg career.

Mendeleev's major accomplishment at the Bureau was the 1899 law that led to the optional introduction of the metric system into the Empire.[105] This law, a highly articulated elaboration of the framework that Glukhov had laid out, capped off a series of stages all designed to bring about the eventual conversion of Russia to the metric system. This metric reform represents the clearest exposition of Mendeleev's political economy in practice, and the instances where it failed likewise provide illuminating points of contact with the unfortunate realities of Imperial administration.

Mendeleev firmly believed in the inevitability of the introduction of the metric system. The nature of his job, he felt, was to make sure that it happened smoothly. Why did Mendeleev believe in this inevitability?[106] There are a few possible explanations. The central conveniences of the system were the primary reasons (decimal accounting, for example), but Mendeleev knew that those conveniences could easily be incorporated into any system of measurement—a decimal breakdown of the *sazhen,* for example. What made it inevitable had nothing to do with the metric system per se or with its relationship to nature. Man-made, arbitrary conventions—such as Russia's adherence to international agreements, or the fact that other nations had converted—made it inevitable. For Mendeleev, the appeal of metric reform had nothing *natural* about it; what he found most persuasive was precisely the *artificial* reasons for its adoption. It should be adopted because it was an appropriate and accepted convention; and its very status as an artificial convention determined the way it should be instituted.

Effective implementation of the metric reform would require that the new conventions be presented as natural standards. Hence the first stage of the reform: The standard exemplars of weights and measures needed to be recalibrated. Russia needed new prototypes not only for Russian units but for metric and British units as well (the three major systems of units used in European spheres of influence). In 1892, before he officially became director, Mendeleev informed his superior, Director of the Department of Trade and Manufactures (under the Ministry of Finances) V. I. Kovalevskii, of "the necessity of renewing them [the Russian prototypes], since all means which touch on the unification of weights and measures in the empire must, by their very essence, depend on the maintenance of prototypes."[107] The wear and tear on the old standards, last renewed in 1835, provided most of the motivation for Russia's recalibration, but European events also played a part. In 1834 a fire in England had destroyed their old exemplars so they had to create new prototypes, which since 1852 had been recalibrated every ten years. Most European nations followed suit. In fact, the example of several nations showed that any prototypes forged earlier than 1850 would have to be renewed.[108] The Russian 90% platinum–10% iridium prototypes were requisitioned from the London firm of Johnson, Matthey, and Co. in 1893. Over the next five years Mendeleev produced a substantial paper trail on this stage's progress, ending upon reception of the prototypes in 1898.[109] Much of Mendeleev's dramatic metrological reform was accomplished in the same fashion—by pushing paper across a desk. The Imperial machinery then converted Mendeleev's sedentary actions into radical transformations.

After the prototypes were created, the reform's next stage was the correlation of Russian prototypes with metric equivalents. In the past, given the exact relation of the British weights to the Russian standards, it was simpler to generate conversion tables between these two. Mendeleev was clearly concerned with the ratios of conversion between the British and the Russian, but he wanted ultimately to relate both systems to the metric. The numbers of the conversion may have seemed messier than with the British system, but that was precisely the point. Treating the metric system as just an independent convention, there was no reason to expect a neat relation, but Mendeleev established a correlation to make the transfer smoother.[110]

The third stage was also the most important for Mendeleev: the establishment of local verification stations (bureaus) dispersed across the Empire.[111] Russian officials had long lamented the lack of a system of verification for weights and measures. The 1835 standardization law had distributed exemplary weights to major centers of trade and industry, but no mechanism required individuals to have their weights verified, and, even where some rudimentary ad hoc process was enforced, the exemplars had so deteriorated as to make any verification counterproductive. Some localities still used weights from the 1830s, and even the Mint and the State Bank had inadequate proto-types (Mendeleev had to renew them as his first act). As V. S. Glukhov, Mendeleev's predecessor, bemoaned in 1889: "It is hard in a few words to imagine the sorry state in which weights and measures are found in Russia."[112] They needed to establish some procedure by which localities could be cali-brated with the center.

A major distinction between Glukhov's and Mendeleev's approaches was that Glukhov preferred importing a Western protocol for local verification, whereas Mendeleev wanted to adapt a Western system to local Russian circum-stances.[113] The first few issues of the Chief Bureau's journal—itself founded by Mendeleev as part of the modernization of the institution—contained reports of inspection trips by Mendeleev's assistants to various parts of European and Siberian Russia, as well as Western Europe, to report on the various mecha-nisms (or lack thereof) for verification of weights and measures.[114] The extent of domestic disorder was extraordinary. In each place, the inspector would first make sure that exemplars existed (often they did not), and then go to shops with the police to determine trade measures. In Russia proper *only* Moscow and Petersburg had special sites for the stamping and verification of measures, and the Petersburg one was handicapped by ancient exemplars. The only well-functioning system in the entire Empire was Riga's. Reform of the verification structure was necessary before any thought of the metric system could be en-tertained: "The immediate introduction of the metric system without prepara-tory verification work and teaching the sense of the metric system in schools [the local official in Irkutsk] considers unthinkable, if one wants to have any kind of order."[115] At least now locals knew the current measures—more or less—and in order to introduce a new system you had to instill trust that the new system would be at least as fair as the old one. Mendeleev concurred.[116]

He first mentioned these new local "establishments" in a series of memo-randa in 1896. Russia was a vast empire, so the presence of accurate metrolog-ical standards in Petersburg (located in one faraway corner) would do nothing for standardization, let alone getting the metric standards distributed to all lo-calities (a proposition amply demonstrated by the status quo). Mendeleev di-vided the country into metrological zones, each with its own verification bureau and team of "verifiers." These individuals would inspect the trade and industrial measures in each region, enforce local standards, and thus proceed to correlate and standardize the Empire.[117] The Bureau did not micromanage and command all the other bureaus directly; rather, its control was constituted by the widespread distribution of the local bureaus—the more separate decen-tralized centers of standardization dispersed, the stronger the Chief Bureau be-came. The Chief Bureau worked beyond its localized site in St. Petersburg only to the extent that it invested each local bureau with part of its authority; these local bureaus returned the favor by making the authority of the Bureau more solid. Mendeleev hoped to expand the number of local verifiers until the day "when the entire empire is covered by a net of local bureaus."[118]

Just what was meant by this "net" proved complicated in practice. A debate erupted about whether to place them in local technical societies, railroad administration offices, or educational institutions.[119] Mendeleev opposed using city councils or chambers of commerce, since they were staffed by the proprietors of the large shops who had the most to gain from inaccurate measures. Instead, he proposed relying on individuals with scientific training, who would be "entirely of good conscience and independent." He hoped, in discussions leading up to the law, "for Russia to decentralize a bit, at least in these matters." After considering using the local rural councils (*zemstva*), Mendeleev opted instead for creating a system within the Chief Bureau's administrative network.[120] The number of intended bureaus grew correspondingly large: Given the 90 districts and zones in Russia, Mendeleev projected a need for at least 100 fully stocked verification bureaus. An additional fifty smaller ones would be partially equipped (with units of weight and length, but not more specialized measures) to accommodate areas with excessive demand. Mendeleev did not want to rush construction. Given the start-up time and the lag in recouping costs, no more than fifteen bureaus were to be constructed in a year.[121]

The bureaus were not just meant to standardize measures; they were also supposed to standardize people, both the verifiers and the verified. The system of regulations was standardized by conferring all control over the local bureaus, originally split under a heterogeneity of jurisdictions, under the Chief Bureau exclusively.[122] Training in Petersburg standardized the inspectors, followed by their distribution to all the corners of the Empire.[123] These local verifiers were then deposited (along with their standards) in the local bureaus. But how was one to standardize the practices from one bureau to the next? Mendeleev had two solutions: First, to foster coordination and to deal with remote areas or very large and immobile equipment, two reserve verifiers per bureau would travel to where demand required; and second, mobile bureaus on train wagons would travel around the country and standardize equipment wherever the tracks led (only one such wagon bureau was actually built, with quite some fanfare).[124] The fabric of the net thus consisted of both stable bureaus and mobile connecting nodes.

Not only did Mendeleev create the supply of verifiers, he also created the demand. Under the 1899 law, all local standards used by tradesmen and industries needed to be verified once every three years. In addition, Mendeleev had his assistants conduct surprise inspections of post offices and banks between the triennial verifications.[125] At one point he even proposed moving to an annual check—so the verifiers would constantly circulate and standardize—and then the levied fees collected would make the system self-financing (and, potentially, these bureaus could collect other taxes as well). Furthermore, the bureaus could use the inspection certificates from annual checkups to gather data on economic activity and thus inform a more accurate industrial policy.[126] The circulatory process of standardization thus fed into the circulation of goods, labor, and capital prompted by Witte and Mendeleev's economic policies.

Mendeleev considered the complete network of local establishments to be a necessary component of full introduction of the metric system, but only a few bureaus were required for the fourth and final stage of Mendeleev's metric reform: the optional adoption of the metric system. Why, one might ask, did Mendeleev not push for mandatory adoption of the metric system? Because this would counter the efficacy of his decentralized mechanism of control. Imposing the metric system, as had been done in France a hundred years earlier, would inspire popular hostility and invite failure. According to a fundamental conservative principle, if one *imposed* unity, it would be unstable and undermine the very unity one desired. If verification establishments were distributed everywhere and if informal metric conventions were adopted, the people would be, so to speak, "metricized," and it would become second nature for them to adopt the system:

> I am a great supporter of the metric system but an even greater supporter of the Russian people and their historically formed conditions. I would like, for my part, that the people themselves, gradually, having the legal right to use the metric system, spoke out in its favor. I know that a state branch could very easily order the usage of some kind of system by a circular. But the issue is how the people will relate to it. It's easy to give an order. Thus it was done in France, an order was given by some sort of still revolutionary government that the metric system be introduced, and the people didn't use it, even after thirty years!
>
> And thus, being a supporter of the metric system and understanding its benefits, I would like that it be voluntarily disseminated in the Russian milieu, and I oppose its immediate and compulsory introduction, I stand for its optional use, and chiefly for the introduction of verification establishments so that cheating in weights and measures will diminish somewhat.[127]

The verification establishments offered a means to calibrate the population, which would then gradually, like a compass in a magnetic field, align itself with the metric system. The important point was not to make the people's bodies metric, but their minds.

All of this is a further articulation of Mendeleev's Imperial Turn. Metrology provided a mechanism for Mendeleev to correlate the behavior of individuals onto the correct path, that of a functional civil society. Although Mendeleev never stated it this way, there is a direct analogy to his conception of the periodic system. The units (elements, meters) are arbitrarily posited but internally consistent under an immutable framework (periodic law, metric system). Although there is something "true" about both systems (periodic and metric), you could only persuade people to use them through their utility, pedagogical or economic. Properly directed, individual initiative would make proper choices, just as it should according to Mendeleev's political-economic model of provoking proper development.

Mendeleev's metric reform was so sweepingly conceived and authoritatively promulgated that it is tempting to view the 1899 optional introduction of the system as the successful conclusion of the affair. That would be a mistake. Mendeleev's metric reform foundered on the shoals of the political events that swept Russia in the first years of the twentieth century. Of the 150 proposed local bureaus, only twenty were built by the 1905 Revolution, at which point the budget crunch caused by the Russo-Japanese War forced a halt in construction.[128] The metric reform met difficulties similar to other efforts in state-building in late Imperial Russia. Just as it was essential to impose some kind of effective governance over the provinces, it was equally important to do so without straining an already stretched budget or alienating localities.[129] Until his death in January 1907, Mendeleev fought as director of the Chief Bureau of Weights and Measures for the completion of the most majestic of his systems. Ironically it would be the loathed Marxists, with their scatter-brained notions of political economy, who would finish the job.

CONCLUSION: VIRTUOUS CIRCLES

This chapter and the one that follow separate the systematic Mendeleev from Mendeleev the misfit. This division is artificial because these twin aspects of his persona came together and reinforced each other. The separation serves the analytic purpose, however, of isolating two different styles of reasoning. At first glance, it seems that Mendeleev's style expressed in this chapter was "rational" or "practical," whereas what follows was more outlandish. Such an assessment reflects a present-day bias toward systematic, black-and-white thinking.[130] On the contrary, much of what I will attribute to the misfit Mendeleev was harmless and amateurish, while a great deal of his systematic thought was recklessly ambitious and utopian, not least the metric reform.

As an example of systematic "overreaching," take Mendeleev's proposal for calendar reform. In the Imperial period, Russia adhered to the "old-style" Julian calendar, which lagged 12 days behind the generally employed European Gregorian calendar in the nineteenth century. Apparently following no higher directives, Mendeleev made reform of the calendar—a problem of correlating two nominal conventions—one of his pet projects. The obvious reform, one would think, would simply be to add the necessary days to the Julian calendar (which is exactly what the Bolsheviks did in January 1918). Mendeleev did not advocate this approach. After carefully examining both systems, he decided they were *both* inaccurate. Instead of retaining either the "old-style" Julian or switching to the "new-style" Gregorian, Mendeleev advocated a "Russian style," which was one day removed from new-style dates.[131] Rather than taking the simplest approach to negotiate between two different conventions, Mendeleev wanted the entire world to conform to his new calendar. Suggestions like these point to something other than rationality behind Mendeleev's systems.

The two different, equally "Mendelevian" styles in fact differ not by utopian qualities vs. rationality, but by the type of *motion* embedded in their structure. In each of the interlocked systems discussed in this chapter, the stability of the network is guaranteed by circulation. Money, labor, data, experts, or standards traveled around the Russian Empire and returned to the point of origin (Petersburg) qualitatively unchanged. (In the case of capital, there might be a gradual quantitative increase as a return on investment.) Their circulation bolstered the uniformity of the whole. This "circular" style has roots in Mendeleev's early views of agriculture and meteorology.

His other, more Romantic notions of the 1880s and beyond were grounded in linear journeys that brought an irreversible qualitative change. These represented travels of adventure and exploration in which the emphasis was on the *traveler,* not the system through which he or she traveled. After the crisis of 1880–1881, Mendeleev began to mine his past experience in order to provide coherence to his plans for a post–Great Reforms program for Imperial modernization. The Romantic undercurrent that he imported alongside his evolutionary systematics, however, would function as a Trojan horse in undermining his hopes for a functioning Imperial Turn.

MAKING NEWTONS

Mendeleev, on left, playing chess with artist Arkhip Kuindzhi. The woman in the back is Mendeleev's second wife, Anna. This photograph is from the late 1880s or early 1890s. From Smirnov, *Mendeleev*, insert 2.

CHAPTER 7

MAKING NEWTONS: ROMANTIC JOURNEYS TOWARD GENIUS

Oh you, whom the fatherland
Expects out of its depths
And wishes to see those,
Like it summons from other lands.
Oh your days are blessed!
Dare now to be encouraged
By your zealousness to show,
That to its very own Platos
And the quick reason of Newtons
The Russian land can give birth.

—M. V. LOMONOSOV[1]

MIKHAIL LOMONOSOV (1711–1765), the natural philosopher and poet often touted as the "father of Russian science," was an inveterate optimist. Mendeleev, no less optimistic in his own way, invoked this Lomonosov stanza in 1901—ten years after leaving St. Petersburg University—in order to disagree with its conclusions: "[I]f I can add the following from my own part, in present times we will, perhaps, get by even without Platos . . . and instead of them it would now be better to wish for Russia a double quantity of 'Newtons'." He continued: "But Plato and Newton were teachers of youth, and thus, considering Russian education, above all we must concern ourselves with teachers." (He jokingly added that if one constructed a proper educational system, one might even be able to harvest a few Platos, but only in the south, "where it's a bit warmer."[2])

Mendeleev was referring to the reform of Russian education. In the wake of his rejection by the Academy of Sciences, he turned to the bureaucracy as a preferred agent of social change; at the same time he began to capitalize on the image of rugged individualism attributed to him by the Petersburg dailies. He also began to think more broadly about his place in Russian culture: "This was a transitional time for me: a lot changed within me; I began to read a lot about religions, about sects, about philosophy, economic articles."[3] His desire to "make Newtons," as well as his desire to fashion himself as Newton's successor (and hence the confirmation of Lomonosov's prediction), dovetailed with his wide-ranging views of social reform.

The educational reform Mendeleev proposed during the last decade of his life encapsulated his fully articulated view of how expertise should be marshaled in Russian culture, and thus stands as one of the best exemplars of Mendeleev's Imperial Turn. This reform contrasted sharply with the decentralized pedagogical approach of the 1863 statute that had structured Mendeleev's vision of the Great Reforms. Now, the first step to improving education would be to ban general examinations, since they discouraged individual innovation. Encouraging mindless regurgitation was no way to make Newtons. Instead, more attention should be paid to training teachers from the elementary school level through university—along the model of the Chief Pedagogical Institute, his own alma mater.[4] The educational reform would be monitored by continually sending inspectors, drawn from the ranks of the most experienced teachers, to schools to ensure equivalent levels of teaching in all corners of the Empire—another instance of Mendeleev's belief in Imperial systems.[5]

He had argued for an earlier variant of this reform in a somewhat rambling letter to the Minister of Finances, Sergei Witte, on 15 October 1895. To counteract the predominance of "Brahmins and Confucians" (by which he meant classicists) in education, he proposed a different model:

Although by its usual essence enlightenment (*prosveshchenie*) comprises familiarizing youth with the means and benefits of what is known, . . . one must consider as its more immediate goals, especially now, in the age of the dominance of the industrial cast of life, the instillation of reasoned (i.e., withstanding the criticism of experience) means of skilled life activity, showing the dependence of success on the quantity and quality of applied labor and the formation of habits which make going on life's path easier.[6]

For Mendeleev, education entailed the installation of proper habits. As in the rest of his Imperial reform proposals, he saw his perpetual motion machines as generating standardized and ultimately interchangeable components. This made his system lawlike: It generated entirely predictable results because all of the elements fit into a larger framework. His system left no room for individuality that would undermine the predictability of the educational "products." Yet this seems to contradict his views about the dangers of standardized examinations as stifling innovation—and indeed it does. This conflict between trying to generate Newtons while not generating too many of them plagued all of his efforts at reform. For his part, Mendeleev papered over these conflicts or tried to pretend that they were not conflicts at all—a temporary solution at best.

One way he thought to accomplish the enhancement of creativity while increasing predictability was by replacing classical education (Greek and Latin) with an emphasis on the natural sciences. He now faulted the statute of 1863 (the open support of which had led to his dismissal from St. Petersburg University) for making students believe they were fully formed adults, thus promoting the hubris instilled by classical education.[7] By emphasizing "form" over "content," such education absorbed the students' and teachers' attention without teaching necessary skills. As a result, students were left on their own to speculate about that content without validating their ideas against experience. Even more dangerous, they were left to this speculation without guidance, thus never experiencing the salubrious subservience one felt toward a mentor, such as the director of a laboratory. The explosion of recent revolutionary activity in Russia demonstrated this danger of classical education: While revolutionaries often claimed they were a scientific vanguard, most of them were in fact trained in classical schools, Mendeleev alleged, and thus had little understanding of the quiescent gradualism that permeated science.

Mendeleev's negative view of classicism reflected the underside of his Imperial Turn, a Romantic elevation of individualists who worked outside the system. This is a vision of conservatism as an aesthetics and a psychology—a vision concurrent with Mendeleev's Russian conservatism as a set of political and economic practices. Mendeleev's entire system of predictability hinged upon the existence of mavericks working outside a regularized system. He could attack Latin and Greek and propose a system to create obedient model students only because he was *not* a product of his own system: He was its Newton, its unique genius. He could not be replaced. For all of his interest in providing expertise to the state, Mendeleev offered no way to produce an expert to marshal all the other experts—that position he took upon himself. His economics relied on an absolute autocrat unfettered by compromise—an irreplaceable maverick whose will kept the system moving. By analogy, the elements in the periodic system could be seen as individuals created separately, but—once posited—they could be organized in a general framework. Much as his economic system can be understood as a circulating motion that left the travelers unchanged, Mendeleev's self-fashioning as Russia's Newton was a linear, transformative journey toward adventure. Mendeleev used four journeys to construct his maverick ideal—two historical, two actual. When the tension between this new ideal began to chafe against the rigors of the inflexible systems Mendeleev himself built, he would be the one caught in the middle.

OUT OF SIBERIA: ROMANTIC BIOGRAPHY

One of the few aspects of Mendeleev's biography that has gained relatively wide currency—in both Russia and the West—is his Siberian origin. He was born in the moderate-sized town of Tobol'sk, Siberia. His grandfather, Pavel Maksimovich Sokolov, served as Orthodox priest of the village of Tikhomandritsy, in the Tver region near Moscow. Pavel Sokolov had four sons, but according to the custom of the clerical estate at the time, he could pass his last name only to one of them—typically the one who remained in the priesthood—and his other sons had to come up with new surnames, often drawn from local geography. The sons were Vasilii Pokrovskii, Timofei Sokolov, Aleksandr Tikhomandritskii, and Ivan Mendeleev—our chemist's father.[8] (Contrary to popular Russian belief, the name has no connection to "Mendel" and is not a Jewish surname.) Educated at the Chief Pedagogical Institute in St. Petersburg (later his son's alma mater), Ivan Mendeleev was assigned to a school in Tobol'sk, where he taught from 1807 to 1818. After transferring to several cities in European Russia, he returned to Tobol'sk in 1827 as director of the local gymnasium until blinded by cataracts the year his son Dmitrii was born.

Dmitrii's mother, Mariia Dmitrievna Kornil'eva, hailed from a local Siberian manufacturing family whose fortunes had declined in recent years. She married Ivan Mendeleev in 1809, bearing seventeen children, of whom eight survived to young adulthood. Mendeleev was the youngest, born on 27 January 1834, when his father was 51 and his mother 41. His mother dominated his childhood. After Ivan's blindness, she reopened her family's glass factory and managed it until it burned down in December 1848; the next year she would take Mendeleev to Moscow and then to Petersburg to enroll him in an institution of higher education. In 1841 the 7-year-old Dmitrii entered the Tobol'sk *gymnasium*, but he had strong memories of only a few of his teachers—indicating the limited influence of these early experiences. He enrolled in the Pedagogical Institute at the age of 16 in the fall of 1850, and his mother died immediately afterward on 21 September (his father had died on 13 October 1847).[9] After graduating he was sent to Simferopol in the Crimea to teach, but the school closed because of wartime conditions in neighboring Sevastopol. Such are the meager facts of his youth.

Siberia remained an occasional theme in Mendeleev's adult work. In his 1906 text, *Towards a Knowledge of Russia,* which analyzed data from the recent Russian census, Mendeleev drew a new demographically informed map that did not separate European and Asiatic Russia, to help Russians develop a better sense of national identity, of understanding the "spirit" of their country.[10] He proposed building a city at Russia's geographical center, which was, unsurprisingly, in Siberia. The most explicit invocation of the Siberian theme, though, came in the summer of 1899, when the chemist returned to his childhood home of Tobol'sk as part of a mission tasked by Sergei Witte. With younger associates—including one from the Chief Bureau of Weights and Measures and another from the Navy Scientific-Technical Laboratory, both institutions he had helped create—Mendeleev toured major cities of iron production across the Urals, concluding with a survey of forests around Tobol'sk. In a chapter of the 1,000-page work that resulted from this journey, he commented:

> I was called to Tobol'sk not only by the business for which we embarked, but also from the attachments of childhood. I was born there and went to *gymnasium* there; some people still live there who remember our family; there at the glass factory, directed by my mother, I received my first impressions of nature, of people, and of industrial affairs. It has been almost exactly 51 years since my mother, having set up almost all of her other children, brought me, the youngest, to Moscow after finishing the gymnasium. Long ago, every year I would plan to return to my homeland, but it never happened, and thus I went with particular emotion, which lasted during my whole stay in Tobol'sk.

Mendeleev concluded the chapter on Tobol'sk with a plea for building a railroad through the town: "Only when a railroad reaches from the center of Russia to Tobol'sk will my native city have the opportunity to show its most excellent position and insistent enterprise of its inhabitants, who preserve the memory of the old force of the ancient capital of Siberia." On 16 December of that same year, Mendeleev was elected a distinguished citizen of his native town.[11]

It is understandable that Mendeleev would maintain an interest in his origins. This knowledge, however, trickled into the wider Petersburg community as a fundamental aspect of the adult Mendeleev's popular image. Indeed, his mother's glass factory assumed a significant role in narratives of his economic thought. But the truth of the matter is that the most important factors in Mendeleev's adult thought—both in and out of science—developed in the course of his time as a student in Petersburg and Heidelberg, and, most importantly, from the 1860s in Petersburg. The image of Mendeleev as the Romantic Siberian upstart, a Horatio Alger of the Slavs, was built in the 1880s upon the foundation of the maverick Mendeleev constructed during the Academy affair. The Siberian aura made Mendeleev even more colorful, given the importance of Siberia as a symbolic motif in Russian culture—the equivalent, in some respects, of the frontier myth in the United States.[12] Although most of the tapestry of "Mendeleev, Siberian" was sewn by others, one of the most well-known threads he spun himself. In 1887, he published the results of years of research as *The Study of Water Solutions by Specific Weight*. He dedicated the volume to his mother:

> This research is dedicated to the memory of my mother by her youngest child. She managed to raise him through her labor alone by directing a factory; she trained by example, corrected with love and, in order to give to science, rode out of Siberia, expending her final means and strength. Dying, she intoned: avoid the self-deception of the Latinist, rely upon labor, and not words, and patiently seek divine or scientific truth, because she understood how often dialectics deceive, how much there still was to be known, and how with the aid of science prejudices, falsehoods, and errors will be dismissed without violence—lovingly, but firmly—and the preservation of achieved truth, the freedom of further development, the general good, and internal success will be achieved.[13]

Although Mendeleev's mother held some of the views expressed here, she almost certainly did not communicate them to her youngest child on her deathbed. (In particular, the reference to Latin seems out of place.) Mendeleev never mentioned this injunction at any point from 1850 until 1887, and its emergence surely had more to do with image than historical veracity.

The book that follows this dedication is also a distinctive product of Mendeleev in the 1880s. After the embarrassing collapse of his gas project in January 1881, Mendeleev retreated from his ambitious goal of isolating the ether in the laboratory (although his ether dreams would later resurface). This failure strained his relations with the local physicists, and he retreated from the border of physics and chemistry to a small-scale, table-top laboratory project investigating the properties of solutions of water, especially of table salt and sulfuric acid. This was in some sense a strategic retreat from the twin peaks of fundamental physics and the Academy of Sciences.

This solutions work stands as Mendeleev's longest sustained research program, yet it was entirely removed from his search for unification in the physical sciences. Doggedly empirical, this work sustained only the faintest superstructure of theory in a fashion quite similar to his early research at Heidelberg on the capillarity of organic solutions, and was related directly to his 1864 doctoral dissertation on alcohol solutions.[14] Here Mendeleev began his defense of a chemical vision of solutions that emphasized the equilibrium of solution and solvent and rejected the now commonplace notion that compounds can break into ionized (electrically charged) component parts when dissolved. This idiosyncratic vision of solutions—the core of contemporary physical chemistry—achieved some currency in England, despite almost irreproducible results, attracting defenses from luminaries such as H. E. Armstrong and P. S. U. Pickering.[15] This acclaim would not last for long, as dissociationist theories became the standard of physical chemistry, and remain so to this day.

Mendeleev retreated on the scientific front to the comfortable zones of the chemistry of his youth, while the rugged hardship implied by the Siberian narrative more than compensated for any wounds sustained in 1880–1881. The out-of-Siberia story proved remarkably popular in building a cultural image of Mendeleev as the Russian avatar of that legendary beast, the "scientific genius." Paul Walden, a physical chemist based in Riga (who later turned down Mendeleev's chair at St. Petersburg University after the latter's death), diffused the Siberia story to the West. In his widely read obituary for Mendeleev in the *Berichte* of the German Chemical Society, Walden—also a historian of chemistry—emphasized Mendeleev's life as a hard-luck narrative that had all the elements of a rags-to-riches story (except that the "riches" in this case were scientific glory). In particular, by stressing the story of the glass factory, he made even Mendeleev's most profit-centric economic work (such as oil) appear to be about knowledge rather than cash. Walden declared that "[Mendeleev's] name and his words often brought forth a fascinating effect, his personality represented a cultural program, to which both broad social classes as well as the state regime paid extensive attention."[16]

For Walden, Mendeleev was even distinctive in his "Russianness"—and what could be more Russian than Siberia? In the middle of the obituary, Walden extended this idea of "the great Russian folk type (*Volkstypus*)" to Mendeleev's physical appearance. Stressing the man's large head, ringed with flowing hair and a famously long beard and set on broad shoulders, he used it as both a metaphor and a cause for Mendeleev's entire character. The chemist's famous gesticulations and staccato expressions were deemed to reflect essential and eternal characters of the Russian people as a whole, who had finally produced a distinctively Slavic representative in the sciences. And, oddly, he characterized Mendeleev, whom he knew quite well, as "an enemy of all phrases and posturing, all popularity-mongering and showiness—and thus he was more popular than hardly anyone else, and he drew everyone into his spell, releasing tension and excitement."[17] (Walden knew perfectly well that Mendeleev craved attention and display, but felt an obituary was not the place to go into it.) Accounts like Walden's (and even recent accounts like that of Oliver Sacks) hailed the image of Mendeleev as a solitary, inspired, daring, loner genius.[18] True, Mendeleev had occasionally been solitary and even daring, but he was certainly not *always* that way. In fact, much of his activity, both before and after the Imperial Turn, was devoted precisely to containing such disruptive misfits.

The construction of Mendeleev as a *Romantic genius* was distinct from the construction of Mendeleev as a *Russian scientist*. The latter process also emerged in the aftermath of Mendeleev's rejection by the Academy of Sciences. Emphasizing the periodic system, the daily papers created the image of Mendeleev as a patriotic Russian and of the modern scientist as a valuable citizen. In stressing the periodic system, these writers remained indifferent to the *process* through which Mendeleev accomplished this feat. The story of genius is different: It is a tale of how inspiration enables a select few to move beyond the herds of humanity. The prediction of the eka-elements, combined with Mendeleev's public efforts of derring-do and seasoned with a Romantic past, offered a heady brew. As his youngest son Ivan recalled:

> I knew, as it were, two Mendeleevs. One was a painstaking gatherer of facts, a detailed empiricist—Goethe's Wagner, for whom the highest reward was the working out of numbers, the gathering of data, the observation of curious individual particularities of phenomena. The other was an audacious Faust, carried off far into the "world of spirits," the world of ideas, the world of general laws and the deeper theories under the flat surface of empirical phenomena. Both of these personalities sometimes warred with each other, threw one another out of some area—and then the work carried a one-sided character and did not give everything that moments of harmony could give. Father achieved the heights of creativity when both personalities flowed into and helped each other.[19]

Mendeleev the Siberian was thus a special kind of original Russian genius who could reconcile inspiration and perspiration. This image was not just a product of the Russian national myth, however; it also combined with the narrative of scientific progress over the centuries. Mendeleev as Romantic Hero had one foot in Asia and the other in the history of science.

RUSSIAN NEWTON: MENDELEEV THE LAWGIVER

"Genius, genius, genius. I work hard, I worked hard all my life, and they say: genius, genius."[20] So Mendeleev would occasionally grumble when feted by the public both at home and abroad. The complaint was somewhat disingenuous, since no one had done more than Mendeleev to endow him with this label. Yes, he had worked hard, but he also devoted part of that hard work to building the reputation of a genius. The conscious recasting of Mendeleev's story as an out-of-Siberia narrative was only one aspect of a larger historical epic. Mendeleev's personal history was one matter; locating himself as a pivotal figure in the history of science was quite another. As he aged, he began to adapt his image as radical maverick into an interpretation of himself as a successor to Sir Isaac Newton. Mendeleev did not just reformulate himself as a seminal figure in *chemistry*; his aims were grander. Both the Siberian and the Newtonian historical journeys unified objects through the passage of one traveler. In the first, the journey unified the Russian people; in the second, science itself.

At the basis of this transformation lay the periodic law. In the years after 1869, Mendeleev rethought his classification of the elements into a natural system, and finally into a law of nature, at times haltingly, at times boldly. By 1871, Mendeleev was convinced that the periodic law was indeed a *law*; the difficulty now was to develop a sense of what laws *meant* in the natural sciences. When the stakes were raised, he turned to an obvious exemplar: Newton's three laws of motion and his law of gravitation, which had enabled physicists for a century and a half to describe the motion of heavenly bodies with astonishing accuracy. They also allowed scientists to predict new planets from aberrations in orbital motion, predictions which were later vindicated.

The Newtonian model became increasingly important over the course of Mendeleev's career. As the discovery of his eka-elements affirmed his confidence (and the confidence of other chemists) in the periodic law, Mendeleev began to elevate the periodic law from an "ordinary" law of nature—such as the gas laws—to a fundamental law like Newton's. One of the clearest ways to observe this shift is to follow the eight editions of *Principles of Chemistry*. Mendeleev heavily revised each edition following the first (1869–1871), often by adding lengthy discursive footnotes. These notes, Mendeleev's signature feature, often elicited comments from reviewers: "The work as a whole appears as a highly idiosyncratic effort. Philosophical postulations, excursions in the areas of astronomy, physics, mineralogy, geology, technology are present in so rich a quantity that one often forgets, for example just reading the footnotes, that a textbook in *chemistry* lies before you."[21] These revisions, layered like geologic strata, record Mendeleev's own changing understanding of the power of laws.[22]

Each edition was revised, but not all revisions were equal. Mendeleev altered the second edition (1873) remarkably little. In the third edition (1877), however, he approached the periodic system quite differently. Just the year before, Lecoq de Boisbaudran had discovered gallium, and Mendeleev had demonstrated the identity of this element with his predicted eka-aluminum. He promptly revised his original plan for *Principles*: "Having become convinced in the truthfulness of the basic principle, I am bringing it into this edition more strictly than it was in the two preceding. But all the same I understand that the true path to the further development of our science is still not discovered, that we should await soon large changes in [our field]."[23] The fourth edition (1881) was little changed, and appeared when it did largely to finance

Mendeleev's divorce.[24] The most substantial revision of *Principles* came with the fifth edition (1889)—also the most widely translated—with heavy attention to Winkler's discovery of germanium (eka-silicon). Mendeleev also reduced the dimensions of the book and moved the small type intended for specialists into massive footnotes in order to make the book more palatable for the introductory student.

The central change in the fifth edition was not Mendeleev's views of the power of periodicity, but his conception of the properties of a *law*. Earlier, he had treated laws as *explainable* regularities; now laws were understood as *invariant* regularities: "The laws of nature do not tolerate exceptions, and this differentiates them from rules and regularities, such as, for example, grammatical ones."[25] Similarly, in his inorganic chemistry lectures of 1889–1890, Mendeleev proclaimed: "Laws can be important in that they give mastery in the place of slavery, they give the possibility to guess what is factually unknown."[26]

Mendeleev now emphasized mass as the source of periodicity and declared that the variation of properties with mass was the law of periodicity, even if he had no explanation of how mass worked. He reiterated this view of mass in the relatively unchanged sixth edition (1895) by analogy with Newton's law of gravity. After all, no one knew how that worked either, yet it *invariably* showed force's relation to *mass* and enabled predictions. This insistence on invariance underlay Mendeleev's continuing discomfort with the inversion of tellurium and iodine in the periodic system: Iodine weighed less than tellurium but followed it in the periodic system. To the end of his life, Mendeleev insisted that one or the other of these atomic weights was incorrectly determined, and they appeared with question marks in every edition of *Principles*.

It is important not to overemphasize the philosophical coherence of Mendeleev's views. For example, the American magazine *The Nation* found his notion of law particularly confusing in the third English edition (based on the seventh Russian edition):

> [The book] is also valuable as expressing with unusual openness all the processes of thought of one of the greatest scientific reasoners that ever lived. It cannot, however, be called a model of judicious and calm logic. Whatever proposition Mendeléeff inclines to, which must be something illuminating his most famous discovery, will be for him "a logical development"; while anything else will be a "hypothesis," regardless of its logical genesis. . . . On many points he is skeptical about the doctrines of the new chemistry, and sometimes his objections have no little force, but they are always exaggerated.[27]

In the seventh edition (1903), Mendeleev began not just to modify the importance of periodicity, but even to alter the history of the discovery itself. He declared that the periodic law had "appeared to me in its entirety exactly in 1869, when I wrote this work"—this was a conscious falsehood.[28] Mendeleev knew that he was bending the truth here, as he had commented in a French article a few years earlier: "As to the evolution of this law, it is very important to take into consideration that it was not recognized by everyone in a single instant, that it met many adversaries, that it was only gradually that it was accepted as rigorous, gradually that it predicted certain facts, and that the conclusions that it extracted were found verified."[29] The eighth edition of *Principles* (1906) in turn emphasized Mendeleev's *historical* position by emphasizing the precursors to the law and its "strengtheners" (*ukrepiteli*) who had discovered the predicted elements. Likewise, Mendeleev's contemporaneous series of articles for the tremendous Brockhaus-Efron encyclopedia, such as "Substance," "The Periodic Law of Chemical Elements," and "Elements (Chemical)" all emphasized this vision of natural law.[30]

The problem with Mendeleev's definition of laws was that periodicity did not meet it. The model for "lawlike" regularity became stronger over time, and was expressed in 1901 in strictly mathematical terms: "Now, a law always expresses a relation between variables, such as fixing their functional dependence in algebra." The real triumph would have been to deduce a mathematical representation that would replicate the core regularities of the periodic law. Mendeleev spent several years on such a rendition before ruling it impossible—as it still is.[31] Not meeting his own standards apparently bothered Mendeleev rather little; instead, he dwelled on the rhetoric of mathematical dependence (from which the term "periodicity" had been derived) and retreated to his general interest in invariance and generality. By 1905 he maintained that the periodic law was lawlike not by its invariance or generality or mathematizability or explanatory power, but by its endurance: "Apparently the periodic law will not be threatened with destruction in the future, but only promises refining and development. . . . I was lucky here, especially with the prediction of the properties of gallium and germanium."[32]

The emphasis on invariance and regularity in the power of laws elides an important transformation in the way Mendeleev understood the development of science. During his gas work of the 1870s, he had explicitly opposed theory as the motive force of scientific change. As he perceived it then, theories generalized from the body of experimental data, but theories did not generate predictions to set tasks for experimenters. The periodic law in its original formulation represented precisely such a generalization and explanation of available evidence about the elements; its power of prediction came later, both historically and conceptually. By 1905 his tune had changed. Observe how he introduced gravity experiments he supervised at the Chief Bureau of Weights and Measures: "I consider it unavoidable, as much as possible, to assist in the establishment I supervise the study of the theoretical side of the matter, so to speak, since, as is generally known, the practical side of matters is fundamentally dependent on theoretical observations which relate to them."[33] The collapse of the gas project and the unparalleled success of his periodic law had shifted Mendeleev's interest toward theory as *primary.*

Mendeleev now needed a new historical model to replace his original model of an organic chemist generalizing from disparate data, a model based on his youthful hero, French chemist Charles Gerhardt. He found him in Sir Isaac Newton. The fascination with Newton was not new. In the late 1850s, Mendeleev taught his mentor Voskresenskii's course on the history of chemistry at St. Petersburg University. In his lecture notes on the biographies of major chemists, Newton merited eight pages, more than any chemist. He also opposed the nascent structure theory of organic molecules and its subsidiary concept, valency, on the grounds that a tetrahedral carbon tetrafluoride atom violated Newton's third law of motion.[34] Yet so far Newton had interested him only as a man who gave science certain theories useful in accounting for data. Mendeleev would embrace Newton as a *personal* model after the discovery of the eka-elements.

In January 1883, writing the president of the Royal Society of London to thank him for the honor of the Davy Medal (received in conjunction with Lothar Meyer), Mendeleev paraphrased the Lomonosov quotation at the head of this chapter: "May future generations of Russians know their own Newtons, Daltons, and Davys!"[35] He would articulate his Newtonian ambitions in two lectures in England in 1889. The first, "An Attempt to Apply to Chemistry one of Newton's Laws of Natural Philosophy," delivered before the Royal Institution on 31 May 1889 (N.S.), tried to connect his work with that of the former president of the Royal Society. He argued that the almost universally accepted structure theory was in opposition to Newtonian dynamics, in particular his third law that every action has an equal and opposite reaction.[36] Much as in his aeronautical work, Mendeleev cast himself as the interpreter of Newton's intentions in the *Principia.* In fact, there was scarcely anyone better qualified than a chemist to safeguard Newton's legacy, since "it is necessary to note that Newton studied chemical experiments for a long time, and, in explaining the questions of celestial mechanics, constantly had in mind the mutual interaction of the worlds of the infinitely small which appear in chemical evolutions."[37] If Newton based his astronomy on chemical models—and not chemistry on astronomical models—then only a chemist (i.e., himself) would be a suitable interpreter. All theories of chemical dynamics had to be mediated not by physicists, but by a general chemist who followed the master's model:

A coming Newton will find the laws of such changes. . . . The achievement of the laws of this harmony in chemical evolutions seems to me possible only under the banner of Newton's dynamics, which for a long time has been fluttering over the domains of mechanics, astronomy, and physics. Calling chemists to this peaceful and universal banner, I think that I am strongly serving scientific unification, by which I explain the great honor, shown to me by the much respected representatives of the Royal Institution, which gave me—a Russian—the possibility to express before Newton's countrymen an attempt at bringing into chemistry one of his immortal principles.[38]

Thus it was the British scientific community that had cast him in the Newtonian role by inviting him to speak. As to who the future Newton might be, Mendeleev feigned no hypotheses.

He treated these themes more abstractly in his Faraday lecture, "The Periodic Law of Chemical Elements," read before the same audience on 4 June 1889 (N.S.). Here Mendeleev did not lecture directly on Newton's laws, but on the nature of his own achievement. He chose to emphasize two aspects of chemistry: the communal effort of chemists to establish frameworks for knowledge, and the necessity of adhering to laws to avoid speculation. Both, he implied, were ideals Newton would support. (The historical Newton's distaste for communal work seems to have been unknown to Mendeleev.) Mendeleev's ideal of cooperation was the Karlsruhe Congress of 1860.[39] In this fashion, he could lionize the entire community for their contribution to his own individual success, much along the model of the Imperial Turn: groups of individuals constantly circulate in an ordered fashion, buttressed by a maverick. In 1860 the maverick had been Stanislao Cannizzaro; in 1869, it was Mendeleev himself. By trusting to a morally ordered framework, Mendeleev could obtain a place in posterity akin to Newton's:

Arising from the virgin soil of newly established facts, knowledge relating to the elements, their masses, and the periodic changes of their properties, has motivated the formation of utopian hypotheses, probably because [the periodic system] could not be foreseen by the aid of any of the various metaphysical systems, but exist[s], like the law of gravitation, as an independent outcome of natural science, requiring the acknowledgment of general laws, which have been established with the same degree of persistency as is indispensable for the acceptance of a thoroughly established fact. . . . It is only by collecting established laws, that is by working at the acquirement of truth, that we can hope gradually to lift the veil which conceals from us the causes of the mysteries of Nature and to discover their mutual dependency.[40]

Mendeleev also admired Newton as Master of the Mint, in which he safeguarded the integrity of English currency against counterfeiters and fraud. Mendeleev could envision his own position as director of the Chief Bureau of Weights and Measures in a similar standardizing light. For Mendeleev, patterning himself on Newton, the essence of any measure was gravity, and thus gravity (like the ether) held the potential to unify all the sciences.[41] Accordingly, Mendeleev began a research program at the Chief Bureau to measure precisely the strength of gravity in St. Petersburg. He had already approached this problem briefly while organizing his gas research, marking in his private notebook measures for g, the local acceleration due to gravity, in Paris and St. Petersburg in order to recalibrate Victor Regnault's pressure results to his Russian lab.[42] In the late 1890s, however, Mendeleev began a full-scale program that measured the precise local g using pendulums, later considered his most important metrological work. He simultaneously sent his associate F. I. Blumbakh to Sèvres, Budapest, and other cities to measure g values throughout Europe. Blumbakh eventually calculated g in St. Petersburg at 9.8193 m/sec^2.[43] In the publications on this topic in the journal of the Chief Bureau, Mendeleev cited Newton (and, to a lesser extent, Galileo) as a source of inspiration for placing *exact measurements* of *mass* at the center of the physical sciences.[44]

Note his emphasis on the physical sciences, as opposed to merely chemistry. In his later years, Mendeleev consistently turned to Newton as his own historical forerunner rather than a more chemical precursor, such as Antoine Lavoisier (1743–1794). Lavoisier would actually seem almost an overdetermined choice for self-modeling. He had redefined chemistry in the late eighteenth century through his notion of "simple substances," an idea Mendeleev would later adapt (via Charles Gerhardt) into the "elements" that structured his periodic system. Both men created tables of elements, and both wrote textbooks (*Traité élémentaire de chimie* and *Principles of Chemistry*) to defend their views. Both men worked extensively on gases—Lavoisier's fundamental work in pneumatic chemistry led to his discovery of oxygen. Both Lavoisier and Mendeleev worked on gunpowder for their governments. Both publicly organized commissions designed to debunk "pseudo-sciences" (Mesmerism and Spiritualism, respectively). And both were trusted state advisors in political economy. With all these similarities, he would have been an obvious choice as a historical exemplar for Mendeleev when the latter searched for a way to define his importance in the history of science.[45]

Yet Mendeleev made very few references to Lavoisier as a model, only reproducing Lavoisier's speech on Mesmerism (*not* on chemistry) in his *Materials for a Judgment about Spiritualism* and his lithograph in *Principles of Chemistry*. Instead of selecting a model that would place his periodic law and himself squarely in the chemical tradition, he opted for Newton, a man with interests in optics, alchemy, mechanics, mathematics, theology, historical chronology, and monetary reform (none of which were Mendeleev's strong suits). Why did Mendeleev insist on Newton? First, although Lavoisier's importance in the history of science cannot be disputed, much of that reputation was solidified late in Mendeleev's life, especially during the centenary commemoration (in the 1890s) of his execution by the Jacobins, whereas Newton had been a representative genius since the days of Voltaire.[46] Second, much of Newton's fame stemmed from his creation of laws that could make predictions (Halley's comet, Uranus, Neptune). In Mendeleev's eyes, Lavoisier predicted the results of specific experiments, not the structure of the universe. Mendeleev's own international reputation was heavily based on his prediction of the three eka-elements, making the analogy with Newton even more appealing. Mendeleev took the Newton exemplar seriously; so should we.

Mendeleev cast his personal journey through history as a transformation of his person and the physical sciences, as a Romantic adventure akin to Newton's reformulation of natural philosophy. The journeys from Siberia and "from history" served as fitting examples of Mendeleev's one-way travels as opposed to the stability of circulatory motion, but still they remained metaphorical. Although he had in fact journeyed from Siberia, his presentation of the journey was a literary *re*-creation. Mendeleev still hungered for actual travel.[47]

NORTHWARD BOUND: THE ARCTIC PROJECT

"You will rejoice to hear that no disaster has accompanied the commencement of an enterprise which you have regarded with such evil forebodings," declared Captain Walton in the opening lines of Mary Shelley's novel *Frankenstein*.[48] Walton referred not to Victor Frankenstein's attempt to generate life in his "workshop of filthy creation," but rather to his own efforts to find the North Pole, a parallel tale of scientific hubris. Interestingly, Walton's story begins in St. Petersburg, before he set off northward from Archangel (Arkhangel'sk). The opening of this novel on a ship destined for vigorous yet futile Arctic exploration, a journey that reveals the far greater adventure of Victor's disastrous experiments, resonates with one of Mendeleev's late pet projects, the northern sea route. First, Walton's journey is one of empire-building through travel northward. Second, Victor and, to a lesser extent, Walton are Romantic heroes who brave great peril for the sake of knowledge. The first similarity encapsulates the dream of the Imperial Turn, the second its Romantic undercurrent.

Arctic exploration had been a passion of Mendeleev's since his days as a graduate student. In 1856, he favorably reviewed a translation from the German of E. Hoffman's travels to the Northern Urals. The young Mendeleev was gripped primarily by the adventure. He contended that "the desire to give a new way of life to the activity of their countrymen . . . elevates the spirits of explorers to the fitful yearning, upon which everything egotistical is forgotten."[49] Exploration elevated the spirit by demonstrating bravery while also yielding useful knowledge.

Mendeleev separated these two functions in later years, emphasizing adventures in his personal life and knowledge-production in his professional life. Despite efforts by Soviet and Russian historians to align all of their cultural heroes, portraying Mendeleev as a lover of the literature of Dostoevsky and Tolstoy, the chemist had little patience for the psychological novel. After reading such works, he cried out: "Torture, such torture, they describe so much! I can't ... I am not in any condition!" *Crime and Punishment* in particular paled before Western shoot-'em-ups between cowboys and Indians: "Here [in Dostoevsky] we have a guy kill a person and [then] two volumes of tortures, and here [in a Western] six are killed on one page and no one feels bad."[50] As he told his family: "I can't stand these psychological analyses. It's better when in the Pampas Indians take scalps from whites, seek out footsteps, fire without missing."[51] His favorites remained picaresque tales of swashbuckling and derring-do, like Dumas's *Three Musketeers* or Jules Verne's novels.

Not that Mendeleev was entirely devoid of ambition to become a highbrow cultural critic. On 13 November 1880, a somewhat unusual article appeared over Mendeleev's signature in a Petersburg daily. For several paragraphs Mendeleev discussed a painting by Arkhip Ivanovich Kuindzhi, a well-known artist of the "Wanderer" school who would become a close friend of Mendeleev.[52] According to Anna Popova, soon to marry Mendeleev, she and the chemist visited Kuindzhi in his studio, where suddenly Mendeleev stood rapt before the painting "Night on the Dniepr" (1880). The painting depicts a mostly dark night, illuminated only by the greenish glow of the moon through an intervening cloud, reflected off the river (Figure 7.1). Mendeleev supposedly delivered a flowing discourse on the nature of artistic landscapes and their parallels in the natural sciences.[53] Inspired by the general admiration for Kuindzhi's unusual style, he wrote up his thoughts and sent them to *The Voice.* Although he may not have composed the letter on the spot, he revealed a deeply philosophical approach to the sciences.

Mendeleev's article centered on the kinds of commentary a *scientist* could uniquely offer on a work of art. (In a curious coincidence, Mendeleev's Spiritualist counterpart N. P. Vagner published an equally laudatory review of this very painting on the same day in a different paper.)[54] In antiquity, Mendeleev claimed, people mostly wanted to represent other people:

In science this was expressed by the fact that its apex was mathematics, logic, metaphysics, politics. In art it was self-adoration, expressed by the fact that only the human form was studied and caused inspiration. I think and write, however, not against mathematics, metaphysics, or classical painting, but for

FIGURE 7.1 "Night on the Dniepr" (1880) by A. I. Kuindzhi. Source: Nevedomskii and Repin, *A. I. Kuindzhi,* opposite page 60.

landscape, for which there was no place in antiquity. Times have changed. People became disenchanted with the autonomous power of human reason, with the possibility of finding a true path only by going deep into oneself, into the human, becoming an ascetic, a metaphysician, or a politician. It was understood that, in directing study to the external, at the same time they began to understand themselves better, to achieve the useful, the peaceful, and the clear, because one could relate to the external more truthfully. They began to study nature, natural science was born, which ancient ages and the Renaissance did not know. Observations and experience, induction of thought, obedience to the inevitable, its study and understanding, soon appeared stronger, newer, and more fruitful than pure, abstract thinking. . . . It became understood that man, his consciousness and reason, were only a fraction of the whole, which is easier to achieve from the exterior rather than the interior of human nature. . . . The apex of knowledge became the experimental, inductive sciences, which used internal and external knowledge, reconciling regal metaphysics and mathematics with modest observation and the request for an answer from nature.

The landscape, Mendeleev concluded, was born along with the transition toward science: "And our century will eventually be characterized by the appearance of the natural scientist in scholarship and landscape in art."[55]

He and Kuindzhi were doubles: They both represented the most progressive strains in their fields; they both tried to unify the internal and external worlds by recourse to "nature"; and they were both public intellectuals feted by the Petersburg elite. Mendeleev already hosted artist friends at Wednesday salons in his apartment—he now made his bid as cultural critic outside of his professed bailiwick of the natural sciences, whose borders he had policed against the Spiritualists with such vigor.[56] However much he enjoyed the possibility of being a public friend of art, the Academy of Sciences debacle soon dampened his enthusiasm at being a "public" anything. He kept his artist friends at home in his salons and his literary views to himself. But he did not abandon his adventure novels. Late in life, while Mendeleev was awaiting surgery to remove blinding cataracts, he had his wife read him favorite passages of Jules Verne, and on his deathbed, he specifically requested Verne's tale of a journey to the North Pole.[57]

In a curious episode, Mendeleev attempted in the 1890s to realize the action of that novel—and was cruelly disappointed. While the dominant wave of Russian geographical exploration in the eighteenth century had been to plunge eastward into Siberia, in the nineteenth century interest expanded to include the frozen wastelands of Russia's Arctic coast. In the second half of the nineteenth century, Siberian industrialists, especially M. K. Sidorov and A. M. Sibiriakov, took the lead in seeking a northern naval route that would enable goods to be moved from the Baltic port of Petersburg or the Arctic port of Arkhangel'sk to the Pacific port of Vladivostok, just as the land route, the Trans-Siberian Railroad, was being constructed. And in 1878–1879, A. E. Nordenscheld completed a circumnavigation of the Arctic Ocean.[58] In the climate of optimism that followed, Mendeleev joined with S. O. Makarov, naval hero of the Russo-Turkish War, in a project to explore both a route to the North Pole and a passage to the Pacific. In 1899 they built an icebreaker, the "Ermak" (named after the man who claimed Siberia for Russia), and in 1901 Mendeleev proposed a journey on this craft to explore the motion of polar icebergs so as to chart a viable northern sea route.[59]

Mendeleev tried to incorporate his daredevil ideas into an economic project. He wrote to Sergei Witte in 1901 that Russia should make a tremendous effort "since no one has such a large coastline on the Arctic Ocean and enormous [Russian] rivers flow into it, draining the majority of the Empire, which part is scarcely developed not so much because of climatic conditions, but rather because of the absence of trade conduits via the Arctic Ocean."[60] Mendeleev sought government funds to find these routes and thus promote industrial exploitation of the North by circulating goods to and from the frozen coast. Once again, circulation formed the heart of the Imperial Turn, but it would first require a linear journey of adventure. This particular plan for the Arctic, however, ended ignominiously. On 8 January 1902, V. I. Kovalevskii told Mendeleev that Witte refused to endorse the plan, supposedly because Prince Aleksandr Mikhailovich, the Tsar's close relation and advisor, refused "to help such an impertinent man as Mendeleev."[61] Even had support been forthcoming, Mendeleev and Makarov fell out acrimoniously, as Witte recalled:

> Mendeleev, a very able but quarrelsome man, had sharp differences with Admiral Makarov over which route to follow to the Far East: he favored a route that would cross the North Pole, while Makarov favored one that followed the Siberian coast, each arguing that his route was less hazardous than the one proposed by the other. After a very bitter argument, in my office, the two never met again.[62]

There was no opportunity to resolve their differences; Makarov departed almost immediately for a military post at the port at Kronstadt. In either case, the venture inspired by Jules Verne foundered on Mendeleev's character: The rigid institutions of the Imperial Turn clashed with the very same brash personality that was supposed to underwrite it.

FULL OF HOT AIR: MENDELEEV, AERONAUT

The Arctic project never came to fruition. The half-steps of recrimination and stonewalling contrast sharply with Mendeleev's experience with bold adventurous travel only ten years earlier—also undertaken with military support—when he fulfilled another Romantic dream that he would later describe as "one of the most remarkable adventures of my life."[63] In the middle of the 1870s, while battling Spiritualists, he became interested in ballooning as a way of exploring gases in the upper atmosphere and as a way to organize meteorology. His publication strategy to raise money to build a scientific aerostat failed to reach his (woefully underestimated) target cost, and he shelved that dream as a dead end.[64] He may have overtly abandoned his interest in aviation, but the sponsors for his foreign travels to study the matter, the Navy, had not lost theirs. Following the German lead, the Russian military grew more interested in balloons in the 1880s—for reconnaissance, weather observation, and bombing—acquiring fifty-two of them by 1907.[65] In 1887, Mendeleev would be allowed to fly in one of these. When he returned, he flaunted the experience in Byronic overtones, reflecting the projection of masculinity in science that he had exercised for many years.

Yet the ascent did not conform to Mendeleev's expectations. He had originally wanted to use a fleet of scientific aerostats to measure meteorological variables (relative humidity, pressure, ambient temperature) in the upper atmosphere. It was to be a communal project, designed to help understand the physics of gases and only secondarily to improve weather forecasting. Such a straightforwardly meteorological ascent had already taken place under the auspices of Russia's chief meteorological establishment, the Chief Physical Observatory, by M. A. Rykachev. Rykachev echoed Mendeleev's refrain that balloon ascents were necessary since "there are no other means in flat and level countries to achieve significant heights, and even where one finds high mountains, observations carried out from mountain ascents can give entirely different temperature results than those taken at the same elevation far from the earth's surface."[66] Rykachev had begun his studies in the 1860s with James Glaisher, the same Victorian balloonist-meteorologist Mendeleev had once criticized, and spent years perfecting measurement techniques. This was no spur-of-the-moment solo venture. The Russian Geographical Society acquired a second-hand balloon from a French sailor in 1873, and Rykachev had equipped it. In his report, he mostly refrained from Romantic descriptions and vaunted his own heroism only briefly:

> Air flight is a purely naval affair. Sailors should take it up, [as] the direction of the balloon demands the same qualities necessary to sailors: the quickness of thought, orderliness, preservation of presence of mind, powers of observation, attention, agility. This is why, being of a naval family, I let myself enter into certain details of our trip and will add also a few words about directing the balloon and its construction.[67]

Here at his most flowery, Rykachev could not compete with even the least theatrical aspects of Mendeleev's later account.

Despite the fact that this ascent was accomplished right outside Petersburg in 1873, and the account (with data analysis) was published in 1882, it drew little public attention, largely because Rykachev failed to project the image of Romantic masculinity and showmanship that would turn Mendeleev's ascent into such a spectacle. Ironically, it was precisely the aspects of Mendeleev's ascent that were unrelated to scientific aims—and thus removed from his original intention in the 1870s—that made his trip such an advertisement for science.

Much of our information about the circumstances of the flight comes from Mendeleev's article, "An Air Flight from Klin during the Eclipse," published in the thick journal *Northern Herald* (*Severnyi Vestnik*), a periodical often identified with the rise of the Symbolist literary movement.[68] A full solar eclipse was expected for 7 August 1887, and since it was rare to observe such an event in a populated area of Russia, it commanded much attention. The eclipse would be full at Klin, near Mendeleev's summer house of Boblovo, at 7 A.M., and would last for roughly two minutes. The weather, however, might be overcast, hindering observations of the sun's corona, the "atmosphere" of the sun visible when the moon blocks the solar disc. Thus, the balloon. On the night of 29 July, Mendeleev received a telegram from M. N. Gersevanov of the Russian Technical Society informing him that they were financing a balloon near Klin, "in case you desire personally to use the ascent for scientific observations." The Technical Society, burned only six years earlier by Mendeleev with his failed gas experiments, turned to him because he lived near Klin and had knowledge of aerostatics. He leapt at the chance:

> If aerostats or flying machines can today battle with weak winds, tomorrow they will defeat even strong ones, and—if they will have the strength to stay in the air a long time and ascend high—they will fly over mountains, flee from storms, and wait them out in the ocean of air where buffeting, the chief reason for the destruction of ships, is very unlikely, because [aerostats] can change the level at which motion occurs, which water does not allow.[69]

Indeed, it almost sounds like something out of Jules Verne.

The eclipse expedition was not devoid of scientific content. Mendeleev did make observations on the corona, the importance of which he explained to the popular audience of the *Northern Herald,* using the analogy of cosmic dust gravitating around a comet to form its tail. By looking at these smaller objects, he contended, it might be possible to understand the more interesting larger objects: "Thus the 'corona,' possibly, is the condensed mass of these small cosmic bodies, which form the sun and support its power."[70] Mendeleev had originally intended to employ balloons for meteorology and not for sporadic one-off eclipses. His shift of topic reveals that, as far as aviation was concerned, he was less interested in the observations than in the mechanism. For example, when V. P. Verkhovskii, pioneer of Russian aviation, wrote to Mendeleev, he treated the latter as an advocate of military, not scientific, aviation:

> In any case, one must hope that the matter won't die and sooner or later an aerostat which satisfies military goals will be built under your leadership. The numbers of proposed costs of which you speak in your letter are so moderate, that no one can raise any objections against them. The cost of balloon and gas, one supposes, also will not exceed the cost of an ironclad. And, finally, if it is brought into existence only once, [and they] realize the power of a military aerostat, then money will not be an obstacle.[71]

Mendeleev had moved rather far from his original interest in decentralized aviation sponsored by scientific societies. The military's money was just as good.

So when Mendeleev received the telegram, he immediately wired back asking for the military's best hydrogen balloon. In direct connection to his new
emphasis on the Empire, Mendeleev wanted to use science to further military
goals (and make himself a Romantic soldier in the process):

> In other words, I wanted to add to the primary goal of observing the eclipse
> another goal—to test existing aerostatic capabilities in peacetime for those
> goals it might serve in wartime. It seemed to me that the War Minister would
> agree to the Technical Society's petition precisely because he saw this side of
> the matter and would consider it useful to test prepared aerostats in coopera-
> tion with scientific goals. . . .

Mendeleev received authorization from the minister for a balloon (called the
"Russian"—*Russkii*) that would easily ascend 700 meters. A military aeronaut,
A. M. Kovan'ko, was ordered by the minister to take the trip with the chemist.
Throughout the event, Mendeleev construed the flight as a military affair, re-
peatedly noting that he took actions that were "exactly as in war."[72]

Mendeleev wanted to ascend five minutes before the scheduled eclipse, just
enough time to reach the desired altitude without being blown off course by
wind. It was at this point, on the morning of 7 August, when the expedition
ceased to be about Empire or Science and became about Mendeleev the Romantic Hero. In keeping with a venerable Victorian tradition of eclipse expeditions, a large crowd had assembled that day, treating the scientific-military
event as a public spectacle (Figure 7.2).[73] However, it appeared that with all
the instruments on board, the ballast would not allow for two to ascend.
Rather than let Kovan'ko make the observations, Mendeleev—a 53-year-old
novice—insisted that he travel alone. In retrospect he portrayed his decision as
self-sacrifice for the greater glory of science, since canceling the expedition
*"would mean to rupture faith in science. After all a balloon is the same kind of sci-
entific instrument as those I am used to working with."*[74] It also directly contra-
vened military orders.

The public adored Mendeleev's heroism. Clemens Winkler, the discoverer of
Mendeleev's predicted eka-silicon (germanium), wrote that he had followed
the balloon flight eagerly in foreign papers. V. A. Giliarovskii, who penned a
newspaper account of the event, presented a breathless image of Mendeleev's
titanic bravery: "As if it were now, I see the enormous figure of the professor,
his wind-tossed hair underneath a hat pulled down over his eyes. . . . Arms
raised up—he arranges the strings. . . . And instantly vanishes. . . . It becomes
totally dark. . . . It became cold and terrifying."[75] Paul Walden's Mendeleev

FIGURE 7.2 Mendeleev's balloon ascent at Klin. This is one of several photographs taken on 7 August 1887. Particularly noteworthy is the scale of the balloon with respect to passenger, and the large crowd that had gathered to enjoy the novelty. Source: *MS*, VII, opposite page 480.

obituary likewise rendered the eclipse trip a centerpiece of his image of the chemist, even though he had not been in attendance:

> Seconds upon seconds were getting lost and each was costly, since the solar eclipse was only supposed to last for a few minutes. Mendeleev suggested to his driver to step out, when the latter demurred, [Mendeleev] made a gesture of throwing him out, quickly gave the final command to cut the line, and— less burdened—the now driverless balloon rose with Mendeleev alone into the air, soon to vanish behind the clouds. . . . Despite the fact that this was his first and last ascent, Mendeleev immediately understood how to master his critical position and to orient himself, so that he made the desired observations and could manage to pull off a nevertheless undamaged landing on earth, with a certain alarm on the part of the local residents.[76]

Upon witnessing Mendeleev's safe landing in the Yaroslavl province, local peasants were supposed to have said: "Dmitrii Ivanovich has flown in a bubble and would have broken through the heavens themselves. . . ." For his part, Mendeleev joked, once safely on the ground: "Before the flight I had no fear; I was just afraid that when I landed the farmers would take me for the devil and thrash me."[77] The perception of danger, however faint, titillated the audience. In England, James Glaisher's near brush with death in an 1862 scientific balloon expedition was the stuff of national legend.[78] As if to emphasize how "ordinary" adventure was in the life of the scientist, Mendeleev traveled to Petersburg three days later and two days after that was already en route to England for a meeting of the British Association for the Advancement of Science, the original patron of scientific ballooning.[79] There remain almost no dispassionate accounts of the expedition: Each rendition is so colored with the imagery of heroic literature that it is hard to extract even the sequence of events. It is noteworthy that Mendeleev's balloon trip is one of the most remarked-upon in all reminiscences and accounts of the life of the scientist, a poignant complement to the common image of the introverted genius laboring over the periodic system.

Military glory and self-fashioning as a Romantic figure were two ingredients of Mendeleev's balloon ascent. There was, however, a third: the construction of the masculine scientist in Imperial Russia.[80] It is no coincidence that so few female figures appear in my account of Mendeleev's travels through Petersburg culture, for in constructing the Russian chemical and bureaucratic community, he and his contemporaries built a male world. Masculinity is an important undercurrent in these narratives. In the formative years of both Imperial Russia and modern science, controlled modes of male sexuality were central issues, especially for the latter in terms of clear demarcations of science as a "proper" vocation for modern men.[81] Similarly, for Mendeleev, modernizing Russia was to be an autocracy built on scientific expertise, and only male rationality seemed stable enough to bear the burden.

The balloon expedition offers only one of the more explicit examples of this emphasis on masculinity. Male expeditions, and especially male *accounts* of those expeditions, established standard narratives of manly behavior.[82] Mendeleev's version highlighted his bravery, his passion for knowledge, and his martial fervor; this vocabulary—both heroic and chauvinistic—seeped into other accounts, especially those of women. In Mendeleev's wife's memoirs, she reproduced numerous snatches from his published account and those of local newspapers, but did not include her own observations, even though she was present. She did relate, however, the prior domestic drama caused by Mendeleev's command for her not to observe the event in case he was harmed—shielding her womanly sensibilities. She defied him and attended, but her brief observations concerned incidentals, like the presence of famous artist Ilya Repin, who intended to paint the flight. (It seems he in fact did not.)[83] According to the hagiographic testimony of a female employee at the Chief Bureau of Weights and Measures, except for his second wife and a few select women, Mendeleev treated women as empty, nervous creatures.[84] His views are typical of his place and time, but they are of interest because of his role in the construction of boundaries between men and women in the domain of the natural sciences.

Mendeleev was, for example, an activist for women's education. While feminist groups in the United States and England were lobbying for suffrage and inheritance rights, in Imperial Russia, where public participation was substantially more limited, education served as the main front in battles for equality.[85] Mendeleev may have distinguished scientific discovery and active exploration as arenas for male mavericks, but he nonetheless thought that women should be educated in the sciences and humanities to the same degree as most men. This set the demarcation line rather cleanly between teachers (men) and students (lesser men and women), and it also made him a thorn in the side of the Ministry of Popular Enlightenment, which opposed higher education for women. Section 103 of the draft of the 1863 university statute explicitly excluded women from courses. In 1862, Mendeleev—then a privatdocent at St. Petersburg University—and some colleagues wanted to amend the wording to "Persons of all estates and both sexes have the right to register as voluntary auditors." The Ministry rejected the suggestion. At the first Congress of Natural Scientists and Physicians in 1867, Mendeleev and his friend A. P. Borodin were among the few to endorse the idea of a women's university. In 1878, women's courses finally opened at St. Petersburg University, and Butlerov, and later Mendeleev, lectured in introductory chemistry throughout the 1880s.[86] This would be the last decade of teaching for Mendeleev to any audience.

THE LIMITS OF ROMANCE:
MENDELEEV LEAVES PETERSBURG UNIVERSITY

Practically since the moment of his return from Heidelberg in 1861, Mendeleev's path through the culture of Petersburg and Imperial Russia was associated with St. Petersburg University in some way. Its campus on Vasil'evskii Island, next to the Academy of Sciences on the banks of the Neva River, was the site of his first lectures, the place where he observed the student rebellions of 1861, and the scene of his greatest professional triumphs as a chemist and a bureaucrat. In 1890, however, Mendeleev's service at the University ended, in circumstances that have endowed him with the aura of a revolutionary (despite his years of loyal service to the Tsar and the principles of autocracy). Mendeleev's departure reflected a conflict between two different modes of self-fashioning—as bureaucrat and as daredevil—and how the university statute of 1884 created an environment that could no longer withstand their tension.

In August 1885 Mendeleev completed thirty years of service for the Ministry of Popular Enlightenment. As per statute, the faculty could vote to extend the tenure of an emeritus professor every five years—Mendeleev had already been reinstated once in 1880—and he was approved again. In keeping with his shifting intellectual interests toward theory, and his financial interests in maintaining high sales for *Principles of Chemistry,* he maintained only his introductory course, lecturing between three and five hours a week throughout the 1880s, for a total of only 84 hours a year (low by contemporary standards).[87] Attendance in his course, a prerequisite for specialization in the natural sciences, remained high, and he was still popular among students, if not among certain members of the faculty or the administration.

In 1885, partially in response to the jolt of the more restrictive 1884 university statute (which replaced the liberalizing 1863 statute), Mendeleev became more involved in University administration. For example, he proposed that topics for examination questions not be determined by statute, but rather by the actual content of that year's lectures. He included a draft of the official proposed revision to the 1884 statute for the physico-mathematical faculty of the University replete with minute (and somewhat catty) marginalia, reminiscent of the denigrating rhetoric of juxtapositions in the 1876 *Materials for a Judgment about Spiritualism.* In a letter to colleague N. A. Menshutkin, he argued from the moral authority of the professoriate:

> University teaching is impossible without the scientific autonomy of professors. Obviously, and also involuntarily, the student who begins in science is first submitted to the scientific influence of his professor, and thus in training students it is necessary to demand knowledge within the bounds and direction of the lectures he has heard. If this demand is absent, then the spiritual connection of professors and students can be subjected to ugly transformations and in place of the authority of the professors, which is subject to general control, he will easily take up as an authority one that is not open, not scientific, [but is] dreamy or crudely materialistic, which is ruinous for youth, who need authority as a connection of the present to the past.[88]

Already here the tensions between Mendeleev the bureaucrat (prescribing standards through regulation) and the individualist (leading through charismatic authority) are clear. Mendeleev had strong personal reasons to hold on to the moral authority of the teacher as a model for peaceful transformation and consensus. It was precisely such reasoned (but still magnetic) argumentation that had allowed Cannizzaro to propose his reform of atomic weights at Karlsruhe in 1860.

The 1884 university statute was not the ideal backdrop for taking a personal stance. Despite assurances during the Great Reforms that allowing educational institutions autonomy would enable professors to control student disturbances effectively, unrest continued throughout the late 1870s and early 1880s, leading to counterreforms in the reign of Alexander III. Structurally, the university statute of 1884 reinstated the conditions of the 1835 statute. Over the objections of the vast majority of professors, Minister D. A. Tolstoi, the architect of the reform, also eliminated subsidies for poor students in an attempt to restrict admission to members of traditionally loyal estates.[89] The reform failed on several levels. Student demonstrations not only continued, they became worse, culminating in the 1899 disturbance that proved a dress rehearsal for the 1905 Revolution. And rather than preventing poor students from attending St. Petersburg University, the lack of subsidies only further impoverished such students and made them more likely to erupt in violence. It also alienated the professors, who were the only allies the Ministry of Popular Enlightenment had on the front lines.[90]

Mendeleev, in particular, made himself the center of student attention during several such disturbances. On 11 December 1887, students gathered in Mendeleev's auditorium—the largest meeting space at the University—and the University was sealed off "so that even Prof. Mendeleev did not get to the auditorium without bumping into the police." Mendeleev insinuated himself as spokesman for the students, telling them: "The University is a whole thing, and each part of it should sacrifice itself for the greater good, therefore I, Professor Mendeleev, will be the herald of your desires before the higher administration."[91] The same month of that initial unrest, Mendeleev interceded with the police after they arrested nine students and were searching for two others on University grounds, arguing (incorrectly) that the police had no right to arrest on the premises.[92] It is important not to misconstrue Mendeleev's role here. He was *not* an advocate of the students' views, frequently nihilist ones of which he strongly disapproved. He acted as their intermediary because it would serve his own cause of University autonomy.[93] Being emeritus, he also had less to lose than many of his peers.

Mendeleev's swan song at the University took place during the ides of March 1890. According to an anonymous police report, unrest was expected on 12 March, but it did not appear until two days later. Between 200 and 250 students occupied the corridors of the University. Professor V. P. Sergeevich appeared and brought the students to the largest auditorium. Sergeevich began to talk the students down from their furor, and had almost succeeded, "when suddenly Professor Mendeleev approached them and said that the students were making noise in vain, and that they should set forth their demands in a special petition, which he was ready to present to the Minister of Popular Enlightenment."[94]

At 9 A.M., on 15 March 1890, Mendeleev handed the following petition to I. D. Delianov at the Ministry:

> The petition of the students of St. Petersburg University. To the Minister of Popular Enlightenment:
>
> On Wednesday, 14 March, we were for the first time given the opportunity to express before the collegium of respected professors with the Rector at the head our urgent needs and burning wishes.
>
> Firmly certain from bitter experience of the necessity of reforms to the university order, we are convinced that our desires are entirely realizable, and we formulate them as the following:
>
> We want that the statute of the universities and other higher educational institutions be founded on the principles of autonomy—that the Rector and professors be selected according to the university statute of 1863, that a university and student court be established, and that student corporations be recognized.
>
> We want that all who have finished middle educational institutions have unrestricted access to university without distinction of creed, social position, and without any hidden characterization from *gymnasium* administration and police.
>
> Finally, we are certain that along with this, our professors can be given freedom in teaching, which existed earlier under the 1863 statute.
>
> Our deep conviction is that all these consequently conducted changes in the sense of our desires will assist the development of student life and only they can set up its normal course.
>
> We insist on the immediate elimination of police functions of inspection, the reduction of pay, and, in particular in relation to our University, to the establishment of a scientific literary society, which existed until 1886, and also a student reading hall.
>
> First using the opportunity to express our wishes, without leaving the bounds of legality, we firmly believe that a similar means of expression of our needs will enter everyday student life. [Signed:] The students of St. Petersburg University.[95]

The next day, Mendeleev received a response from Delianov reminding him that by the oath of Imperial service, no servant of the Tsar was allowed to receive or transmit such papers, an oath that bound both Delianov and Mendeleev. By this technicality, he returned the document to Mendeleev.[96] Delianov was, of course, correct, but Mendeleev took it as an insult and resigned in protest—of the insult, *not* the denial of the students' claims. When students heard that Mendeleev was retiring because of the petition, rumors abounded that he was fired by the Ministry, inspiring public protest and a demonstration. Roughly 360 students were detained in consequence.[97]

This additional unrest must have been painful for Mendeleev. After all, the whole purpose of the petition was to prove that students could function—under the tutelage of professors—within the limits of the 1863 statute. The reaction undermined Mendeleev's faith in the Russian people's ability to take a gradualist path toward change, although it reinforced his belief in the *need* for such gradualism. As he had told the students on 14 March: "It is impossible to force any kind of event—this is impossible in this world—nothing, nowhere, never—all history, all life, all experience of life shows this. And there will come a time—you must believe this, you must have the vision that it will come, it will come: After all everything moves gradually forward—and you must understand this and act accordingly."[98] Resignation was not evidence of Mendeleev's radicalism, both because the student petition was hardly radical, and also because he would later move on to work for both the Navy and the Chief Bureau of Weights and Measures, hardly fitting posts for a revolutionary. Instead, it was the action of a maverick who felt that he could transcend rules so as to bring about a better order. This was the Mendeleev of the balloon flight, of Arctic exploration, hoisted by the petard of the other Mendeleev, the one who helped insulate the regime from hotheads like himself.

The story of Mendeleev's departure from the University was subjected to substantial revision, much as his Siberian origins were. Understandably, the event was painful and he wished to cast it in a positive light. In 1905, in the midst of renewed (and much more political) student unrest, Mendeleev recalled in a diary entry: "I left the university defending its authority and the students. I am not bitter about it, but have only a clear enmity to the regime. . . ."[99] This version fails to correspond with Mendeleev's attitudes to student unrest around the time of his resignation. For example, he recalled that "[i]n 1887 I was so fed up with university disturbances, that I wanted to leave the university." The University pacified him instead with a trip to the Don region to investigate coal. He made similar allusions to the irritations of student unrest in his 1887 solutions text.[100] In a letter to Witte composed 1903 (sent in 1906), Mendeleev declared that after thirty-five years at the University, "I decided to leave it entirely, moreover, since the renewed student unrest simply impinged on my weak health, and the new university statute went into effect, already beginning to snuff out the bright sides of our scientific activity that had begun not long ago, and lowering the influence of pure science on the youth." Witte for his part recalled that "because of his exceedingly quarrelsome nature he had left the university as soon as he had become eligible for a pension." (Mendeleev already had his pension in 1885.)[101]

The day of his final lecture, Mendeleev, still furious at the perceived insult, indulged his Romantic side even more by contemplating the creation of a daily newspaper with artists A. I. Kuindzhi and I. I. Shishkin. Originally, he wanted to call it *Rus'*, but that title had already been taken by Ivan Aksakov (in the very paper that had published Butlerov's defense of Mendeleev against the Academy). He then wanted *Foundation* (*Osnova*) or *Motherland* (*Rodina*), but both of those were also already in print. He settled—fitting for the balloonist—on *Ascent* (*Pod"em*). Mendeleev had to petition the same Ministry of Popular Enlightenment, specifically its Chief Administration of Printing Affairs, for permission to publish the periodical. He wrote on 29 March 1890: "I declare as the chief goal of the proposed newspaper the impartial discussion of various questions relating to the industrial development of Russia, because there I see the natural and reliable path to the further success of the Fatherland and a means to the improvement of the well-being of all strata of the people."[102] Delianov, still irritated by the troublesome professor, resolved that Mendeleev could publish the journal only if he eliminated all discussion of politics and literature and only discussed industry, *and* if he submitted to the preliminary censorship reserved for politically sensitive cases.[103] Mendeleev refused this kind of oversight, and commented to a friend in a fit of sour grapes: "Delianov didn't allow it. And I'm glad. This isn't my cup of tea: after all with this there wouldn't be peace day or night."[104]

Despite the intrinsic tension between Mendeleev's bureaucratic systems and his Romantic journeys, the combination rested on two basic pillars that seemed to be perpetually stable. As long as autocracy continued to be unconstrained and committed to modernization, he was confident that technical expertise would have a prominent role in the future of the Russian state. For its part, the Romantic self-image was buttressed by the immutable status of the periodic law, seemingly incontestable after the discovery of germanium in 1886. The twin journeys—of circulation and adventure—would meet unexpected challenges in the early years of the twentieth century, however, and Mendeleev would be hard pressed to maintain his shaky equilibrium. The periodic law would be threatened by the rise of a new chemistry that refuted Mendeleev's fundamental conception of his own achievement, and autocracy would shake under the impact of revolution in 1905. Not even a Russian Newton could save it.

DISINTEGRATION

Mendeleev at his working desk in 1904, from Smirnov, *Mendeleev*, insert 2.

DISINTEGRATION:
FIGHTING REVOLUTIONS WITH FAITH

My God! it is a melancholy thing
For such a man, who would full fain preserve
His soul in calmness, yet perforce must feel
For all his human brethren—O my God!
It weighs upon the heart, that he must think
What uproar and what strife may now be stirring
This way or that way o'er these silent hills . . .
—SAMUEL TAYLOR COLERIDGE, "FEARS IN SOLITUDE"
(1798)

THE FIRST SIX YEARS OF THE TWENTIETH CENTURY—the last six of Mendeleev's life—were difficult for both the chemist and the Empire he served. Beginning with widespread student rebellion in St. Petersburg in 1899, the Tsarist regime met with increasingly vocal (and often violent) opposition from a broader spectrum of society than ever before. International currency fluctuations made it harder to acquire necessary financing, while transformations in military technology demanded heavy state expenditures. Nevertheless, Russian ministers and Nicholas II took aggressive action to expand the Empire eastward via the Trans-Siberian Railroad, antagonizing the nascent military forces of a modernized Japan. On 26 January 1904, the Japanese launched a preemptive attack on Port Arthur, Korea, and the ensuing Russo-Japanese War ended in Russian humiliation and the almost complete devastation of the flower of the Russian Navy.

In January 1905, rising discontent with the regime led to a protest at Palace Square in Petersburg that—through incompetence and failure of nerve—erupted into Bloody Sunday, as soldiers fired on unarmed civilians. The erosion of the goodwill of the populace and the regime's financial reserves forced Nicholas II to renounce autocracy and submit, in the famous October Manifesto, to restrictions on his absolute power and an elected parliament. The Russian state was now financially weakened, militarily embarrassed, and—most importantly—no longer unfettered. The late Imperial period was coming to a close.

These events alone, which undermined so much of Mendeleev's project to restructure a modernized, but still autocratic, Russia, would have been enough to devastate the aging chemist without the personal woes that dominated his final years. After leaving St. Petersburg University in 1890, he worked for four years as a consultant to the Navy developing a new form of smokeless gunpowder. Although his new variant showed tremendous promise, the Navy opted to adapt the Army's alternative. Mendeleev resigned and turned his time to his post as director of the Chief Bureau of Weights and Measures.[1] The work at the Bureau was personally satisfying, but he was unable to see his metric reform to completion, largely as a result of the budget crisis consequent to the Russo-Japanese War, and it proved a great disappointment. In late December 1898 his eldest son Vladimir died suddenly. Mendeleev contracted cataracts a few years later—the same affliction that had blinded his father the year he was born—and had to undergo surgery to restore his vision. Furthermore, a series of ill-advised investments dealt a severe blow to his family's financial security. As if this were not enough, newspapers constantly attacked him for his economic policies. He applied the balm of self-pity in a solitary diary entry dated 10 July 1905: "For the future, I confess: I would like that the traces of my efforts in life would remain stable, of course not for centuries, but for a long time after my nearing death. There are only two categories of my efforts in life I consider stable: [my] children and my scientific works."[2] At the peak of those scientific works was the periodic law. It may all have been worth it if that could remain stable.

Yet now, in his final years, Mendeleev confronted a series of chemical "attacks" to his periodic law. He sought salvation in his age-old Holy Grail—the ether—and attempted to save his system by appropriating certain threats to tame others. This strategy, while temporarily effective in defending the periodic system, would fail to safeguard his vision of an unfettered, rational autocracy, the axis of the Imperial Turn. Mendeleev's last years were lived in the shadow of disintegrating systems, scientific and political, to which he had devoted his entire career.

CHEMISTRY UNDER ATTACK: DISINTEGRATION IN FIN-DE-SIÈCLE PHYSICAL SCIENCES

At the end of the nineteenth century, Mendeleev's understanding of the natural world was in peril. While his views on chemical and physical laws had undergone occasional revision by the 1890s, his beliefs as to what matter *was* and how it behaved were fairly set. This understanding was heavily conditioned by the periodic law itself. Matter, according to Mendeleev, had three essential properties: It was "atomic" (each atom was integral—it had no substructure); it was immutable (each specific element had fixed mass and could not "become" any other element); and each element possessed a specified valency (a numerical "charge" that determined how a given atom would combine with others, a concept to which Mendeleev eventually reconciled himself). Thus, each element in the system was placed as an atomic *individual* (in the literal sense of being "without divisions"), according to its *mass,* in a periodic relation marked by a recurrence of *valency.* Mendeleev considered these three properties of integrality, immutability, and valency to be of a piece. They were simply what it meant to be a "chemical element."

Beginning in 1894, a new phenomenon had emerged to assault directly each of these qualities of matter, and these assaults threatened both the borders of chemical knowledge and the stability of the entire discipline. Mendeleev felt he had to preserve the integrity of the chemical worldview to which his periodic system had contributed so substantially. The three phenomena were the discovery of noble gases, radioactivity, and the electron (and its relation to William Prout's hypothesis that elements were formed by the grouping of subparticles). The integrity of Mendeleev's chemical vision was at stake in each.

Despite Mendeleev's famed predictions of the properties of empty spaces in his periodic system, he was taken by surprise in 1894 by William Ramsay's announcement of a new chemical element, tentatively dubbed argon—the inert one. Ramsay's research itself was not shocking. The two had been in correspondence for some years concerning Mendeleev's moribund gas expansion research. On a recent trip to England connected with his gunpowder work, Mendeleev had given Ramsay a copy of *On the Elasticity of Gases,* but the latter was unable to read Russian, and he wrote Mendeleev (in French) in January 1892 to see if there were any translations or summaries of the text in Western journals, or, if not, whether the Russian could send him the chief results.[3] Mendeleev, recall, considered his gas work a failure since he had been unable to locate the ether in the deviations from the Boyle-Mariotte and Gay-Lussac laws, and thus had not made a greater effort to publicize his failure in Western Europe. He responded to Ramsay's request by sending another copy of the text and some brochures, to which Ramsay responded with gratitude and a repetition of his inability to read Russian, "but I see in the text some figures which guide me, and I will do my best to understand your beautiful work. I was quite surprised to see the immensity of your work."[4]

Ramsay was after fundamental prey. Together with Lord Rayleigh, he had embarked on pneumatic experiments to review Victor Regnault's gas work—just as Mendeleev had two decades earlier. While Ramsay was not trying to find the ether in the interstices in the data, he hoped to be able to explain the margins of error of the deviations. He wrote Mendeleev another sympathetic letter saying that he had met similar difficulties as Mendeleev had.[5] However, late in 1893, one of Ramsay's students discovered a problem with Mendeleev's measurement of volumes by displacement of mercury, and Ramsay tactfully pointed out that a small factor needed to be added. Afraid of angering his colleague, and perhaps aware that he had unwittingly exposed a sensitive nerve, he continued: "Perhaps I am wrong, and in that case I apologize. But instead of publishing my criticisms which may very well be wrong, I turn to you in hope that you can correct me if I have not followed your explanations due to my ignorance of your language."[6] Mendeleev's error turned out to be Ramsay's gain, for it was in the correct measurement of the volume displacement that he was able to postulate in 1894 a new constituent of the atmosphere to make up the shortfall exactly. This was to become argon.

Mendeleev's reaction was mixed. While he had greeted the validating discoveries of gallium, scandium, and germanium with pleasure, argon was the first announced element that he had no empty space for in the periodic system. It had a measured atomic weight of 40, which would place it between chlorine and potassium (where there was no empty space), and it seemed to be completely unable to bond with other elements (meaning it had a valency of zero, inconceivable for an element by Mendeleev's definition). He immediately telegrammed Ramsay (in French): "Delighted at the discovery of argon. Think molecules contain three nitrogen bound together by heat."[7] Here we see Mendeleev resisting novel discoveries that could be interpreted as violating his periodic law. The threat was not just in Mendeleev's head. After reviewing the properties of the inert gases discovered soon after argon, an American chemist remarked: "The appearance of so many new elements at one time will no doubt prove embarrassing with the present arrangement of the Periodic System, and attempts will probably be made to rearrange the system to conform to these new discoveries."[8]

The first strategy, favored by Mendeleev, was to deny argon elemental status. In Chapter 5 of the sixth edition of *Principles of Chemistry* (1895), Mendeleev included a supplement on argon, "the new component part of air," where he claimed that it was too soon to consider argon an element. It was more likely a compound or simple body, either of which would explain why it did not react with any other elements. Mendeleev suggested that, in analog to ozone (O_3), argon could in reality be N_3. (James Dewar of England had proposed this solution even earlier, explicitly in order to save the periodic law.[9]) Mendeleev worked out the details of his view in a meeting of the Russian Chemical Society on 2 March 1895. In a fashion harkening back to Cannizzaro's atomic weight determinations from Karlsruhe, he discussed the likelihood that argon was composed of from one to six component parts, concluding that a triatomic molecule was the most likely option.[10]

He soon changed his tune. In 1903, he considered Ramsay's findings "some of the most brilliant experimental discoveries of the end of the 19th century," and admitted that his early hypothesis of triatomic nitrogen was incorrect. What changed his mind? Mendeleev cited five pieces of evidence that swayed him: the finding that argon's density was just barely greater than 19, while N_3 would have been around 21; Ramsay's discovery of helium in 1895, which also displayed chemical inertness; the later discoveries of the other inert gases neon and krypton; the uniqueness of their spectra; and Ramsay's proof of the constancy of chemical features when correlated with density.[11] Mendeleev now became a proud partisan of the idea that the inert gases should be considered a zero-valency 0-group, to be placed on the far left of a periodic system (see Figure 8.1) (and not the far right, as in modern representations). This way, he argued, the system would be organized from least reactive (the inert gases) to most reactive (the halogens).[12] By the seventh edition of *Principles* (1903), Mendeleev had fully endorsed the "argon group" and considered Ramsay an affirmer (*ukrepitel'*) of the periodic law. He had abandoned one of his essential views of matter that underlay periodicity: valency.

(Mendeleev's failure to predict the inert gases was later referred to in the debates by the Nobel Committee as a reason not to award him the 1906 Prize in chemistry. Ostensibly, because Mendeleev's periodic law had not yielded any recent fruit—given that he had not predicted argon and its brethren—the Committee voted instead to award the prize to Ferdinand-Frédéric-Henri Moissan, who had isolated fluorine and invented the chemically useful electric furnace. In fact, Mendeleev had won support from the chemistry section of the Swedish Academy, but violent opposition from Svante Arrhenius, whose theory of solutions Mendeleev had criticized in the 1880s, derailed the nomination. Defeat was snatched from the jaws of victory, a stinging reminder of the Academy of Sciences debacle.[13])

Argon was something of a sideshow compared to radioactivity, which was perhaps the most controversial topic in the physical sciences at the turn of the century—it elicited claims of the disintegration of the elements, spiritual forces, revolutions in medicine, and the reemergence of the alchemists' dream of transmuting elements.[14] In 1896, in an effort to demonstrate that the phenomena of X-rays (discovered by Wilhelm Conrad Röntgen the year before) were related to fluorescence, French physicist Henri Becquerel undertook a series of experiments on uranium. By accident, he discovered that uranium would cloud photographic plates; a series of further experiments led him to conclude that uranium spontaneously emitted energy. In 1898, Pierre and Marie Curie, in their Paris laboratory, discovered the new elements polonium and radium, which emitted energy of extreme intensity—dubbed "radioactivity" by Marie. Radioactivity fast became one of the most vigorous fields of research in the physical sciences, and research along these lines in Russia proceeded apace. In April 1897, I. I. Borgman conducted the first Russian scientific work on radioactivity, followed shortly by N. A. Gezekhus at the Technological Institute in St. Petersburg. Both, incidentally, had been on Mendeleev's Commission on Spiritualism. Borgman and others, including N. N. Beketov, successor to Butlerov's chair at the Academy of Sciences, read a series of popular lectures on radioactivity. Their work culminated in the establishment of a radiological laboratory in Odessa in March 1910.[15]

Most of this initial Russian work was physical rather than chemical in nature, since the latter required larger samples of scarce radium. Nevertheless, Mendeleev quickly began speculating in opposition to the French interpretations of radioactivity as disintegration of the elements. The first announcement of radioactivity generated substantial stir in the Imperial capital, and Mendeleev's views, as the dominant scientist in the Empire, were in heavy demand. For example, on Thursday, 25 January 1896, there was to be a display of Röntgen (X-ray) experiments at a private citizen's house with both the Ministers of War and of Finances (Witte) in attendance. Mendeleev's presence was requested to add an air of scholarship to what was otherwise a sensational display of novelty.[16] After this, Mendeleev seemed to let the topic rest for several years.

He later began a research program at the Chief Bureau of Weights and Measures to investigate radioactivity, part of his efforts to meld practical standardization with original scientific research. In 1903–1904, he directed one of his assistants, M. V. Ivanov, to perform experiments on the "strength" of radioactivity using what was essentially a large capacitor. A radioactive sample was placed in between two plates, and then the voltage drop caused by the radioactive emissions was measured across a wide variety of temperature and pressure conditions. Mendeleev was already deeply aware of preexisting Western literature on radioactivity.[17] Organizing such work was difficult, as samples of radium were hard to find in Russia. Mendeleev ended up having to solicit a personal connection in Berlin, W. F. Giesel, for samples of radium from Braunschweig. Giesel responded promptly with a milligram of radium bromide (and apologies that he was unable to isolate pure radium). While Mendeleev sent the request for this radioactivity exchange under the letterhead of the Chief Bureau of Weights and Measures as an official letter, he kept

Giesel's response in his personal papers. This was not an idle mis-filing; it demonstrates the constant slippage between Mendeleev's standardization work and his chemical worldview.[18]

Mendeleev's most salient exposure to radioactivity, and the genesis of most of his hostile views of the phenomenon, was his visit to the Curies' laboratory in Paris in 1902. What he saw evoked similar worries as the Spiritualists had. He wrote in his Paris notebook: ". . . [M]ust one admit whether there is spirit in matter and forces? Radio-active substances, spiritualism? What are individualization, association, general or individual eternity???"[19] Mendeleev included an extended footnote on uranium's supposed radioactivity in the seventh edition of *Principles* (1903), arguing that one should not attribute radioactivity as a property of an element, but rather as a phenomenon that occurred *to* the element, like magnetism.[20] He wrote to a friend, A. Ia. Bogorodskii, author of the anti-radioactivity *Materials on the Electrochemistry of Inorganic Compounds in the Fiery-Liquid State* (1905), that he was sure that "[success] will come, if the current scientific superstition does not seduce you."[21] In accordance with his conservative orientation, Mendeleev preferred innovation when it was built on long-standing tradition, such as the periodic system. He remarked to a friend: "Tell me, please, are there a lot of radium salts in the whole earth? A couple of grams! And on such shaky foundations they want to destroy all our usual conceptions of the nature of substance!"[22] One of the conceptions that would be destabilized was the immutability of the elements; in particular, his conviction that elements could not "transmute" into each other—a modern alchemy. In 1906, the Parisian *Revue Scientifique* asked Mendeleev, along with other scholars, to write their opinions for a special issue on how radium could be "reconciled with the grand principles considered until now immutable, for example conservation of energy and the fundamental hypotheses placed at the base of the physical sciences (the hypothesis of the ether, the mechanism for the transmission of radiation, atomic theory)."[23]

Transmutation was no idle threat to Mendeleev's system. Consider, for example, the "Emmens incident," which Mendeleev saw as an instantiation of the potential for a "new alchemy." Stephen Emmens was an American who called himself a doctor (it has proven impossible to trace any doctoral degree in his name). In the 1890s he claimed to be able to transmute Mexican silver dollars into gold through a secret chemical and mechanical process that he tried to patent under his own syndicate. Emmens announced his results in a privately published investment prospectus. Throughout this document, he attempted to legitimize his new silver–gold allotrope by locating it as a new element "Argentaurum (Ar)" in the periodic system.[24] (The symbol "Ar" is used today for argon.) Although Emmens undertook a series of other scholarly efforts that received little support—like his disproof of Newton's theory of gravity—scientists like silver expert M. Carey Lea and even distinguished English chemist William Crookes expressed interest. Emmens eventually alienated Crookes by refusing to answer his correspondence in the open fashion expected in the scientific community, yet for years afterward he still received support from investors in what appears to have been an elaborate confidence scheme.[25]

This rather minor affair particularly irritated Mendeleev, as it encapsulated his fears for the integrity of expertise in the scientific world. Although Emmens's claim was a forgettable incident in the then scientific backwater of America, Mendeleev attacked it directly in his high-profile 1895 encyclopedia article on the periodic law. He continued in his 1898 article "Silver from Gold," one of the lengthiest contemporary critiques of Emmens, which began in almost Marxist fashion, reducing the affair to financial concerns:

> At the very same time that the capitalist and profiteering world—with feverish activity—is worried about the question of a gold currency, of mono- and bimetallism, when in America, as with us, as in all countries, many tomes are written and many unexpected—even political—combinations are accomplished concerning the dominant significance of gold and the fall of the value of silver . . . a great commotion has occurred from the fact that in the USA Doctor Stephen H. Emmens in the spring of 1897 announced, not in any scientific society, but under the title of a special syndicate "Argentaurum" (silver–gold), news that he had achieved the transformation of the silver of Mexican dollars into real gold. . . . [26]

Mendeleev inveighed against the publication of scientific findings, however dubious, in commercial venues, allowing individuals to hide behind corporate secrecy defenses to prevent replication of their findings. While Mendeleev claimed he would have preferred to let the claim pass in silence, people had been asking his opinion, and he thought it was beneath the dignity of the scientific community to respond to Emmens in a scientific journal, the very forum the latter had disdained. He continued: "In the republic of science all are 'Barons,' and the freedom for fantasy is not subjected to any limits besides the scope of journal articles and that there are subjects [such as a perpetual motion machine, he stated in a footnote] that are not allowed in the annals of science, and [therefore] there are writings which must hack their own special paths to publication."[27] Mendeleev accused Emmens of violating the code of science on four grounds: He had revitalized an old superstition (transmutation); he did so in secret; he advocated theories that were clearly in his commercial interest; and he had poor experimental support. For these reasons, Mendeleev explicitly compared him to the alchemists. In the end, as he commented in a French article in 1899, the major problem with Emmens was his poor philosophy of science: "Even if the validity of the American chemist's observations come to be demonstrated, and if *argentaurum,* recalling the epoch of the alchemists, does not exist only in the imagination and was not destined to serve purely material interests, even in these conditions, I say, it would be impossible to deduce any general law from this isolated fact."[28]

Mendeleev's anger about Emmens—a provincial charlatan—derived from deeper worries about the chemical community, and in particular widespread sloppy thinking about "mass." For Mendeleev, mass was not merely a secondary characteristic of an element's properties as, say, its crystalline structure; it constituted the very identity of an atom. It was how one knew an oxygen atom to be different from a cobalt atom. Mass—to Mendeleev—was the most fundamental discriminator. This view stands in sharp contrast to today's understanding of matter, in which each atom is composed of a definite number of protons, neutrons, and electrons, and any given proton in a cobalt atom is identical to any in an oxygen atom. In other words, Mendeleev firmly rejected any notion that atoms were *composite,* or complex constructions out of one or several types of "primary matter." His view was not rare for nineteenth-century chemists, many of whom were united in opposition to the hypothesis of British chemist William Prout (1785–1850).

"Prout's hypothesis" was actually two different hypotheses. In 1815–1816, Prout had proposed, first, that all atomic weights were integral numbers of hydrogen's and, second, that all elements were composed of some form of primary matter—often dubbed a "protyle"—and that this primary matter was hydrogen. So, for example, the fact that a volume of oxygen weighed sixteen times that of hydrogen, carbon twelve times, and sulfur thirty-two times was explained by an atom of sulfur really being just thirty-two individual hydrogen atoms glommed together, and so on.[29] When Jean-Servais Stas definitively measured several atomic weights in the early 1860s to be nonintegral multiples of hydrogen, including that of chlorine (35.5), chemists widely discarded Prout's first hypothesis. Many still saved the second by shrinking the postulated protyle, believing elements to be composed of yet smaller particles than hydrogen.[30]

Mendeleev consistently opposed Prout's hypothesis as antithetical to both laboratory evidence and the metaphysics of chemistry. In a theoretical chemistry lecture in 1864, Mendeleev entertained Prout's ideas for a bit before rejecting them with Stas's data, but his denunciations of Prout would only rise with time, and become more and more sophisticated.[31] In 1886, after his three predicted elements had been discovered and several chemists proposed Prout's hypothesis as an excellent metaphysical explanation for why the periodic law worked, Mendeleev was careful to state that "now, as upon establishment of the periodic law, [I] sooner tend to see in it the induction of the recognition of an independent autonomy (individuality, heterogeneity) of elements, under the sovereignty of a general law."[32] Mendeleev attributed the recurrent interest in Prout's hypothesis, despite contrary experimental evidence, to chemists' desire to make order of their field: to clarify the "murky" concept of mass with a "quantity of matter." However, just as one had to work with conservation of mass as an unprovable postulate, so it was safer not to postulate complexity of elements. In the seventh edition of his *Principles* (1903), he bluntly asserted that all speculations on primary matter "relate to the area of fantasy and not science, and I don't recommend to persons beginning to study chemistry (for whom this book is written) to fall into this area."[33]

The rise of domestic advocates of Prout was only one reason why Mendeleev became yet more impassioned about immutability in the late 1890s.[34] Comparing "chemical individuals"—modern atoms—with ancient Greek (Democritean) atoms, Mendeleev noted that:

> [J]ust most recently, especially in connection with radioactive substances, some have begun to recognize the splitting of chemical atoms into smaller "electrons," and this logically wouldn't be possible if "atoms" were recognized as mechanically indivisible. The chemical worldview can be expressed in an exemplary fashion, comparing the atoms of the chemists with heavenly bodies: stars, the sun, planets, satellites, comets, etc. Just as from these separate entities (individuals) systems are formed like the solar system or binary star systems, or certain nebulae, etc., so one can conceive the formation of entire molecules from atoms, and of bodies and substances from molecules.[35]

Atoms were no more reducible to one primary matter than Jupiter and Venus were made of a certain number of "moon units." Mendeleev's worries about Prout's hypothesis became more acute with the discovery of the electron in 1897. J. J. Thomson, discoverer of a constant charge-to-mass ratio in the emanations from cathode-ray tubes (interpreted as a charge-carrying particle), considered himself a chemist, and chemists began to appropriate the electron even while physicists were still skeptical. A few even proclaimed it a new element, including Ramsay, the discoverer of argon.[36] Mendeleev was not among the electron's enthusiasts; he considered it most likely an epiphenomenon of atomic interactions. He came from a chemical tradition starting with Newton, and moving through John Dalton, Jean-Baptiste Dumas, and Charles Gerhardt, all of whom emphasized integral treatments of molecules and were hostile to the electrochemical explanations of the rival tradition that also began with Newton but emphasized the role of electricity and bonding, as continued by Humphry Davy, Jacob Berzelius, and Michael Faraday. His opposition to this supposed *element* of chemical charge is thus easy to understand.[37] This new claim may not have been as egregious as that of Emmens, but he considered it erroneous all the same.

Mendeleev could not let such transgressions against his fundamental conception of matter and, even more importantly, his periodic law pass unanswered. Interpreting the situation in fin-de-siècle physical sciences as chemistry under attack by superstition and sloppy reasoning, and exasperated by people letting their irrational preferences dissuade them from proper scientific method, Mendeleev undertook a chemical interpretation of the ether that would harness the inert gases to stave off the twin dangers of radioactivity and Prout.

PONDERING THE IMPONDERABLE: THE CHEMICAL ETHER

In the seventh edition of his *Principles of Chemistry* (1903), Mendeleev reflected on the set of difficulties confronting contemporary physics and chemistry:

> The root of the inadequacies of contemporary atomism, in my opinion, should be sought in the lack of clarity of the understanding of the "ether," which fills both interplanetary and interatomic space. . . . Contemporary natural science strives, but still does not know how, to come to terms with a material, but not ponderable, chemically active . . . ether with the necessary clarity. This is one of the tasks the 19th century dedicates to science.[38]

This was not an idiosyncratic belief. Physical scientists at the turn of the century still considered the ether an absolute necessity for the explanation of not just light undulations, but also Newton's gravity. Since Mendeleev's first attempts to isolate the ether in the 1870s with his gas experiments, theories of the ether had become ever more central to the physical sciences, and mathematical modeling of them proceeded apace. As a result, a panoply of models continued to proliferate, even after Albert Einstein's 1905 Special Theory of Relativity supposedly banished the ether as "superfluous." The ether's chemistry was usually overlooked, however, primarily because the function of the ether was typically understood to be *physical* and not *chemical*, meaning that the theories concentrated on the mechanics of the ether and paid little or no attention to its composition.[39] Mendeleev, who proposed his model of the chemical ether in 1903, was one of the very few exceptions. Scientifically, the work reflected his almost complete obliviousness to the extensive mathematical and physical requirements of the ether developed in the West, especially in England. But mathematics was not important for Mendeleev here; he was not after equations and structure, he was after substance.

Mendeleev's ether project has been interpreted—by both his contemporaries and historians—as an attempt to stop the encroachment of modern physics on chemistry.[40] This is a misreading. First, such interpretations are based on a view of chemistry and physics in which the dividing line between the two sciences is static or at least unproblematic. Mendeleev was highly troubled by the exact placement of such a line, or even whether such a line existed, and he did his best throughout his career not to place himself too firmly on either side.[41] Second, Mendeleev's first instinct was *always* to try "physics" first, as evidenced by his experimental ether investigations of the 1870s. Only when that failed, and Mendeleev had recovered from the disappointment, did he move to reasoning from chemical periodicity. Third, in his ether theory, he admitted that the periodic law would only produce partial results, and thus one had to invoke kinetic theory—the cornerstone of contemporary *physics*—to complete his predictions. In truth, the central issue at play in Mendeleev's ether speculations was not physics versus chemistry, but rather experimentation versus theory. It was about the stability of knowledge in the physical sciences.

Mendeleev had long considered the ether to be an essential component of the physico-chemical universe, but the ether began to take on new functions as he transitioned from teaching to his later bureaucratic career. In his last year of lectures (1889–1890), he stressed the *unity* the ether could bring to the physical sciences: "The newly discovered connection [between electricity, magnetism, and optics by Maxwell] shows only that unity of the forces of nature . . . will be intermediate in the ether and that all material, weighted, chemical relations will be forgotten."[42] In his earlier Gerhardtian ether theories, Mendeleev had divided the world into two types of atoms, corporeal and ethereal. The former composed all brute matter, and the latter, through its interactions with corporeal atoms, generated the phenomena of heat and light. Mendeleev still retained elements of this view, as late as 1895, in his account of incandescence. Taking a specific problem, such as why zinc oxide glowed blue-green when heated to 500–600° while most other bodies glowed cherry-red, he suggested: "The reason for such a particularity is somewhat understood, because illumination depends, in essence, on the vibration of ethereal atoms, and they vibrate only under the influence of the motion proper to corporeal atoms and molecules, the nature of which is different in different bodies, and the type of motion generated by their incandescence can thus not be identical."[43]

In 1903, by contrast, Mendeleev saw the ether as of a piece with the rest of the elements. While on a medical rest at Aix-les-Bains in 1905, Mendeleev had noticed ripples in the tank of water and used this phenomenon to explain how the undulatory theory of light could be reconciled with molecular phenomena only through "the necessary materiality and expansiveness of the luminiferous or world ether."[44] The ether, "originally proposed exclusively to explain optical phenomena," could be expanded to include other forces, such as gravity.[45] Mendeleev sought a unification in the ether that was distinct from disintegration and reckless homogenization. The ether was not meant to be a substrate that *composed* everything, à la Prout; it was rather a *medium* to reconcile the interactions of nature.

There were very few *chemical* ether theories to serve as models for Mendeleev's attempt. An important forerunner, however, was Charles Schinz's *Attempt at a New Chemical Theory* (1841), which Mendeleev owned and read.[46] In his personal library, this was the only text classed under his "Ether" heading to predate the 1880s, and was probably purchased while he was at Heidelberg in the late 1850s. Schinz's theory bears striking intellectual similarities with Mendeleev's. Schinz claimed at the beginning of his slender volume to provide a coherent means of unifying heat, light, electricity, and magnetism, through a chemical element called "ether." This element, similar to Gerhardt's ethereal atoms, served as a substrate for these forces and possessed the properties of matter: extension, impenetrability, divisibility, porosity, and weight. This last property posed some complications, as "the finest of our balances" could not yet weigh this substance, but Schinz insisted that the ether did have weight, but "to so weak a degree that we cannot detect it with our instruments."[47] Mendeleev would later use an almost identical argument about the "weightlessness" of his ether. A passage Mendeleev marked in his copy likewise hints at his own later solution to radioactivity:

> It is possible that there exist some combinations where the quantity of the ether contains an equal amount to the sum of quantities which are contained in primitive bodies; but these would be only exceptions, and we are authorized by experience to establish that there are combinations [which are] part ether, and it is this part which produces the phenomena of heat and light, according to the intensity with which the combination is made.

However, "the ether does not play, in the end, any other role than all other elementary bodies."[48] Schinz stands in marked contrast to James C. Maxwell, William Thomson, and other ether theorists and cosmologists. Mendeleev owned many of their works (often in Russian translation), but his personal copies were barely marked, and never in the mathematical sections, which suggests that he did not read them. His attention was directed at the qualitative role the ether could play, and not the mathematics of the mechanics (with which he had always been uncomfortable).[49]

Mendeleev had the requisite tools, then, to draft his ether project for a long time: Schinz's and Gerhardt's ether theories, as well as his own past research in gas–ether connections. What inspired Mendeleev's ether pamphlet at the turn of the twentieth century? The answer lies partly in his perception that chemistry was threatened, but the sudden desire for publicity almost certainly stemmed from something much more personal. On 19 December 1898 Mendeleev's eldest son from his first marriage, Vladimir, died of illness while in Navy service. Mendeleev spent much time disconsolate in his quarters at the Chief Bureau of Weights and Measures. One letter of 8 March 1899, though, "especially revived me," as he wrote in its margins. This letter was from L. N. Shishkov, a chemist at the Artillery Academy and also a noted expert on explosives, who offered condolences on his son's death. Then Shishkov moved to issues of molecular motion:

> It has obviously become time to study the laws of this motion, for which the separation of sugar into alcohol and CO_2 comprises an evident illustration. In all probability, the universal motion of stars, just as molecular internal [motion] have one and the same sense and cause, and precisely by means of this motion yields from heterogeneity a certain relative homogeneity, the degree of which is the measure of energy present in matter, which reveals itself in electrical and like phenomena.[50]

Shishkov handed Mendeleev a means not just of unifying the physical sciences under the ether, but also of using particulate motion to connect to his earlier gas research. Mendeleev's work on his ether concept began directly after the Shishkov letter.

Thus primed, in 1901 Mendeleev was approached by the editors of a new journal, the *Herald and Library of Self-Education,* to write an article on the state of contemporary science for the first issue. The journal's publishers were Brockhaus and Efron, who were simultaneously publishing the monumental encyclopedia for which Mendeleev edited the articles on technology and industry. This new magazine was the perfect venue to work out his ether views, since "the subject touched on many areas of the natural sciences, and seemed to me amenable for popularization."[51] The editors apparently agreed. Mendeleev saw these articles as a scientific complement to his economic summa, *Cherished Thoughts.* He had originally intended to write an appendix to that text that would express his philosophical worldview, but he did not want to rush into print without contemplation during such revolutionary times. He could not make his thoughts cohere to his satisfaction with so many distractions, and so he opted not to publish half-digested principles that would express his ideals only incompletely.[52] Mendeleev disengaged his philosophical concerns about the natural world from politics and economics. Instead he would compose a unified vision of nature connected by the chemical ether.

The piece fared surprisingly well, published in four installments in the *Herald and Library* and then as an independent pamphlet. Mendeleev distributed complimentary copies widely to various scientists, and it was also repeatedly translated.[53] He preferred the German translation over the English, which lacked his precious philosophical opening, and he fretted that "such an omission removes from the entire article the real significance which I wanted to give it in trying to introduce the ether into the system of elements."[54] Mendeleev was especially amused by the work's translation into Esperanto for the journal *Internacia Scienca Revuo,* which his correspondent claimed would give Mendeleev a wider audience: "[W]ith the help of Esperanto, all will be clear to everyone as a clear day."[55]

The pamphlet, much like Esperanto, was intended to unify different communities. The essence of the ether project was to locate the ether in the periodic system of elements and then use interpolation techniques to predict its necessary properties—just like the prediction of the three eka-elements in 1871.[56] He began his quest for the ether with the issue of weight. Typical descriptions of the ether in Mendeleev's time described it as "imponderable," as having no weight. For Mendeleev the idea of a substance without weight was ridiculous. The only way we could know matter was through its set of measurable properties. So if the ether were to exist for Mendeleev, not only must it have some mass (for it must be made *of* something), but it had to be a definite quantity (assuming it was a pure simple substance) and could thus be located in the periodic system. The reason the ether *seemed* to have no weight was that all substances were permeated by this ether. Just as one could not weigh air before the advent of air pumps, he argued, the ether's weight could not be determined without some kind of fictive ether pump.[57]

Mendeleev explicitly labeled his philosophical position here "realism," by which he did not mean simply the belief in an external reality. He began with a discussion of his meaning:

> Like a fish frozen in the ice of ages ago there has beat in the thought of wise men their striving for unity in everything, i.e., searching for a "beginning of all beginnings," but only to the point necessary in order to recognize the indivisible, yet not fusible trinity of the eternal and original: substance (matter), force (energy), and spirit, although to distinguish them to the end is impossible without blatant mysticism.[58]

Mendeleev's proposal for unification through the ether, then, was not so much a reduction to a unity as a general organization under three unifying principles. This philosophical introduction was important for two reasons: First, Mendeleev repeated these phrases constantly in his later writings; and, second, they exemplified his strategy of unification through heterogeneity. To make explicit the implicit (and oddly Catholic) religious metaphor, recognition of the trinity would redeem the physical sciences from the corruption of Prout and radioactivity.

In his updated ether theory published in the *Herald and Library*, Mendeleev conceived of the ether as a gas—specifically, a noble gas. This confounded his earlier understanding in *On the Elasticity of Gases* (1875) of the ether as a combination of rarefied gases. In the 1903 pamphlet, he explained the change, reflecting on his gas work of the 1870s: "I was silent because I was not satisfied with what appeared at first glance. Now my answer is different, but it still doesn't completely satisfy me. And I would still remain guardedly silent, but I no longer have years before me for thinking and experimental trials, and thus decided to set forth the subject in its immature form, presuming that to keep silent is also not right."[59] Such a rarefied-gas view was no longer possible because it denied the fundamental need for *homogeneity* in the ether.[60] The position here is subtle but not unstable: The ontology of the world was fundamentally heterogeneous, whether into the broad categories of matter, force, and spirit, or within matter into the nontransmutable elements; yet properties ascribed to a particular body, like the ether, had to belong to a single, homogeneous body. Heterogeneity, after all, is merely a collection of individualized homogeneities.

The core of this new principle of organization was the group of inert gases, elevating what was once the albatross of chemical inactivity to a virtue. Ether was the lightest element (lighter than hydrogen), and at the top of the 0-group (above another postulated element, coronium).[61] (See Figure 8.1.) Note that the 0-group of inert gases is not on the right but on the left, the standard placement before today's electronic interpretation of the periodic law. This position left two blank spaces above helium, and led Mendeleev to some of its properties:

> Thus *the world ether can be conceived*, like helium and argon, *as incapable of chemical combination.* . . . When we recognize the ether as a gas this means, above all, that we strive to relate its concept with the ordinary, real concept of the states of matter: gas, liquid, and solid. There is no need here to admit, as Crookes does, a peculiar fourth state of matter, removed from the real understanding of the nature of things. The mysterious, almost spiritual support is removed from the ether with this provision. . . . If ether is a gas, this means that it is ponderable, it has its own weight. We must ascribe to this if we are not to discard on its behalf the entire conception of the natural sciences which takes its origin from Galileo, Newton, and Lavoisier. But if ether has such a highly developed power of penetration that it goes through all envelopes, then it is impossible to think about experimentally finding its mass in a given quantity of other bodies, or the weight of its specific volume under given conditions, and thus one should speak not of the imponderable ether, but of the impossibility of weighing it.[62]

Series	Zero Group	Group I	Group II	Group III	Group IV	Group V	Group VI	Group VII	Group VIII
0	x								
1	y	Hydrogen H=1·008							
2	Helium He=4·0	Lithium Li=7·03	Beryllium Be=9·1	Boron B=11·0	Carbon C=12·0	Nitrogen N=14·04	Oxygen O=16·00	Fluorine F=19·0	
3	Neon Ne=19·9	Sodium Na=23·05	Magnesium Mg=24·1	Aluminium Al=27·0	Silicon Si=28·4	Phosphorus P=31·0	Sulphur S=32·06	Chlorine Cl=35·45	
4	Argon Ar=38	Potassium K=39·1	Calcium Ca=40·1	Scandium Sc=44·1	Titanium Ti=48·1	Vanadium V=51·4	Chromium Cr=52·1	Manganese Ma=55·0	Iron Fe=55·9 Cobalt Co=59 Nickel Ni=59 (Cu)
5		Copper Cu=63·6	Zinc Zn=65·4	Gallium Ga=70·0	Germanium Ge=72·3	Arsenic As=75·0	Selenium Se=79	Bromine Br=79·95	
6	Krypton Kr=81·8	Rubidium Rb=85·4	Strontium Sr=87·6	Yttrium Y=89·0	Zirconium Zr=90·6	Niobium Nb=94·0	Molybdenum Mo=96·0	—	Ruthenium Ru=101·7 Rhodium Rh=103·0 Palladium Pd=106·5 (Ag)
7		Silver Ag=107·9	Cadmium Cd=112·4	Indium In=114·0	Tin Sn=119·0	Antimony Sb=120·0	Tellurium Te=127	Iodine I=127	
8	Xenon Xe=128	Caesium Cs=132·9	Barium Ba=137·4	Lanthanum La=139	Cerium Ce=140	—	—	—	— — — (—)
9		—	—	—	—	—	—	—	
10	—	—	—	Ytterbium Yb=173		Tantalum Ta=183	Tungsten W=184	—	Osmium Os=191 Iridium Ir=193 Platinum Pt=194·9 (Au)
11		Gold Au=197·2	Mercury Hg=200·0	Thallium Tl=204·1	Lead Pb=206·9	Bismuth Bi=208	—	—	
12	—		Radium Rd=224	—	Thorium Th=232	—	Uranium U=239	—	

FIGURE 8.1   Mendeleev's periodic system with the chemical ether. The ether is the box at the upper left labeled *x*, and the element below it, *y*, is coronium. Source: Mendeléef [Mendeleev], *An Attempt towards a Chemical Conception of the Ether*, 26.

While the ether could not be weighed, its weight could be determined—just not experimentally. The properties of the ether had to be deduced through the periodic law, where the ether was construed as a noble gas. The periodic law gave only an upper cap for what the element x, in row 0 and group 0 of the periodic system, should weigh ($x = 0.17$, with H = 1). To find a more exact prediction, one invoked physics, specifically the kinetic theory of gases, computing what the average weight must be for the gas to escape planetary atmospheres. Upon a simple calculation using Newton's law of gravitation, Mendeleev argued that x had to be less than 0.038 to escape earth's atmosphere, and 0.000013 to escape the sun's. He then scaled up to a larger star, γ-Virginis, which had 32.7 times the sun's mass. His final result was $0.00000096 > x > 0.000000000053$. Interestingly, even though mass canceled out of all the escape velocity equations, he did not cancel the term in order to make the calculation more "visualizable." This stemmed from Mendeleev's fixation on Newton's concept of mass as the centerpiece of the physical sciences. He finally calculated that the ether must weigh nearly one-millionth of an atom of hydrogen, and must move at about 2,250 kilometers per second. This

ether penetrated everything and produced observable effects when it interacted slightly with elements.[63]

These were the theories of the new Newton: Mendeleev assimilated this
project for a chemical ether seamlessly with his new self-presentation. His bold
formulations of the periodic law were on a par with Newton's laws of gravity,
and now he sought to dedicate the latest fruit of that research to the Master
himself. In the ether pamphlet, he added as a brief footnote: "I would like preliminarily to call it 'newtonium'—in honor of the immortal Newton." In an
early draft, scrawled illegibly on both sides of a flimsy scrap of paper, he emphasized this Newtonian aspect even more, concluding: "[The ether is] the
lightest elementary gas which penetrates everything (row 0, group 0), which I
would like preliminarily to call newtonium, since the thoughts of Newton
penetrate all parts of mechanics, physics, and chemistry."[64]

Mendeleev considered his central contribution not the prediction of the
weight of element x, but rather the ether's ascription to the family of inert
gases. The ether as noble gas had two central properties: First, it was "the lightest—in this sense the limiting—gas, which has a great degree of penetrating
power," taking up the mantle of the most "typical" element from hydrogen;
and, second, it could dissolve in other substances without combining with
them, just as argon could dissolve in air or water.[65] This property enabled
Mendeleev to save his system:

> And secondly, recently people have begun to speak often and a great deal
> about the smashing of atoms into tinier electrons, and it seems to me that
> these should not be considered so much metaphysical as metachemical repre-
> sentations, which stem from the absence of any kind of definite notions re-
> lated to the chemism of the ether, and I wanted in the place of these kinds of
> confused ideas to set up a more real representation of the chemical nature of
> the ether, thus, at least until something shows either the transmutation of an
> ordinary substance into the ether and back, or the transmutation of one ele-
> ment into another, any representation of the breaking of atoms should be
> considered, in my opinion, contrary to contemporary scientific discipline,
> and those phenomena in which one recognizes the breaking up of atoms
> could be understood as the emission of atoms of the ether, which penetrates
> everywhere and is recognized by everyone.[66]

Mendeleev noted that the chiefly radioactive elements (uranium, thorium,
radium, etc.) were the heaviest ones, and thus they must attract a large proportion of lighter matter, just as the sun attracted planets and cosmic dust. Naturally, uranium would be surrounded by a great cloud of attracted ether that
dissolved and intercalated with the uranium mass itself. At some critical point,
too much ether penetrated the uranium and certain chemical processes, of

whose exact nature we were ignorant, caused quantities of ether to be ejected from the sample. Radioactive energy was just the reaction energy produced by the minute and highly diffusive ether. Ether atoms, and not a "decayed" part of the primary atom, were ejected. There was no transmutation, no primary matter from which all elements were constructed, and the periodic law was preserved in all its integrity.[67]

The ether also maintained Mendeleev's intricate worldview by disarming the other threat: Prout. In a draft of the ether pamphlet Mendeleev wrote: "Considering possible the existence of even one of the 'pre-hydrogen' elements. . . predicted by the periodic law, I think that the confirmation of this would serve greatly for a new unification and strengthening of such fundamental real knowledge as mechanics, physics, and chemistry—instead of the doctrine of 'primary matter.'"[68] In the final analysis, for Mendeleev all of the various threats to his vision of the unity of the physical sciences (except for noble gases, which had been domesticated and appropriated to that worldview) stemmed from Prout. Clearly, he argued, there was little to no evidence for Prout's view that matter was composed out of bits of homogeneous primary matter; what made so many chemists adhere to this view was the *unity* it promised. The ether, on the other hand, not only dealt with radioactivity, but satisfied this yearning for unification without caving to what Mendeleev considered "metaphysics":

. . . [W]ithout the development of individuality it is absolutely impossible to admit any kind of generality (*obshchnosti*). In a word, I see no kind of goal in the prosecution of the thought of the unity of substance, and I see a clear goal both in the necessity of admitting the world ether's unity, and in the realization of the conception of it as the last residue of that process, by which all other atoms of the elements were formed, and from them all substance. For me this type of unity speaks much more to real thinking than the concept of the formation of elements from a single (*edinoi*) primary matter.[69]

Mendeleev was not opposed to adding elements to the ontology of the natural world; he objected to the protyle as a *reduction* of matter that would suggest transmutation. (Mendeleev was even willing to admit the possible existence of elements with atomic weights between hydrogen and helium, such as a very light halogen.[70])

The reactions to Mendeleev's pamphlet at home and abroad were, in general, rather positive, as the overwhelming translation efforts would indicate. Among the Russian audience, Mendeleev's status seemed to have reinvigorated belief in the reality of the ether, as noted by student Andrei Litkin: "Some say that if [William] Thomson calculated its density and you have begun to study its chemical properties, then obviously the existence of the ether has already been entirely proven, since neither you nor Thomson would begin to study the physical and chemical properties of problematic bodies."[71] Foreign reviewers were equally enthusiastic about the potential of Mendeleev's theory as a unifying hypothesis, although they were somewhat more skeptical about how much Mendeleev ignored physical theory. While one reviewer of the English translation felt that "[a]ll chemists and physicists will find this pamphlet interesting and suggestive," the reviewer from *Nature* pointed out that Maxwell had proven that if a particulate ether existed, then all energy in the universe would already have been transferred to it, and Mendeleev had said nothing to counter this point.[72]

An American review of the English translation of the seventh edition of Mendeleev's *Principles of Chemistry* (1905) considered the ether prediction as "[p]erhaps the most remarkable thing in the book," but cautioned that "[i]nasmuch as this conception is largely the result of extrapolation over a long range, the conclusions are correspondingly hazardous."[73] The French review—from a prominent journal on radioactivity studies—was supportive of its spirit of unification. Likewise, Mendeleev was informed by a friend at the German standards bureau in Charlottenburg that the German translation in the journal *Prometheus* was making quite a splash.[74] The enthusiasm was short-lived, and soon his theory faded from the international scene. Mendeleev bemoaned to Clemens Winkler, the discoverer of germanium, who was likewise skeptical of the dissociation of radium into daughter elements, that his cataract surgery prevented him from attending the World's Fair in St. Louis, "although precisely there I intended to put forth my opinion about the semi-spiritualist state, into which they [radioactivity researchers] are now trying to enmesh our science. It behooves us to stop it while we can still act."[75]

Mendeleev's attempt to preserve his chemical worldview soon vanished entirely amid the rising nuclear model of the atom and the general acceptance of the electron, which was essentially complete by Mendeleev's death in 1907, although he never admitted it. Mendeleev increasingly separated himself from chemistry at large, seeking to bring the fold back to him with pronouncements of the reality of his chemical ether. In an interview with a Petersburg paper in January 1904, Mendeleev stated that his current scientific projects "are directed exclusively towards the confirmation of the theory, or rather, attempt, of a chemical understanding of the world ether which I established last year."[76] His assistant M. Ivanov attempted, under Mendeleev's instructions, to perform observations of the sun's corona to evaluate the density of coronium (and, by extension, of the ether), but these efforts were soon discontinued.[77]

Mendeleev's attempt to let theory guide experimental investigation into the core concepts of matter and energy, however, illustrates the transition in his views on the power of theory that is intimately tied to his reinterpretation both of the periodic system as a law of nature, and of himself as a disciple of Newton. Most of all, though, the ether pamphlet, along with Mendeleev's worries about the fate of chemistry, reflected underlying uncertainties in his philosophical system, a tension between system and inspiration that bedeviled his thought after the 1880s. His attempt to reconcile them philosophically was equally fraught.

TRIPARTITE METAPHYSICS: MENDELEEV IN THE ABSTRACT

Mendeleev had always held that science could not progress without a deliberate choice of fundamental assumptions to structure our knowledge. For example, atomism could be accepted or rejected, but at some point we had to accept fundamentals about mass and measurement if science were to exist. In a rare formal statement of his view of the scientific method, Mendeleev insisted on the need for scientists to make conscious choices: "It is manifest that it is only possible to carry out these investigations when we have taken as a basis something which is incontestable and self-evident to our understanding; such, for instance, as number, time, space, matter, form, motion, or mass."[78] A philosophical project thus had to coexist with any attempt to synthesize results in the sciences, or even to find results that were amenable to synthesis.

This was nowhere more evident than in the ether project, which was explicitly couched as an exercise in "realism." For Mendeleev, realism formed the bedrock not just of the periodic law, but of all his mature thought. Mendeleev had a metaphysics, and it was a philosophy of his own making. As he put it in a footnote to the chemical ether pamphlet: "It is now better to compose new rubbish than to repeat the old, which leads to instability both in thinking and in social relations."[79] He began his doctrine with an attack on dichotomies. In the first paragraph of *Cherished Thoughts,* he insisted that the binary division between materialism and idealism was too simplistic. He rejected both of these as extremes caused by devotion to classical philosophy; he opted instead for "realism" as a middle pass: "True idealism and true materialism are the products of antiquity, realism is a new affair compared with the length of historical epochs." This "realism" was ideally suited to Russians, who were geographically between Europe and Asia, and were thus "a real people, a people with real concepts."[80] Mendeleev had stated a similar thought as early as 1881 in a letter to his second wife. He refused to let himself become trapped by the standard dichotomy between Slavophiles and Westernizers:

. . . I will begin to write those results that I consider true and good, those foundations of a party to which I would join, with which I would act. Let us call it *popular* (*narodnaia*). Its foundations are clear to me even now. It isn't Slavophilism, it isn't Westernism, it isn't prostration to the people, it isn't the exaltation of rights which do not call forth obligations. And it will be that which will not be achieved: a union of education (*obrazovannosti*) with popular foundations, with its immediate needs and not by fashionable ideas, but by the simple feeling for popular truth, with labor, with freedom of thought, with freedom of industry, with science in life, [which is] necessary to achieve not some kind of utopia, but a possible, attainable, peaceful [world] that can develop healthily.[81]

Mendeleev's philosophy of realism was bound up with his view of the proper form for the Russian state, and thus also the necessary form of the Russian economy. This was literally a "worldview," the implications of which he drew out in an unpublished manuscript of that name.

Mendeleev had originally written his "Worldview" piece as a conclusion for *Cherished Thoughts,* to offer the metaphysics that undergirded his economics and politics, to demonstrate the relation between "matter, force, and spirit; instinct, reason, and will; freedom, labor, and duty." He omitted it from the final publication because he felt he lacked the artistry to convey what he meant beyond the level of caricature—and so he resigned it to the proverbial sock drawer.[82] It never saw publication in his lifetime. The manuscript, however, clearly demonstrates Mendeleev's philosophical commitments. He argued that the problems of contemporary philosophy stemmed from a tendency—derived from the speculative philosophers of antiquity—to search for *one* unifying principle, a "beginning of all beginnings." Instead, Mendeleev argued, there were three basic components of nature: matter (substance), force (energy), and spirit (soul). Everything was composed of all three in some measure, and one could not reduce any one category to another. For example, an oxygen atom was composed of matter (measured by its atomic weight), energy (valency, its potential of chemical combination), and "spirit" (what made it an essential oxygen atom, which could not transmute to any other atom). Likewise, Russia was a nation with its quantity of matter (people), energy (economy), and national spirit, and thus was not reducible to Germany, England, or any other country. "Spirit" reflects what we today would call "essentialism"; it manifests that which is irreducibly proper to the object in question. This notion of spirit/soul is something like an Aristotelian final cause—the teleological purpose of a subject's existence—but Mendeleev would never have identified it with a classical concept. The fact that this is clearly metaphysics removes Mendeleev from the companionship of positivists. Instead, "Worldview" was a text about humility, about confessing the limits of our knowledge, rather than attempting the hubris of the Greeks.[83]

"Worldview," composed in September 1905, represented a substantial development of views that had been brewing for some time before the Imperial Turn. In his entire voluminous corpus of writings, Mendeleev only wrote one article, entitled "The Unit," under a pseudonym (D. Popov, his future wife's maiden name). Mendeleev claimed that he "avoided signing my name in this instance only because in those times it was considered inappropriate for a professor-naturalist to meddle in questions of a more or less philosophical-social character, and even more from the purely popular side."[84] Considering this was written directly after Mendeleev's multiple meddlings in the philosophical-social question of Spiritualism, this seems implausible. Rather, since it appeared in the journal *Svet* (meaning *Light* or *World*), edited by his Spiritualist counterpart, N. P. Vagner, Mendeleev was most likely trying to keep a low profile. The choice of pseudonym was probably designed to impress the young woman he was courting.[85] Although he wrote this piece in 1877, he reprinted it in the footnotes of *Cherished Thoughts,* which indicates that he still held to these views, and he explicitly endorsed them in private notes in 1899.[86]

The piece was officially a response to an article in *Svet* by V. Alenitsin, which argued that there was no such thing as a "zero" in nature, that the concept was an artificial introduction by philosophers.[87] Mendeleev argued that the "unit" was likewise artificial, as were attempts to reduce everything to composites of a single unit. Thinking about the world through elementary individuals was destructive: "Individualism, or all the essence of our education, is the ripe and even rotting fruit of the concept of a unity which exists independently in nature. . . . Your *individual* is only zoological, animal, and all that is *human* . . . is from others, with others." In other words, the collective gives meaning to the individual, just as the state gives meaning to the entrepreneur by allowing him to understand where his true interests lay. Alenitsin had attacked the wrong concept: "The idea of zero, in my opinion, was harmless, but the concept of unity is the beginning of much that is bad." Like all other words and ideas, the "unit" was just a convention: "The unity of measures, weight, time, all attractive forces—in a word, all unities used in science—are known to be conventional. They do not exist, we think them up ourselves, that is, they are fictitious." Only when the fact that a unit was conventional, that not even our most cherished concepts of unity, be they scientific, political, or religious, were realized to be conventions would we understand the world.[88]

This discussion of metaphysics and unities raises the question of Mendeleev's views of religion. It is striking how rarely he mentioned this topic in his vast oeuvre. In marked contrast to contemporary Victorian naturalists, such as both the proponents and opponents of Spiritualism, Mendeleev seemed to have very few theological commitments. This was not for lack of exposure. His upbringing was actually heavily religious, and his mother—by far the dominating force in his youth—was exceptionally devout. One of his sisters even joined a fanatical religious sect for a time.[89] Despite, or perhaps because of, this background, Mendeleev withheld comment on religious affairs for most of his life, reserving his few words for anti-clerical witticisms. (This is the same man, recall, who bribed a priest to allow for his second marriage to A. I. Popova.) Mendeleev's son Ivan later vehemently denied claims that his father was devoutly Orthodox: "I have also heard the view of my father's 'church religiosity'—and I must reject this categorically. From his earliest years Father practically split from the church—and if he tolerated certain simple everyday rites, then only as an innocent national tradition, similar to Easter cakes, which he didn't consider worth fighting against."[90] It would be tempting to explain this apathy as the fault of the Orthodox Church as an appendage of the state disconnected from social action and without an ability to instill passion. This, however, would be contrary to the active and independent life of the Orthodox Church in nineteenth-century Russia, where it was far from a simple "handmaiden to the state," and instead formed a powerful source of both criticism and, occasionally, support.[91] Mendeleev did not see the Church as corrupt, but simply as incapable of meeting his demands.

Mendeleev's opposition to traditional Orthodoxy was not due to either atheism or a scientific materialism. Rather, he held to a form of romanticized deism. On 19 March 1884, he wrote an exhortation to his children from his first marriage (who lived with their mother after the explosive divorce) about religious belief: "The first and most important thing in life is work for others, but you have to set it up so that you can live for yourself too. It is necessary to live in order to fill the task of nature, the task of God. And its highest point is the society of people. One by himself is nothing. You must remember this."[92] Instead of relying on any established religion (other than following Orthodox customs for the sake of conformity), Mendeleev hinted to his niece of a "new religion" consistent with his metaphysical realism:

> Christ taught about the *internal world of a person,* Socrates about the relation of a person *to the state,* and New Religion will teach *about the relation of a person to society.*
>
> The chief principle of New Religion is the following: man *alone* is zero, a molecule, a cell. One man is part of an organism, and the organism is society, and thus a person should consciously live for society, and this is the chief rational connection of a separate person with society. Of course, each person works for himself, as a cell can live for itself and not consciously for the person, but at the same time serve the person and be nothing without him. Only a person at a low state of development lives for himself alone, like a microcosm, like an individual.[93]

It is difficult to generalize these few statements into a coherent religious position, but this New Religion was in important senses a spiritual analog to his
evolutionary reformism. Much as economies evolved from the tribal to the artisanal to the industrial, religions evolved from man's relation with God toward man's relation with society, providing a theological justification for the
Imperial Turn. At the same time, Mendeleev focused on Jesus and Socrates,
prophets who embodied their doctrines in their individuality, much as he
viewed himself as a Newton for a modernizing Russia. Like the Spiritualism he
opposed, then, New Religion was an attempt to harmonize theology with the
contemporary world. Mendeleev's new religion, much like his economic policies, would allow people to function as a complicated and interconnected industrial organism. In the end, he saw philosophy and religion as the only ways
to stabilize his threatened universe. They were the ether of the social world.

THINGS FALL APART: THE REVOLUTION OF 1905

In 1904 an article in *New Times,* commenting on some of Russia's leading intellectuals, ventured that "Mendeleev is not a wise man and still less a prophet,
although he did manage to predict the eternal facts of nature, the discovery of
new bodies."[94] Mendeleev's reputation as a leading Russian thinker hinged on
his ability to marshal his periodic system to make predictions, and predictions
into a glimmering of prophecy. The chemical ether was his last such effort,
born not of a desire to establish a new science of chemistry but in order to defend that science from threatening newcomers. Mendeleev did not evolve from
"progressive" to "reactionary"; he built up an ideal and then defended it as the
best of all possible systems. Circumstances made his views seem old-fashioned;
the views themselves had not changed. Unfortunately, the vast majority of his
peers in chemistry did not hold to the metaphysical worldview that Mendeleev
wanted to protect, did not see what was so awful about granting the disinte-
grations of radioactivity status as natural phenomena. Mendeleev's predictive
powers fell on deaf ears; he was a chemical prophet no more.

His abilities to forecast fared no better for the Imperial system he had
helped construct in St. Petersburg. Mendeleev's program for political and social reform could adapt willingly to changing circumstances—at least in theory. He often tailored his vision of the proper path for Russian modernization
(cultural, political, economic) in response to personal and external pressures.
The most salient of these shifts came in 1880–1881, upon the final collapse of
the Great Reforms with the assassination of Alexander II, the accession of
Alexander III, and the end of the dream of a decentralized model of scientific
societies upon Mendeleev's rejection by the Academy of Sciences. But his evo-

lutionary economics had weathered the transition from Alexander III to
Nicholas II in 1894 rather well, and Mendeleev's advice was still in demand in
the first years of the twentieth century as the esteemed Director of the Chief
Bureau of Weights and Measures. He believed that he had finally reached the
ear of power and was able to implement a system that was not contingent on
the fluctuations of local Petersburg politics. It was, however, contingent on the
existence of unfettered autocracy. The collapse of that crucial rampart also
struck the failed prophet by surprise.

Western Europe knew Mendeleev as a leading representative of the conservative elite of Imperial Russia, a reputation partially derived from his scientific
achievements (and rather heavily based on Mendeleev's Romantic posturing).
A series of high-profile interviews by foreign journalists while on travels
abroad (usually for metrological conventions) display not only his willingness
to pronounce on the fate of Russia, but also his blindness in the face of the cataclysm that would strike in 1905. In an interview with *Figaro* as late as March
1905—two months after the events of Bloody Sunday and several years after
the onset of increasingly aggressive student protests—Mendeleev blamed the
unrest on residual classicism in the educational system and German *agents
provocateurs*.

As for the possibility of a republic in Russia, he rejected it out of hand. At
one point, he claimed, referring to medieval Novgorod, Russia had indeed had
a republic, "[b]ut the people understood that it was a game of ambitions,
where everyone pushed themselves in order to have offices and to profit, and
everything was swept away. It was replaced by autocracy. The autocracy is perfect! Since the autocracy owns everything, there is no one to rob, right?" Citing his expertise concerning the student mind-set based on his almost forty
years teaching at St. Petersburg University, he dismissed any possibility of revolution: "We are Russians, and not at all prepared for violent demonstrations.
We treat politics like our private affairs: we have whims, caprices, ephemera,
and we play at Revolution because it amuses us. At the base, this is not serious.
The Russian is very calm. . . . The true people are very tranquil."[95] Seven
months later, Russia had a constitution and centuries of unrestrained autocracy were over. Little in Russian history since then has supported Mendeleev's
characterization of his countrymen.

In a poetic historical coincidence, Mendeleev's political blindness corresponded with personal blindness, and as he was undergoing cataract surgery,
Russia's autocratic system imploded. Persistent student rebellions since 1899
and Russia's disastrous rout in the Russo-Japanese War (1904–1905) precipi-
tated a climate of dissatisfaction with the reign of Tsar Nicholas II. Liberal intellectuals and the rising strata of professionals (lawyers, physicians,
pharmacists, journalists, teachers, and others), often organized within the
structures of the rural councils (*zemstva*) that had been created as a Great Re-

form in 1864, agitated for a constitution and an elected parliament. Radical leaders called for the overthrow of autocracy, and the years immediately preceding 1905 saw a rise of political terrorism and assassinations of high ministers reminiscent of 1881. The crisis peaked in October 1905, when a general strike brought the nation to a halt—largely through the very effective striking of railroad workers—and forced Nicholas II to abandon his principled opposition to parliaments and constitutions.[96] Although over the next several years the parliament (Duma) was dissolved twice and reelected with a more restricted and conservative franchise, and although Nicholas repeatedly acted without consulting elected officials, the Revolution of 1905 successfully overturned the old political system. Autocracy was now in principle constrained by the Tsar's own Manifesto of 17 October 1905.

Mendeleev would not live to see the monarchy's retrenchment. He died in 1907 horrified by the specter of parliamentary democracy. The Tsar's October Manifesto appeared just as Mendeleev was finishing *Cherished Thoughts*—which had been gradually serialized for two years—and he expressed pleasure at the actions of the "good, great-spirited Tsar." This, however, flatly contradicted his more articulate views earlier in the text that "[t]he union and unification of Russia, its spiritual and intellectual enlightenment, its external and internal forces, and even the rudiments of the industrial and progressive stratum have been influentially defined by monarchs, and that not only now, but in the foreseeable future Russia is and will be a monarchical country."[97] The very loyalty to the crown that he felt Nicholas had betrayed prevented Mendeleev from censuring Nicholas's actions. With the Tsar subordinated to the laws of the land (the Fundamental State Laws of 1906), the entire framework of the Imperial Turn, itself a mutation of Mendeleev's faith in the Great Reforms, had collapsed.

Mendeleev did not just feel intellectually betrayed by the events of 1905, he felt personally stung by the actions of his friend Sergei Witte, once Minister of Finances, in bringing about the October Manifesto. Witte's interests in Russia, like Mendeleev's, lay in fiscal and political stability; the two men differed only in what they were willing to countenance to achieve them. While some members of Nicholas's government counseled repression, Witte argued for moderation and eventually became the first Prime Minister, consolidating the post-October regime. Witte, whom Mendeleev had helped to build the modern industrial economy that was then leading Russia to extraordinary rates of economic growth, had—in Mendeleev's eyes—forsaken the very autocracy that had made such efforts possible. Witte was no more a "liberal democrat" than Mendeleev, but Witte's conceptual and practical system allowed for more flexibility in the structure of the Russian state.

Tension between Witte and Mendeleev had been building for some time, as exemplified by the events of Bloody Sunday. In 1901, in an effort to limit the ability of radicals to manipulate labor unrest, police official S. V. Zubatov was granted permission to organize labor unions under police surveillance. The idea was to channel the rising pressure into constructive and limited labor groups. On Sunday, 9 January 1905, a cleric named Father Gapon, one of Zubatov's union organizers, led striking workers to the Winter Palace to present the Tsar with a petition to intervene in their case, markedly outside of the original intention of the police-union idea. Carrying crosses and pictures of the Tsar, they marched peacefully toward the Palace, but guards panicked and fired. The massacre was dubbed Bloody Sunday. As fate would have it, the Tsar was not home to receive the petition, and this event became a trigger for the 1905 Revolution.

On that Sunday, Mendeleev was distraught by the rumors that were circulating in Petersburg. He went to Witte's house on the day of what he called the "Gapon riot." Upon his return home, he removed Witte's portrait, turned it so the painting faced the wall, and barked "Never speak to me of that man again." Witte's own admittedly selective memoirs of Bloody Sunday do not mention Mendeleev at all.[98]

There remains considerable speculation of what transpired between the two men on that day. They still met on occasion afterward, during Mendeleev's last year, but their relations were quite strained. The most common interpretation, favored by Mendeleev's family and Soviet scholars, was that Mendeleev tried to convince Witte to negotiate with the workers and stop the bloodshed. This is unlikely. Mendeleev sneered at Gapon's "riot," and he would have known that Witte could not have done anything anyway.

The conflict between the men stemmed from two sources, political and personal. Mendeleev generally felt that Witte was too quick to leap on the nearest fashionable bandwagon. He remarked later that he thought even Witte's protectionist measures were not part of a philosophy of industrialization but instead were opportunistic ad hoc stopgaps.[99] In contrast, Mendeleev was a true believer who demanded sincerity among his fellows. On a personal level, Mendeleev felt that his former friend failed to support him when his fortunes went south. In August 1903, he wrote an autobiographical letter that enumerated his services to Russia, spanning many of the pedagogical, political, and cultural efforts discussed in the preceding chapters. He had the letter sealed to be given to Witte upon the chemist's death, but mailed it in 1906 (after living beyond his expectations). The purpose of this letter, revealed at its close, was a plea for Witte to bail out his family from a real estate deal that had gone south.[100] The succor seems never to have materialized. Mendeleev needed the money, and Witte was not there to help.

The last year of Mendeleev's life was filled with disappointment. Although in 1906 he was still the most decorated and influential scientist in Russian history and enjoyed the respect and admiration of younger generations of Russian chemists—generations he had largely formed through the Russian Chemical Society and *Principles of Chemistry*—he saw his political dream in shambles, not to mention the failure of his chemical ether project to win acceptance. He even came to question his most fundamental political presumptions, such as his confidence in the justice of Emancipation of the serfs, that most vaunted of the Great Reforms. As he told Vladimir Visokovatov after the conflagrations of 1905: "All those who think a great deal of themselves as saviors of our motherland, who want to reshape it in the Western style, are very mistaken. By forcibly putting a Western caftan on our peasant, our superficial political and economic triflers cannot in any way understand that he will not crawl into it and that it will dangle helplessly on our peasant."[101]

The program of the reformers of the 1860s instituted communal land ownership with Emancipation, but without education and an investment of capital, this only led to the impoverishment of the countryside. The Great Reforms, which had earlier seemed to Mendeleev an ideal template for reform, now appeared tragically impotent. Too much advice, too much bickering had been taken into account, and Mendeleev no longer offered his own solutions. He confronted a state and a public that had lost patience with the elements of tradition he considered absolutely indispensable to stability. It was as if "stability" were no longer a goal of "modernization."

Bitter and disillusioned, he retreated into his position at the Chief Bureau until his death in January 1907. A German obituary summed up his last years concisely: "Then the terrible war of Russia with Japan broke out. As an ardent patriot, who knew well the unrefined treasures and powers of his fatherland, who believed firmly in its fortunes in war, he could be pained by the upsets and the collapse so much more heavily. Then came the much more horrible, inner enemy, revolution. . . . Disappointment, hopelessness took hold of his soul, shattered his faith in the greatness of his fatherland, broke his will to live."[102] His dream for Imperial Russia died with him.

CONCLUSION

Mendeleev sketched by his second wife, Anna, in the early 1890s,
from Dobrotin et al., *Letopis' zhizni i deiatel'nosti D. I. Mendeleeva,* 530.

CONCLUSION:
THE MANY MENDELEEVS

A phrase (it often happened when he was exhausted) kept cycling round and round, preconsciously, just under the threshold of lip and tongue movement: "Events seem to be ordered into an ominous logic."

—THOMAS PYNCHON[1]

MENDELEEV WAS NEVER BORED IN IMPERIAL PETERSBURG. From his first days back in the city after studying abroad in Heidelberg, he was enmeshed in projects to reform chemistry, the state, and society that ranged from the most mundane aspects of chemists' professional lives to the most utopian visions of Russia's future. At the base of all of these—the battling of Spiritualists, the organization of a gas laboratory, the tariff—was the same man, Dmitrii I. Mendeleev, attempting to come to terms with a culture in tremendous flux. These contours of possibility and anxiety defined Russia at the end of the nineteenth century, and Mendeleev's attempts to navigate conflicts between systems and the misfits (including himself) that threatened them eventually defined his life. Throughout both his most technical chemical work and his highly involved social activities, he concentrated on the different ways one might control such misfits, drawing solutions from his cultural context in Petersburg.

The importance of *Mendeleev* as an individual cannot be overemphasized. One could in principle similarly follow the paths of many figures in Imperial Russia or in nineteenth-century science—or in fact in almost any place or time. Yet Mendeleev opens a particularly valuable perspective on the history of both Russia and chemistry. The educated elite in Imperial Petersburg was quite small, and individuals who were prominent in several groups—such as Sergei Witte or Feodor Dostoevsky—were able to imprint their concepts deeply in Russia's state or its culture. Mendeleev, on the other hand, unified artists, writers, scientists, and bureaucrats while preserving their traces in his sizable personal archive; his life illustrates what it was like to live and work in St. Petersburg. He considered himself a *Russian* and a *scientist*, both European concepts. His career testifies well to the degree of openness of his European peers, who accepted this man from the boundaries, albeit with some reservations. In one sense, this is not unusual, since Petersburg was one of the political centers of Europe in the nineteenth century. But Russia was not just a metropole; it was also provincial, and many of Mendeleev's interactions with his "peers" placed him in the position of a supplicant, not an equal. Mendeleev's chemical ideas, therefore, demonstrate how European science functioned, as well as how barriers of language and culture placed constraints on scientific attempts at attaining universality.

This story has been biographical, but it has not treated Mendeleev as its exclusive subject. It is almost impossible to look at Mendeleev and not be distracted by—or swept up in—the world around him. In fact, it is at times perplexing to reflect that these are the activities of only one man. Faced with the predictor of the eka-elements on the one hand and an architect of Arctic exploration on the other, it is hard to conceive that one person occupied all the roles this man played. In the end, though, this entire story is structured around how Mendeleev and others *remembered* his role in Imperial Russian culture.

Memory is a tricky matter, hard to pin down. One can hardly propose to analyze all attempts to commemorate the life and achievements of this remarkable chemist. Each remembrance has imparted its own nuances onto Mendeleev's heterogeneous life—a heterogeneity I wish to investigate in all of its contradictions rather than explain away. Memorials, however, are typically constrained to more unitary representations, which, in Mendeleev's case, divide mainly into two categories: the isolated genius and the adept public servant.

The purest expressions of each of these two positions are embodied in memorials crafted by those who knew and respected him. The first—Mendeleev the lonely sage—is captured by the painting "In the Wild North . . . " (1891) by Ivan Ivanovich Shishkin (Figure 9.1). Shishkin gave Mendeleev a copy of this painting, which still hangs over his desk in the Mendeleev Museum-Archive in St. Petersburg. The painting depicts a lonely pine tree looming over

FIGURE 9.1 "In the Wild North . . ." by I. I. Shishkin (1891). A copy of this painting was given by the artist to Mendeleev, who hung it over his desk, where it remains. Source: *Ivan Ivanovich Shishkin.*

a valley, majestic although mournful in its isolation. It was originally one of two illustrations for a commemorative collection of poems by the early nineteenth-century Romantic poet Mikhail Lermontov; it accompanied an untitled poem, based on an original by Heinrich Heine:

> *In the wild north a pine stands*
> *Alone on the naked heights,*
> *And she sleeps, shaking, clothed*
> *In free-flowing snow as a raiment.*
> *And she dreams that in a distant wasteland*
> *In that region where the sun rises,*
> *Alone and sad on a burning cliff*
> *A beautiful palm grows.*[2]

This mournful isolation corresponded to Shishkin's image of Mendeleev. Shishkin and other painters of the Wanderer group—including the famous Repin, Kramskoi, Surikov, and Kuindzhi—attended Mendeleev's Wednesday salons for many years; these weekly gatherings became such an institution that art stores would send albums of other artists' work to them for previewing.[3] It is understandable that Shishkin wanted to see this side of Mendeleev: The lone genius did derive from Romantic-era conceptions of the artist. The Wanderers knew of Mendeleev's social projects, but they preferred to understand him as a soul who could not be constrained by the norms of the broader milieu. That which made him great, his genius, was not connected to external social influences. This depiction correlates well with popular myths about the creation of the periodic law, his rejection from the Academy of Sciences, and his maverick behavior against the Spiritualists.

At the time of Mendeleev's death, however, a different image was prominent, especially among bureaucratic and chemical circles. To these observers, Mendeleev may have been a genius, but he was far from isolated; rather, he was centrally involved in the major upheavals of his world. He was a civil servant, and a distinguished one. A few years after his death, a monument was erected to him in St. Petersburg, but not, as one might expect, at the site of his solitary contemplations of the periodic law, on the campus of St. Petersburg University along the street that bears his name. (Instead, a statue of M. V. Lomonosov was recently placed across from Mendeleev's apartments.) Rather, this statue was erected in the frontyard of the Chief Bureau of Weights and Measures, where Mendeleev died. This was the institution that allowed him to combine his desire to investigate gravity with the demand to standardize the Russian Empire. It was a scientific institution that bore the imprint of his personality. The monument suggests multiple readings, each quite divergent from Shishkin's lone pine (Figure 9.2).

FIGURE 9.2 Monument to Mendeleev, located on Moskovskii Prospect in St. Petersburg in front of the former Chief Bureau of Weights and Measures, erected shortly after his death. Source: Courtesy of I. S. Dmitriev.

The bronze Mendeleev stares calmly across Moskovskii Prospect at the Technological Institute, the first institution to hire him as a professor, but his mind is not on chemistry. He sits turning the pages of the *Chronicle of the Chief Bu-*

reau of Weights and Measures, the periodical he founded to chart his metric reform, smoking one of his ever-present cigarettes. He looks down upon the origin of the unnervingly straight road that connects Petersburg to Moscow, the
artery that linked Russia's two metropoles. This road, more than Mendeleev
Line by the University, represents the link between economics and bureaucracy
that was to be Russia's salvation. Less a snowy pine than a real man (with real
nicotine cravings), this Mendeleev was grounded in politics. This was the
Mendeleev his colleagues chose to memorialize in alloy. On a wall far behind
and above Mendeleev, partially obscured by an enormous tree, is the periodic
system in its canonical final form (it includes the noble gases), rather than the
1869–1871 creation. It was added about two decades later—an afterthought.

These opposed Mendeleevs have almost completely dominated the various
interpretations of his life. To illustrate this point, consider two of the most intellectual and sophisticated interpretations of Mendeleev. Both are exceptionally clever, yet demonstrate the persistence of the Lonely Pine and the Bronze
Lounger. First, the collected observations of Mendeleev's son-in-law, the famous avant-garde poet Aleksandr Blok (1880–1921). Given his importance in
the development of modern Russian poetry and his support of Leftist politics,
Blok has drawn much more attention from scholars than his father-in-law.

Blok was the grandson of A. N. Beketov, rector of St. Petersburg University
(1876–1883), brother to chemist Nikolai Beketov, and a close friend of D. I.
Mendeleev. Mendeleev convinced A. N. Beketov to buy a summer estate near
his own estate, Boblovo, which the rector christened Shakhmatovo. This was
the site of Blok's happiest childhood memories; it was also where he met Liubov' Dmitrievna Mendeleeva, the eldest child from the chemist's second marriage and the great love of Blok's life. Therefore, Boblovo and its owner play an
important supporting role in the drama of the nascent visionary Blok. In the
few cases where Blok's biographers do mention Mendeleev, he is characterized
(with astonishing understatement) as "a well known chemist of his day."[4] Others draw attention to Mendeleev's appearance as "an Old Testament prophet"
who "had been known to take off from his lofty rural stronghold on the wings
of the wind—in an air balloon."[5] They stress the Romantic Mendeleev, the
unconventional genius—like his son-in-law. Alternatively, he is passed over as
a Jules Verne aficionado who did not appreciate Blok's "decadent" poetry. He is
a romanticized sideshow.

The poet himself proved rather more sophisticated in his observations of his
father-in-law. In 1908, partly in reaction to Mendeleev's death, Blok began
reading some of Mendeleev's works, in particular *Cherished Thoughts* and *Materials for a Judgment about Spiritualism.*[6] In a series of letters, book reviews,
and diary entries, Blok crafted an image of the role his father-in-law played in
Imperial Russia. There are two main elements of Blok's portrait. The first was
that Mendeleev was a lone genius who, importantly, was disconnected from

the Russian intelligentsia by his superior understanding of the Russian people (*narod*). Blok was no lover of the established intelligentsia, those who "had left Dostoevsky to die in poverty [and] related to Mendeleev with open and secret hatred."[7] Because he was close to the people, Mendeleev understood Russians more clearly. As Blok wrote to Liubov' in 1903: "Your father is like this: he knows everything that happens in the world already for a long time. He has entered into everything. Nothing is concealed from him. His knowledge is most complete. It is formed from geniality, and simple people don't have that."[8] Mendeleev here was a weapon used to indict effete intellectuals who had stopped acting in the people's interests, thereby losing their *raison d'être.*

Blok's Mendeleev was not a happy figure. As much as Blok loved his wife, he despised his mother-in-law, Anna Mendeleeva. In Blok's depiction, Mendeleev was a pawn caught between his own goodness and her wretchedness, a figure of pathos rather than admiration. In his diary of 1911, Blok penned a "Theme for a novel. A scientific genius falls violently in love with a cute, feminine, and vapid Swede" who "falls in love with his temperament and not him." In this barely disguised portrait of the Mendeleev family (the fictional family has two children, Liubov' and Ivan), the couple grows apart, but the woman stays with the increasingly isolated man only to exploit his money and social connections.[9] For Blok, Mendeleev's private life reflected his public life: a well-meaning genius detached by malicious manipulation from all that he loved.

Blok's second theme treated Mendeleev as counterpoint to novelist Leo Tolstoy. For Blok, Tolstoy in his later writings and Mendeleev in *Cherished Thoughts* offered contradictory optimistic visions of Russia's future, the former building on faith in the Russian peasant and agriculture, the latter on faith in the Russian entrepreneur and industry. The inability to decide between the two was symptomatic of Russia's tragedy: "This tragedy was recently expressed most starkly in the irreconcilability of two principles—the Mendelevian and the Tolstoyan; this opposition is even sharper and more alarming than Merezhkovskii's posited opposition between Tolstoy and Dostoevsky."[10] Blok continued in his notebooks: "*Mendeleev and Tolstoy.* The sharpest doubt (contradiction). I cannot choose: not two abysses, god and the devil, [but] two paths of goodness. . . ."[11] Blok chose neither the rural utopia of the novelist nor the capitalist utopia of the chemist. He found the future in socialism. In his diary entry of 31 January 1919, he ended his tutelage to the image he had crafted of his father-in-law: "Symbolic action: on the Soviet New Year I smashed Mendeleev's desk."[12] Blok consigned Mendeleev to a specifically Imperial past.

Blok's Mendeleev was indeed a lone genius, but not by choice. Precisely *because* Mendeleev was brilliant and able to interpret the plight of the Russian people, the intelligentsia ostracized him. This is merely a social interpretation of the Romantic image of Mendeleev the loner, an explanation of how the genius became abandoned in the first place. Blok may have included some of Mendeleev's context, but this was still the artist's image of Mendeleev the Creator. Blok did not emphasize Mendeleev's service to the state, but he could not ignore it either. Mendeleev's desk, the scene of that isolated thinking, was destroyed in an endorsement of the new Soviet Russia—Mendeleev was thoroughly a creation of his old context, one that needed to be repudiated for Russia to become the Workers' Paradise.

A different interpretation emphasized Mendeleev the public figure and discoverer of periodicity in a spirited defense of a social philosophy. This was Leon Trotsky's vision of Mendeleev, the best of a long series of Sovietizations of the chemist. There were many Mendeleevs used by Soviet historians and philosophers, amidst substantial pressure to assimilate him for the greater glory of the Soviet Union. The most common avatar of Mendeleev was as a dialectical materialist. Dialectical materialism was the official philosophy of science of the Soviet Union.[13] The most important tenet of the theory with respect to Mendeleev was the principle of the transformation of Quantity into Quality. Drawing on Marx's observations on political economy, where an accumulation of quantity (capital) would on occasion qualitatively transform the relations of production (from feudalism into capitalism), Soviet philosophers of science pointed to the periodic law as the scientific equivalent. That is, quantitative increases in atomic weight led to qualitative differences in elements: You add a bit of mass to hydrogen and you produce helium; with a bit more, you get lithium; and so on up the periodic system. Perhaps the most succinct formulation was Joseph Stalin's in his only piece on the philosophy of science, "Anarchism or Socialism?" written in the year of Mendeleev's death: "As to what concerns forms of motion, what concerns how, in accord with dialectics, small *quantitative* changes in the end lead to large *qualitative* changes—this law has force in equal measure in the history of nature. The Mendeleev 'periodic system of elements' clearly shows what a large significance the emergence of qualitative changes from quantitative changes has in the history of nature."[14] This quotation was the obligatory opening of articles on Mendeleev during Stalin's rule, and was later replaced during de-Stalinization by a more politically appropriate endorsement from Friedrich Engels. Eventually, more philosophically coherent and interesting dialectical materialist interpretations emerged, many of which also used Mendeleev's bureaucratic work to defend the validity of dialectical materialism.[15]

Trotsky was thus not alone when on 17 September 1925 he delivered a speech on Mendeleev entitled "Dialectical Materialism and Science." Trotsky's strategy was to take the life and thought of the chemist from various angles to illustrate different aspects of Marxism. He began by using Mendeleev to defend dialectical materialism. The discovery of periodicity pointed to Marxist methodology in two ways: first, by the transition from Quantity into Quality; and second, by demonstrating proper scientific method, as Marxism did: "There is no less difference between the Marxist method of social analysis and the theories against which it fought than there is between Mendeleev's Period[ic] Table with all its latest modifications on the one side and the mumbo-jumbo of the alchemists on the other."[16] He continued:

Chemistry is a school of revolutionary thought, not because of the existence of a chemistry of explosives (explosives are far from always being revolutionary), but because chemistry is, above all, the science of the transmutation of elements; it is hostile to every kind of absolute or conservative thinking cast in immobile categories.

It is very instructive that Mendeleev, obviously under the pressure of conservative public opinion, defended the principle of stability and immutability in the great processes of chemical transformation.[17]

This recognition of Mendeleev's conservatism is important, and worth postponing for a moment.

For man does not live by philosophy alone. Other Soviet writers also used Mendeleev to endorse the Soviet state, despite the chemist's frequent comments about the errors of communism. Here, Mendeleev's legacy was preserved not because of its chemical or philosophical advantages, but because of the reflected glory he would give to *Russia* (blurred during World War II with the Soviet Union) as part of a nationalist propaganda campaign.[18] Through distorting and selective quotations from his economic writings, writers painted Mendeleev as a supporter of forced industrialization and collectivization of agriculture, and popularized some of his more utopian economic schemes. Such efforts were eased along by Mendeleev's optimistic language, eerily echoing a famous catchphrase of Stalin's: Mendeleev wanted Russia "not just to remain behind other states, but to catch up (*dognat*) [and] even to overtake (*peregnat*) where possible."[19] He was also occasionally invoked as an implicit forefather of the Soviet regime's technological prowess. For example, in March 1946 the Supreme Soviet characterized him as "the greatest chemist of the world, who discovered the periodic law—the basic law of chemistry—which to the present time helps scientists discover the secret of atomic energy."[20]

Trotsky's version was unique in combining these tropes with a recognition and repudiation of Mendeleev's conservative political stance, rather than pretending those beliefs did not exist. For Trotsky, Mendeleev presented a typical case of a man who had the necessary evidence to understand the ravages of capitalism, but did not: "Mendeleev did not have a finished philosophical system. Perhaps he lacked even a desire for one, because it would have brought him into inevitable conflict with his own conservative habits and sympathies."[21] Mendeleev was a "spontaneous" dialectical materialist, following traditional Leninist terminology: He acted unwittingly *as if* he adhered to the principles of dialectical materialism.

Trotsky's best example of Mendeleev's conservatism was his advocacy of the old treaty system of international relations, hoping for disarmament through accords negotiated between foreign states. Empirically, these treaties failed to promote disarmament and were considered by some (including Lenin and Trotsky) as proximate causes of World War I. Trotsky was not pointing to Mendeleev's Imperial and conservative errors to malign the man: "But permit me to state that the major miscalculations of this great man contain an important lesson for students. From the field of chemistry itself there are no *direct* and *immediate* outlets to social perspectives. The objective method of social science is necessary. Marxism is such a method."[22] In other words, Mendeleev was held back by his conservative social climate. Even though he performed valuable economic services for his country—Trotsky cited alcohol–water mixtures, smokeless gunpowder, oil, the tariff, and the quest for northern sea routes—Mendeleev was bound by his time to anti-progressive beliefs. His failures as much as his successes pointed to Marxism as proper method.

Trotsky simply expanded on the Mendeleev of the monument. The bronze statue, the figure of the man and its embedding in context, deserves our attention since it shows us the importance of treating the scientist as part of his culture. For Trotsky, the social world of Moskovskii Prospect determined the limits of one's understanding of the science. Mendeleev was so didactically useful precisely because he did not have and could not provide a unified framework. Mendeleev suffered because he could not offer a unified system; Trotsky could. The matter was similar for Blok: Because Mendeleev was close to the people but cut off from his peers, he was unable to persuade anyone of the vitality of his vision, or to combine his approach with Tolstoy's. Blok provided the unification by rejecting Mendeleev and turning to a Soviet future.

The appeal of these images of Mendeleev is their unitary simplicity, and that is also their greatest defect. Blok, Trotsky, Shishkin, and the monument all show us what comes of endowing a single life with a unified interpretation. They offer us systems that handle their misfits by simply excising them. There are better ways. Instead of just emphasizing the individual in the center of various cultural and social currents, one should also emphasize the currents, to look at the systems and the misfits in one glance. Think of Mendeleev as a packet of tracer dye in a turbulent stream, and then concentrate on what the consequent patterns can tell us about the stream rather than the dye. On occasion, of course, one might take the opportunity to see what the stream's effects can tell us about this particular dye. This excises the binary vision of Mendeleev as Romantic and Mendeleev as Civil Servant, a dynamic that brought the man himself considerable grief. Instead of cobbling together a Frankenstein's Mendeleev out of component parts, we find instead a way of examining the history of chemistry, Imperial Russian culture, and a particular individual in their mutual interconnection.

For chemistry, the narrative begins with the formation of the periodic system of chemical elements. Beginning with the Karlsruhe Congress in 1860, a series of theories began to coalesce in the physical sciences that proved remarkably consistent. Thermodynamics and statistical mechanics, electromagnetism, structure theory in organic chemistry, and the standardization of atomic weights all offered a coherent vision of matter and its dynamics. The periodic law formed no small part of this constellation of unifying theories, correlating a vast amount of chemical and physical knowledge in an ordered framework. Periodicity was even more valuable because it enabled different fields to communicate while remaining agnostic about fundamental entities (such as atoms). Such, at least, was the role that the periodic system came to assume. In its early inception at Mendeleev's desk, the system was an attempt to solve a pedagogical demand under pressure from his publishers—a stopgap measure meant to be merely "good enough." It was only after its formulation that Mendeleev began to explore its possibilities as a "law of nature."

Mendeleev gradually built a chemical worldview around his periodic law. First, he came to believe that all matter is to some degree atomic. This belief led him to abandon research on the periodic system in 1871 and to attempt to find the luminiferous ether in the laboratory through deviations in gas laws. In this rejection of "chemistry" for "physics," as we would now put it, Mendeleev merely traversed what he saw as a blurry boundary between similar approaches to the same material. In the organization of his laboratory and his forays into meteorology, aviation, and the popularization of science, Mendeleev envisioned theory and experiment as a cooperative pair.

The stunning failure of his experimental efforts—the first of many formative failures—did not diminish his ardor for unification. The three essential elements of the atomic chemical worldview were so intuitive to him (and to many other chemists and physicists) that it is impossible to locate their origin clearly. All atoms had to be valent—that is, they had some propensity to combine with other elements—or else chemistry as the science of combinations of matter would be nonsensical. That atoms had no substructure was encoded in the Greek meaning of the very word "atom," and emphasized in Mendeleev's preferred latinized variant: "individuals." And, finally, atoms had to be immutable. Any reference to transmutation—the changing of one element into another—smelled rather too strongly of alchemy. After the discovery of gallium, scandium, and germanium, Mendeleev raised the stakes of his lawlike discovery and came to understand the chemical worldview as inextricable from—in fact identical to—his periodic system.

Mendeleev's vision of a unified physical science grounded in the periodic law became besieged in the final years of the nineteenth century, and he spent the remainder of his life battling against incursions into his theory. First the noble gases, then the electron, and finally radioactivity threw into question the three components of the worldview: valency, integrity, and immutability. Mendeleev's response, the chemical ether, is striking precisely because of how deeply it was embedded in nineteenth-century chemical thought just as the physical sciences were developing into new frameworks. Within a decade, special relativity, quantum theory, and radioactivity would make Mendeleev's reasoning seem oddly antiquated. But he had no way of knowing that. Building on the success of his interpolative predictions of the eka-elements from the early 1870s and his quest for unity in the gaseous ether of the later 1870s, in 1903 Mendeleev used the periodic law and kinetic gas theory to predict an extremely light inert gas that possessed the key properties of the ether. Domesticating one threat (that to valency) in order to combat the more sinister danger of radioactivity, Mendeleev argued boldly for a vision of a unified science that maintained crucial components of the stabilizing worldview he had begun to construct over three decades earlier.

Ironically, Mendeleev's historical journey through the physical sciences and his efforts to unite them reveals those sciences as primarily disunified. Underneath the rhetoric of a gradually closing worldview that would encompass all knowledge of the natural world lay a carefully negotiated mixture of theories that sat in a more or less unstable juxtaposition. This is not to say that electromagnetism, thermodynamics, the periodic law, or any of the other "unifications" of the nineteenth century were individually unstable; rather, it is to point out that, to the historical actors, there were multiple ways in which the various components could be reconciled. Some theories could be twisted, others ignored. The disunity that would pop into relief with the advent of relativity theory and quantum theory in the early twentieth century was already evident in the nineteenth century. The dream of Karlsruhe, to remove disagreement by the communal negotiation of foundational concepts in chemistry (or any science), turned out to have severe limits.

Mendeleev's experience of the tensions within the modern physical sciences mirrors the deep historical rifts that beset Imperial Russia. In the late Imperial period, Russia experienced three severe political shocks: the Emancipation of the serfs (1861), the assassination of their Emancipator, Alexander II (1881), and the first Russian Revolution (1905). The political and social consequences of these events have often been investigated, but not their cultural implications. What did these events *mean* for Russians? This question, at the heart of cultural history, is rather difficult to answer; but Mendeleev's extremely wide-ranging activities can at least tell us what they meant for *him,* as well as partially outlining the questions, hopes, and fears prompted in the broader Russian public. The Emancipation, though greeted by many as long overdue, was fraught with questions about order. Now that there was no longer a system of direct servitude of serfs to nobility, nobility to Tsar, how were Russians to understand their place? Were they subjects or citizens? Did they have "rights"? How much could autocracy allow them to participate without destroying itself? How could the "unity" of Russian culture—variously understood—be preserved? Mendeleev had no magic answer to these questions. The important point is that he *asked* them along with Dostoevsky, the Tsar, and thousands of other Petersburgers.

Among the welter of these cultural issues lay the Russian state's concerns about the proper disposition of expertise. The Great Reforms transformed Russia by eliminating serfdom and easing censorship; but they also created the need for lawyers, jurists, agricultural experts, municipal administrators, and professors in the sciences. This expertise was unfamiliar and the problem of organizing it was daunting. Mendeleev's move from local groupings (the Russian Chemical Society, the Russian Technical Society, his gas laboratory, the Commission against the Spiritualists, the Petersburg Academy of Sciences) to Imperial ordered societies (the Academy of Sciences again, the Ministry of Finances, the Chief Bureau of Weights and Measures) reformulated the problem of expertise in late Imperial Russia. Mendeleev's credentials as chemist, economist, bureaucrat, and metrologist positioned him to serve as an expert among experts who could solve the general problems of expertise in a modernizing autocratic state. As a prominent professor at St. Petersburg University, Mendeleev was intimately involved in issues of educating the next generation of specialists, and pedagogical worries abutted directly on political ambivalence toward expertise. He was not the only advocate of systems to preserve the Russian social and cultural order; Butlerov, Aksakov, Witte, Tolstoy, and others proposed alternatives. Yet he was an exemplar of a peculiar breed, the "liberal in the name of autocracy," a conservative reformer precipitously poised between scientists and bureaucrats.

Mendeleev pursued his systems through a series of models based on scientific exemplars. He began by organizing people into small, independent societies that would provide local order. As student unrest, scheming Baltic Germans, and raving Spiritualists continued to threaten this order, Mendeleev experimented with new systems until he lost his faith in the power of purely local ordering to control the misfits. His rejection by the Academy of Sciences in November 1880—his second great failure—definitively changed his ideal approach to creating an ordered society. Now he would move from the decentralized to the bureaucratic, from local Petersburg to Imperial Petersburg. The assassination of Alexander II on 1 March 1881 similarly renewed the urgent need for unification. The Tsar proved vulnerable, autocracy had been weakened by the Great Reforms, and yet somehow Russia had to be governed. Mendeleev's economic consulting, the tariff of 1891, the metric system, Arctic exploration—these were all attempts to provide administrative unity in a culture that was starving for answers to the questions of law and order.

These sets of questions, this life path, of course also provide insights into Mendeleev the man, whose status within Russia is almost unequaled. He remains the most recognized Russian scientific name both at home and abroad. (The competition for second place is fierce.) Many Russian cities have a Mendeleev street or a monument, Moscow even has a subway station. Yet as a person he remains shrouded in historical fog.

When he returned to Petersburg in 1861, Mendeleev was poor and desperate for money; when he died, the city went into mourning. This transformation was the result of much hard work by a man talented in the art of self-promotion; but even more, it was the result of several lucky breaks, the most important of which was seen at the time as a misfortune of gigantic proportions. Mendeleev's rejection by the Academy of Sciences may have been the best thing that ever happened to him. Prior to that debacle, Mendeleev had been a moderately well-known local personality. The newspaper adulation after the Academy rejection made him a star and drew popular attention to what would remain the single greatest success of his life: the periodic system of chemical elements. Yet at the same time his greatest quarry—the ether—eluded him. When he withdrew from his gas project in January 1881, his marriage and professional life were in shambles, and his quest for unification seemed permanently derailed.

But Mendeleev was given a rare opportunity in the early 1880s: the chance to reinvent himself. And reinvent himself he did. For the most part, his wide array of Imperial ventures were attempts to build circulatory perpetual motion machines, systems that would strengthen the Russian Empire through constant internal movement—of capital, metersticks, or labor. Each system depended centrally, however, upon a misfit, a *deus ex machina,* who could conceive and implement these circulations while remaining immune to their standardizing force. Mendeleev reinvented the misfit, too. This other figure, the Romantic Mendeleev of adventurous journeys, was a direct product of his self-fashioning as a maverick, initiated by the Petersburg newspapers after the Academy affair. It was this Mendeleev who crafted himself as Newton's successor, a Siberian hero, the conqueror of the North and the air. It was also this Mendeleev who, as a misfit in the system of education he himself had a part in establishing, was induced to resign in 1890. The student unrest he inadvertently stoked would eventually bubble into the movements that led to the 1905 Revolution, an event that shocked Mendeleev's lifelong faith in Russia. For that Revolution compromised all of the systems Mendeleev had worked so hard to create by disabling the most important Misfit, the Tsar, from his unfettered autocratic ability to implement progressive reform. Mendeleev was a liberal in the name of autocracy, and when autocracy was removed, what remained was barely liberal. The grizzled chemist and bureaucrat died angry but successful, a long way from his Siberian roots.

The story of this book is not merely that of a man who fashioned his own place in Russian culture. Mendeleev certainly had a chance to make his own history, but he did not always have the opportunity to make it as he pleased. In the 1860s, he approached a culture rich in possibilities and mobilized segments of it—pedagogical, chemical, political—to secure a foothold for himself in competitive Petersburg. In the 1870s, he was occasionally able to continue this active strategy of self-promotion—for instance, against the Spiritualists—but he also came up against resistances that he could not surmount. He was not able to make the ether exist just by wishing it so, and he was not able to make the Russian Technical Society underwrite whatever research he wanted. Likewise, in his rejection by the Academy, Mendeleev did almost nothing to generate the public image that proved so valuable to him later. These failures are important, even vital. Sometimes the fault was his, sometimes—and these cases are often more interesting—it was not. It is precisely the points where Mendeleev failed—with gases, the Academy, and the chemical ether, for example—that we see where ambition and wishful thinking collide with greater powers. Failure shows the resistance, the border, the friction of history. Mendeleev repeatedly pushed the bounds of what was acceptable or even possible in both his science and his politics; the limits he encountered bounded his culture. Past studies of Mendeleev have stressed his successful work, most notably the periodic law. In the final analysis, this was his only unqualified success, and it deserves to be treated as such—as a special case. To tell Mendeleev's story by embellishing one moment is both inaccurate and unfair.

It is crucial to shy away from this "periodicity-mania." Mendeleev was an icon across the sciences, but not just there. The preceding story has been in part a plea for reintegrating Mendeleev and, more importantly, the scientific community back into the history of plans for Russia's future. Whether looking at the community of physical scientists, Russian bureaucrats, or Mr. Mendeleev of St. Petersburg, one encounters an intense interest in adapting past knowledge to present circumstances. This modernizing project captured the goal of all the various systems we have encountered. The inescapable entwining of science, culture, and modernity remains the enduring legacy of Mendeleev's life. No matter how wide the seams and how deep the flaws, Mendeleev always returned to the system in order to chart his course. It was the best hope for making the world predictable, and that was a dream too dear to relinquish.

ACKNOWLEDGMENTS

My involvement with Mendeleev began just shy of a decade ago under the sponsorship of three mentors: Peter Galison, Loren Graham, and Mario Biagioli. Thanks to their criticism and encouragement, it has now evolved into a work that bears but faint resemblance to its incoherent origins. All of their exertions are much appreciated. Many different people have read all or part of the manuscript and each provided invaluable suggestions. It would overtax these pages to thank them as they deserve, so instead I will merely list them with a very grateful nod: Daniel Alexandrov, Katherine Anderson, Elizabeth Baker, Jonathan Bolton, Robert Brain, W. H. Brock, Nathan Brooks, Angela Creager, Amy Finkelstein, Adam Fleisher, Karl Hall, John Heilbron, David Kaiser, Masanori Kaji, Nick King, Alexei Kojevnikov, Ilana Kurshan, Adriaan Lanni, John LeDonne, Elizabeth Lee, Theresa Levitt, Eric Lohr, Donald MacKenzie, Ishani Maitra, Julia Mannherz, Ian McNeely, Everett Mendelsohn, Sharrona Pearl, Denise Phillips, Eric Scerri, Sam Schweber, Rena Selya, William Todd, Jing Tsu, and Norton Wise. Two individuals in particular performed services beyond the call of friendship or duty. Olga Litvak's detailed reading of a complete draft immeasurably improved the manuscript, and at times transformed it. Last but not least, Matthew Jones, more than any other single person, has been with this project since the beginning, offering astute comments and pointed criticism to help endow a set of diffuse ideas with form and substance.

Igor S. Dmitriev and Natasha Iu. Pavlova were my hosts and guides in St. Petersburg on several visits, and I cannot begin to express my heartfelt gratitude for all of their kindness. They provided not only a place to rest and think, but, as gifted historians, sharpened many of my notions about Mendeleev and Imperial Russia. This project would have foundered early on without the extensive assistance of the staff of several libraries, especially the Widener Library of Harvard University and, in St. Petersburg, the Library of the Academy of Sciences and the Russian National Library. I would like to thank the D. I. Mendeleev Museum-Archive (ADIM), the Russian State Historical Archive (RGIA), the Institute for Russian Literature of the Russian Academy of Sciences (Pushkin House; PD), and the Petersburg Branch of the Archive of the Russian Academy of Sciences (PFARAN) for permission to reprint material from their collections. I also gratefully acknowledge the innumerable efforts of the staff at each of these institutions—particularly Nina G. Karpilo at ADIM—in furthering my research.

Without financial support, none of the Mendeleev research would have been possible. Research for this book was supported in part by a 1999 grant from the International Research & Exchanges Board (IREX) with funds provided by the National Endowment for the Humanities, and the United States Department of State, which administers the Title VIII Program. An earlier trip to St. Petersburg was made possible by a travel grant from Harvard University's Davis Center for Russian Studies. Finally, a Graduate Research Fellowship from the National Science Foundation allowed me to keep body and spirit together while working in the United States. None of these organizations is responsible for the views expressed. Finally, I owe much to the Harvard Society of Fellows, which gave me the resources, the companionship, and—most valuable of all—the time to bring this work to completion.

Some of the material in this book is drawn from articles I have previously published, and I extend my appreciation to the Society for the History of Alchemy and Chemistry, the American Association for the Advancement of Slavic Studies, and the Regents of the University of California for permission to reproduce selected extracts from, respectively: "Making Newtons: Mendeleev, Metrology, and the Chemical Ether," *Ambix* 45 (1998): 96–115; "Loose and Baggy Spirits: Reading Dostoevskii and Mendeleev," *Slavic Review* 60 (2001): 756–780; and "The Organic Roots of Mendeleev's Periodic Law," *Historical Studies in the Physical and Biological Sciences* 32 (2002): 263–290.

Basic Books has been very helpful both in shaping the project and in expediting its completion. The Basic Prize Committee, who selected this manuscript for publication, deserves particular thanks: Norton Wise, Mary Jo Nye, David Lindberg, Robert Richards, and Ron Numbers. William Frucht and Rich Lane shepherded the work through the rapids of the publication process, and David Shoemaker has been an invaluable editor, improving the text in ways I did not think possible.

Finally, I would like to thank my family for their love and support. My parents and my brothers, Oren and Jonathan, have endured my obsession with Mendeleev without judging me too harshly for it. For their help with everything, they deserve this book more than anyone else. It is for them.

INDEX OF FRONTISPIECE IMAGES

Preface: Rough draft of the first periodic system, dated 17 February 1869, from Smirnov, *Mendeleev,* insert 1.

Chapter 1: Mendeleev in 1869, from Trirogova-Mendeleev, *Mendeleev i ego sem'ia,* facing page 9.

Chapter 2: Mendeleev and his first wife, Feozva Nikitchna, in 1862, from Dobrotin et al., *Letopis' zhizni I deiatel'nosti D. I. Mendeleeva,* 95.

Chapter 3: Mendeleev, painted by I. N. Kramskoi, 1878, from Dobrotin et al., *Letopis' zhizni I deiatel'nosti D. I. Mendeleeva,* 526.

Chapter 4: A group of professors and teachers from the physico-mathematical faculty of St. Petersburg University, 1875. Mendeleev is in the center, leaning to the side. From Smirnov, *Mendeleev,* insert 1.

Chapter 5: This caricature of Mendeleev appeared on 7 December 1880. It features the chemist dreaming underneath a tableau of scheming academicians, none of whom appears to represent an actual member of the Academy of Sciences. The cartoon is entitled "Daydream" and the caption reads: "D. I. Mendeleev. Will they elect him? . . . Won't they? . . . They didn't! . . . What is this: a dream that is similar to reality, or reality that doesn't differ from a dream?. . . " From "Son na iavu," *Strekoza,* 7 December 1880, #49: 1. A copy may be found in A. M. Butlerov's personal files in PFARAN f. 22, op. 1, d. 38, l. 10.

Chapter 6: Mendeleev, on left, with metrological assistants at the Eiffel Tower, 1895, from Makarenia, Filimonova, and Karpilo, *D. I. Mendeleev v vospominaniiakh sovremennikov,* 2d. ed., insert.

Chapter 7: Mendeleev, on left, playing chess with artist Arkhip Kuindzhi. The woman in the back is Mendeleev's second wife, Anna. This photograph is from the late 1880s or early 1890s. From Smirnov, *Mendeleev,* insert 2.

Chapter 8: Mendeleev at his working desk in 1904, from Smirnov, *Mendeleev,* insert 2.

Chapter 9: Mendeleev sketched by his second wife, Anna, in the early 1890s, from Dobrotin et al., *Letopis' zhizni i deiatel'nosti D. I. Mendeleeva,* 530.

NOTES

CHAPTER 1

1 Bulgakov, "Sobach'e serdtse," in *Sobranie sochinenii*, 164.

2 See DeMilt, "The Congress at Karlsruhe." For a history of the formation of the Congress, see Stock, *Der internationale Chemiker-Kongreß Karlsruhe.*

3 The official statement of the Congress resonated well with this ideal: "We may be of differing ethnic origin and speak different languages, but we are related by professional specialty, are bound by scientific interest, and are united by the same design. We are assembled for the specific goal of attempting to initiate unification around points of vital concern for our beautiful science." "Account of the Sessions of the International Congress of Chemists in Karlsruhe," 9.

4 As Mendeleev's German obituary put it: "Mendeleev's attendance at this Karlsruhe Congress is doubtless the greatest spiritual windfall of his foreign studies; the impressions received from the wrestling over clarity exerted a lasting effect: they formed the capstone of his education and development. His years at school and abroad both closed at this moment." Walden, "Dmitri Iwanowitsch Mendelejeff," 4729.

For the traditional interpretation of Karlsruhe, see "The Congress of Chemists at Carlsruhe"; and Ihde, "The Karlsruhe Congress." The conventional understanding of Karlsruhe as a crucial moment in establishing modern atomic weights has been disputed by historian Alan Rocke, who argues that it only codified an earlier transformation in organic chemistry. Even without the Congress, he claims, the new atomic weights would have become commonplace in a few years. See Rocke, *Chemical Atomism in the Nineteenth Century*, 295; and idem, *The Quiet Revolution.*

Avogadro's hypothesis is typically cited as the claim that equal volumes of gases at the same temperature and pressure contain the same number of molecules, but he also noted that some gases must form diatomic molecules in the gaseous state. Historians have debated for some time why this incredibly useful hypothesis, which cleared up the inconsistencies in determining atomic weights, was ignored for so long. There were actually some good technical reasons for its neglect (isomerism, for one). See Knight, *Atoms and Elements*, 100; Rocke, "Gay-Lussac and Dumas"; and Mauskopf, "The Atomic Structural Theories of Ampère and Gaudin." For a methodological approach, see Fisher, "Avogadro, the Chemists, and Historians of Chemistry"; and Brooke, "Avogadro's Hypothesis and Its Fate."

5 Mendeleev to Voskresenskii, 7 September 1860, published as "Khimicheskii kongress v Karlsrue" in *S.-Peterburgskie Vedomosti*, 2 November 1860, #238. The quoted passages are from the reprint in *MS*, XV, 165–174.

6 Mendeleev, "Khimicheskii kongress v Karlsrue," *MS*, XV, 172–174.

7 Pipes, *Karamzin's Memoir on Ancient and Modern Russia.* Given the tendency of thinkers to adhere to different parts of the historical tradition, it is hard to conceive of a nonbiographical approach to this subject until we develop a better picture of the conservative landscape. See Thaden, *Conservative Nationalism in Nineteenth-Century Russia*; Sinel, *The Classroom and the Chancellery*; Gerstein, *Nikolai Strakhov*; Lukashevich, *Ivan Aksakov*; and Dowler, *Dostoevsky, Grigor'ev, and Native Soil Conservatism.* For an attempt to generalize an "ideology" of conser-

vatism, see Pipes, "Russian Conservatism in the Second Half of the Nineteenth Century."

8 This belief lasted until the end of the regime. On its prevalence in the court of Nicholas II, see Verner, *The Crisis of Russian Autocracy*, Chapter 3.

9 Raeff, "The Russian Autocracy and Its Officials," 90. See also idem, "The Bureaucratic Phenomena of Imperial Russia." The inefficient debates in the post–1905 Duma seemed to confirm the conservatives' worries. See, for example, Gorlin, "Problems of Tax Reform in Imperial Russia."

10 Wirtschafter, *Structures of Society*. This was not a "class," but more of a "social stratum," a more appropriate term for reasons articulated in Freeze, "The *Soslovie* (Estate) Paradigm and Russian Social History."

11 Balzer, *Russia's Missing Middle Class*; idem, "The Problem of Professions in Imperial Russia"; Hutchinson, *Politics and Public Health in Revolutionary Russia*; Frieden, *Russian Physicians in an Era of Reform and Revolution*, esp. 107 and 160; Rieber, "The Rise of Engineers in Russia"; idem, "Interest-Group Politics in the Era of the Great Reforms," esp. 69–71; idem, "Bureaucratic Politics in Imperial Russia"; Confino, "On Intellectuals and Intellectual Traditions in Eighteenth- and Nineteenth-Century Russia," on 139–141; Zelnik, *Labor and Society in Tsarist Russia*, 83; Ruane, *Gender, Class, and the Professionalization of Russian City Teachers*; and Wortman, *The Development of a Russian Legal Consciousness*, 261.

12 D. Mendeleev, "Ob ekspertize v sudebnykh delakh," *Sudebnyi Vestnik*, 29 October 1870, reprinted in *MS*, XXV, 613–617, on 617.

13 Lincoln, *In the Vanguard of Reform*; idem, *Nikolai Miliutin*; Pintner, "The Russian Higher Civil Service on the Eve of the 'Great Reforms'"; idem, "The Social Characteristics of the Early Nineteenth-Century Russian Bureaucracy"; Wortman, *The Development of a Russian Legal Consciousness*; Suny, "Rehabilitating Tsarism"; Orlovsky, "Recent Studies on the Russian Bureaucracy"; Rowney and Pintner, "Officialdom and Bureaucratization"; and Pintner, "The Evolution of Civil Officialdom."

14 Lincoln, *The Great Reforms*. His position draws on micro-studies of particular bureaucrats' visions of a reformed Russia: idem, "Russia on the Eve of Reform"; and idem, "Reform and Reaction in Russia." Lincoln actually only discusses *six* Great Reforms. I include the university statute of 1863 both because it very nicely exemplifies Lincoln's general framework and because of its centrality for Mendeleev. For more on Alexander, see Rieber, "Alexander II"; idem, *The Politics of Autocracy*. Alexander's views are crucial to understanding his supposed turn toward Counter-Reform in the mid–1870s, which Rieber successfully interprets as a reevaluation of how best to achieve the cardinal goals of fiscal and military stability. For Alexander's "scenario of love," see Wortman, *Scenarios of Power. Volume 2*, 46 and 71.

15 For a thoughtful analysis of novelist Ivan Turgenev as a liberal trapped into endorsing a radical Left he abhorred—the "liberal dilemma"—see Berlin, *Russian Thinkers*, 261–303. The problem for conservatives was nicely posed thirty-five years ago in Schapiro, *Rationalism and Nationalism in Russian Nineteenth-Century Political Thought*. The branches of the bureaucracy closest to the naked power of the Tsar—the Ministry of the Internal Affairs and the various organs of secret police—were least able to accommodate the spirit of bureaucratic reformism. See Orlovsky, *The Limits of Reform*; Daly, *Autocracy under Siege*; and Lieven, "The Security Police, Civil Rights, and the Fate of the Russian Empire." For the traditional view of the Great Reforms as a "revolution from above," see Mosse, *Alexander II and the Modernization of Russia*. For other interpretations, see Eklof, "Introduction"; and Zakharova, "Autocracy and the Reforms of 1861–1874 in Russia." Soviet historiography tended to emphasize Emancipation to the exclusion of all other reforms. See Field, "The Reforms of the 1860s"; and Gleason, "The Great Reforms and the Historians Since Stalin."

16 The view of a "culture of servitude" suggested here is close to "hegemonic" in the sense proposed by Gramsci, *Selections from the Prison Notebooks*, 210; and Lears, "The Concept of

Cultural Hegemony." "Civil society" is an immensely controversial category for the Imperial period, as it has important implications for the 1917 Revolution. See Kassow, West, and Clowes, *Between Tsar and People*; Rieber, "The Sedimentary Society"; Wirtschafter, *Structures of Society*, 15; and Haimson, "The Problem of Social Identities in Early Twentieth Century Russia." On precursors to civil societies in the eighteenth century, see Smith, *Working the Rough Stone*, Chapter 2. The contemporary anxiety about social forms can also be observed in the proliferation of social-scientific theories. See Vucinich, *Social Thought in Tsarist Russia*.

17 On liberals: Wortman, "Property Rights, Populism, and Russian Political Culture"; Wagner, "The Trojan Mare." On radicals: Naimark, *Terrorists and Social Democrats*; Wortman, *The Crisis of Russian Populism*. On reactionaries: Rosenthal, "The Search for a Russian Orthodox Work Ethic"; Field, *Rebels in the Name of the Tsar*; Rieber, *Merchants and Entrepreneurs in Imperial Russia*, 139; Hamburg, *Politics of the Russian Nobility*.

18 Vucinich, *Science in Russian Culture*, 136. For a more institutional approach, see Solov'ev, *Istoriia khimii v Rossii*, especially Chapter 11. Chemistry's position as exemplifying public knowledge dated at least to the seventeenth century. Levere, "The Rich Economy of Nature"; Golinski, *Science as Public Culture*; Donovan, *Philosophical Chemistry in the Scottish Enlightenment*; Anderson, *Between the Library and the Laboratory*; Riskin, "Rival Idioms for a Revolutionized Science and a Republican Citizenry"; Beretta, *The Enlightenment of Matter*; and Smith, *The Business of Alchemy*.

19 Lincoln, "The Daily Life of St. Petersburg Officials in the Mid Nineteenth Century." There are good studies of the artistic culture and economic development of Petersburg. These studies pay no attention to the role of chemistry in either culture or in the late Imperial economic boom: Volkov, *St. Petersburg*; Bater, *St. Petersburg*; and Brower, *The Russian City between Tradition and Modernity*.

20 Although "laws"—from the periodic law to the legal mechanisms of the state—were of particular import to Mendeleev, he would vacillate over the course of his life about what exactly constituted the essential property of a law of nature. Whatever they were, though, he was certain that he, at the very least, had discovered one. In his definitions of laws of nature, he first emphasized regularity, or generality, or the potential for expressing phenomena in mathematical form, but over the course of the 1860s he settled on a specific property: predictability. Predictability, Mendeleev maintained, was one of the most impressive and persuasive aspects of modern science. (One could argue, following this view, that what makes modern meteorology scientific is our ability to predict the weather accurately in advance. In turn, what makes meteorology somewhat less scientific than, say, astronomy, is that our powers of prediction function much better in the latter.) What made a law of nature "lawlike" for Mendeleev was the fact that one could use it to forecast the properties of nature. This is not a particularly controversial view. But for Mendeleev, the metaphor of the law of nature became central as a model for the organization of knowledge in economics, politics, and culture.

There is some debate among historians and philosophers of science as to whether prediction is any more persuasive to scientists than retrodiction (accounting for past data). For an exemplary study of this question, see Brush, "Prediction and Theory Evaluation in Physics and Astronomy." One of the striking transformations in early modern Europe was the emergence of prediction as a factor in decision-making. Before the sixteenth century, it was quite rare to predict the consequences of one's actions, whether in economics, agriculture, ballistics, or in other areas of applied knowledge. Emerging views of natural philosophy placed a premium on doctrines that enabled prediction, granting individuals a central responsibility over their own fates. See Jennings, "The Consequences of Prediction." Since this period, the inability to account adequately for either the rhetorical or epistemological power of prediction has provoked some of the more vigorous debates in the philosophy of science. For example, Hans Reichenbach split with the mainstream of logical positivism largely because of its inability to account for the temporal asymmetry given to the *future* in scientific prediction. González, "Reichenbach's Concept of Prediction."

By today's definition, laws of nature possess the following properties: They are independent of us and not arbitrary; one has no choice but to obey them; and they are described by experts—scientists—through whom nature speaks. These scientists discovered these laws to be laws through successful prediction. As the discoverer of a law of nature, therefore, Mendeleev presented himself as an individual who understood a form of nonarbitrary, fixed laws, and thus could provide a metaphysics for the newly created importance of law in the Russian Empire of the Great Reforms. Compare the series of studies on the importance of law in the development of early modern science: Shapiro, *A Culture of Fact*; and idem, *Probability and Certainty in Seventeenth-Century England*. See also Shapin and Schaffer, *Leviathan and the Air-Pump*; Sargent, *The Diffident Naturalist*, Chapter 2; and Cromartie, *Sir Matthew Hale*.

21 Sacks, *Uncle Tungsten*.

22 Levi, *The Periodic Table*.

23 Fortunately, unlike many twentieth-century scientists (Russian or otherwise), Mendeleev left a surprisingly complete archival record, permitting a local approach exceedingly difficult for Soviet historical events and persons. He preserved the vast majority of his correspondence and owned an extensive library, which he began to catalog into separate groupings himself. Mendeleev, organizer of the chemical elements, did a much poorer job with his own files, only a small portion of which did he succeed in cataloging. Yet the very fact that he attempted to organize his papers shows a consciousness of his own reputation as a public and historical figure. In another piece of good fortune, the very Soviet regime that obfuscated Mendeleev historiography did its part to maintain this collection. On 21 December 1911, a museum was opened in Mendeleev's former apartment on St. Petersburg University's campus, where it remains to this day. In 1952 the Council of Ministers of the Soviet Union ordered that copies of all Mendeleev's materials in other archives, often composed in his egregious handwriting, be deposited in this archive. Krotikov, "The Mendeleev Archives and Museum of Leningrad University"; Dobrotin and Karpilo, *Biblioteka D. I. Mendeleeva*, 7–8, 26–28; Volkova, "D. I. Mendeleev i kniga"; Ionidi, "O mendeleevskom nasledstve." Mendeleev's notoriously atrocious handwriting is still difficult to make out. On 19 January 1888, chemist Wilhelm Ostwald begged Mendeleev to have his correspondence printed over in a clean hand so they might be legible. ADIM 2/196, quoted in Mendeleev, *Nauchnyi arkhiv: Rastvory*, 126. Regrettably, Mendeleev did not do this for all his letters, and one must rely on the "key" provided by his secretary: Skvortsov, "O priemakh rasshifrovki rukopisei D. I. Mendeleeva."

The Soviet regime heavily controlled interpretations of cultural heroes like Mendeleev, as discussed in Chapter 9. Mendeleev, perhaps the best known Russian scientist at home and abroad, could not be described as the anti-Marxist economist and loyal subject to the Tsar that he clearly was. Historians resolved, in many works, either to whitewash those aspects or to pass over them in silence. The official Soviet interpretation was essentially embodied in the work of B. M. Kedrov. For an English-language summary of this position, see Kedrov, "Mendeleev, Dmitry Ivanovich." For other biographical studies that suffer an official ideological gloss, see Ionidi, *Mirovozzrenie D. I. Mendeleeva*; Zabrodskii, *Mirovozzrenie D. I. Mendeleeva*; Fritsman, "D. I. Mendeleev"; Dobrotin and Kerova, "Logicheskii analiz tvorcheskogo puti D. I. Mendeleeva"; Chugaev, *Dmitrii Ivanovich Mendeleev*; and Pisarzhevsky, *Dmitry Ivanovich Mendeleyev*. The best studies either treat the public figure without a detailed investigation of the chemical research, or commit the opposite sin of focusing too narrowly on a particular research program: Figurovskii, *Dmitrii Ivanovich Mendeleev*; Makarenia, *D. I. Mendeleev i fiziko-khimicheskie nauki*. One is unsure quite what to make of the bizarre fictionalization by Posin, *Mendeleyev*.

Published collections have also helped make Mendeleev's correspondence, manuscripts, and publications available. Regrettably, his collected works were bowdlerized by Stalinist editors whenever ideologically sensitive material presented itself, and I have often turned to original sources to uncover a conspicuously un-Soviet Mendeleev. The editors who began the compilation of the collected works clearly did not realize the size of the project. The original ten volumes were quickly expanded to the twenty-five described in the Bibliography. For the ten-volume breakdown, see Blokh, *Iubileinomu mendeleevskomu s"ezdu*, 4. Still, much of the most revealing material—chemical and political—remains unpublished. On some of the chemical omissions, see Kedrov, "Ser'eznye oshibki i upushcheniia."

24 This approach is similar to other quasi-biographical efforts in the history of science: Smith and Wise, *Energy and Empire*; Biagioli, *Galileo, Courtier*; Geison, *The Private Science of Louis Pasteur*; Ginzburg, *The Cheese and the Worms*; Desmond and Moore, *Darwin*; Rocke, *The Quiet Revolution*; and idem, *Nationalizing Science*. Particularly influential for me has been a biographical study from a rather different time and field: Brown, *Augustine of Hippo*.

CHAPTER 2

1 Shchukarev, "Zakony Prirody i zakony Obshchestva," 86.

2 Mendeleev, "Dnevnik 1861 g.," 131. Mendeleev only kept diaries from early 1861 until early 1862. For more on these, see Figurovskii, "Dnevniki D. I. Mendeleeva 1861 i 1862 gg."

3 Figurovskii, *Dmitrii Ivanovich Mendeleev*, 56; Brooks, "The Formation of a Community of Chemists in Russia," 402.

4 The literature on the periodic law overemphasizes this single episode in Mendeleev's life. As a result, almost all relevant material for an account of the periodic law has been reprinted with commentary. Along with *MS*, II, the best collections of original documents are: Mendeleev, *Periodicheskii zakon. Klassiki nauki*; idem, *Periodicheskii zakon. Dopolnitel'nye materialy*; and idem, *Nauchnyi arkhiv, t. I: Periodicheskii zakon*. For chronologies, see Kedrov, comp., "Kratkie svedeniia o zhizni i nauchnoi deiatel'nosti D. I. Mendeleeva i o ego rabote nad periodicheskim zakonom"; and Menshutkin, "Glavnye momenty v razvitii periodicheskoi sistemy elementov." For bibliographies of primary and secondary literature, see Semishin, *Periodicheskii zakon i periodicheskaia sistema khimicheskikh elementov D. I. Mendeleeva v rabotakh russkikh uchenykh*; and idem, *Literatura po periodicheskomu zakonu D. I. Mendeleeva*.

5 Throughout this book, I shall use either "periodic system of chemical elements" or "periodic law," with "system" referring to the visual representation in tabular form. English-speakers are accustomed to the phrasing "periodic table," which I shall avoid for the reasons articulated in the Preface. For a useful catalog of the various graphical forms of the periodic system, see Mazurs, *Graphic Representations of the Periodic System during One Hundred Years*.

6 *Kratkoe istoricheskoe obozrenie deistviia Glavnago pedagogicheskago instituta*, 2–5. On the establishment of the Institute, see Rozhdestvenskii, *Istoricheskii obzor deiatel'nosti Ministerstva narodnogo prosveshcheniia*, 194.

7 Late in his life, Mendeleev would advocate resurrecting this type of teaching establishment. Mendeleev to Witte, 15 October 1895, RGIA f. 560, op. 26, d. 29, ll. 21–44ob., on l. 25. See also Tishchenko, "Dmitrii Ivanovich Mendeleev," 13–14. For an amusing 1823 satirical poem on the Institute by Griboedov, see Shchukarev, "D. I. Mendeleev i Leningradskii gosudarstvennyi universitet," 149.

8 *MS*, I, 7–137. On isomorphism in general, see Melhado, "Mitscherlich's Discovery of Isomorphism"; and Schütt, *Eilhard Mitscherlich*. This concern for internal–external connections is also evident in the materials for his dissertation on specific volumes (1856), ADIM II-A–17–3–3; and more generally in *MS*, I, 139–311. On the importance of Gerhardt as an intellectual resource for Mendeleev, see Gordin, "The Organic Roots of Mendeleev's Periodic Law."

9 Dobrotin, "Rannii period nauchnoi deiatel'nosti D. I. Mendeleeva kak etap na puti k

otkrytiiu periodicheskogo zakona"; and Shchukarev and Dobrotin, "Pervye nauchnye raboty D. I. Mendeleeva kak etap na puti k otkrytiiu periodicheskogo zakona." Recently, Japanese historian of science Masanori Kaji has revived a less technical variant of the Dobrotin thesis: "Mendeleev's Discovery of the Periodic Law of the Chemical Elements."

10 Quotation from A. P. Borodin to Avdot'ia Konstantinovna Kleineka, 5 November 1859, in Borodin, *Pis'ma*, 34. Details of Mendeleev's stay in Baden are drawn from Mendeleeva, "Novye materialy o zhizni i tvorchestve D. I. Mendeleeva v nachale 60-kh godov," 90–92.

11 Sechenov, *Avtobiograficheskie zapiski*, 96–97. Many other émigré travelers were extremely impressed with Mendeleev. See, for example, Passek, "Vospominaniia"; Romanovich-Slavatinskii, "Moia zhizn' i akademicheskaia deiatel'nost'"; and Iunge, *Vospominaniia*, 285–288.

12 See the correspondence of both with Mendeleev: Volkova, "Perepiska I. M. Sechenova s D. I. Mendeleevym"; and idem, "Pis'ma A. P. Borodina k D. I. Mendeleevu." For more on Borodin's chemistry, see the biography by Figurovskii and Solov'ev, *Aleksandr Porfir'evich Borodin*.

13 Mendeleev's textbook came out in two editions before being entirely eclipsed in 1864 by A. M. Butlerov's structure-theory textbook, *Introduction to a Complete Study of Organic Chemistry*. Butlerov's text is still readable as a clear exposition of the fundamentals of organic theory. The text only came out in two editions during his lifetime: the Russian original in Kazan (1864–1866), and a German translation in Leipzig (1867–1868). The original edition has been reprinted as Butlerov, *Sochineniia*, II. On this text, see Bykov, "Materialy k istorii trekh pervykh izdanii 'Vvedeniia k polnomu izucheniiu organicheskoi khimii' A. M. Butlerova."

14 "Razbor sochineniia D. I. Mendeleeva 'Organicheskaia khimiia,' sostavlennyi akademikami Iu. F. Fritsshe i N. N. Zininym," 25 May 1862, reprinted in Kniazev, "D. I. Mendeleev i tsarskaia Akademiia nauk," 302.

15 Quoted in Menshutkin, *Zhizn' i deiatel'nost' Nikolaia Aleksandrovicha Menshutkina*, 7. K. A. Timiriazev similarly praised the text in 1939: "In the beginning of the sixties [Mendeleev] was primarily an organic chemist; his excellently clear and simply expounded textbook, *Organic Chemistry*, was peerless in European literature, and who knows precisely how much this book facilitated the next generation of Russian chemists [in] moving forward chiefly in precisely this direction." Reprinted in Makarenia, Filimonova, and Karpilo, *D. I. Mendeleev v vospominaniiakh sovremennikov*, 24. See also Mendeleev's later reflections on the importance of this text: "Spisok moikh sochinenii (1899)," in Shchukarev and Valk, *Arkhiv D. I. Mendeleeva*, 50. For more on *Organic Chemistry*, see Gordin, "The Organic Roots of Mendeleev's Periodic Law."

16 For a survey of Mendeleev's pedagogical appointments until his tenure at St. Petersburg University, see the useful article by Krotikov and Filimonova, "Ocherk pedagogicheskoi deiatel'nosti D. I. Mendeleeva v Peterburgskom universitete (1856–1867 gg.)."

17 Voskresenskii was widely know as the "grandfather of Russian chemistry." All biographical accounts of Voskresenskii highlight his role as Mendeleev's mentor, such as: Figurovskii and Elagina, "Aleksandr Abramovich Voskresenskii"; Pletner, *Dedushka russkoi khimii*; and Makarenia, "A. A. Voskresenskii i ego nauchnaia shkola." Mendeleev wrote an obituary for Voskresenskii in the 23 January 1880 edition of the Petersburg daily *Golos* and an 1892 encyclopedia entry on his life: *MS*, XV, 335 and 625, respectively.

18 Alston, "The Dynamics of Educational Expansion in Russia." On the 1835 statute, see Whittaker, *The Origins of Modern Russian Education*. For a schematic account of the eighteenth-century Russian university, see McClelland, "Diversification in Russian-Soviet Education," 180–183.

19 Pushkin, "*Raznochintsy* in the University," 36; and Alston, "The Dynamics of Educational Expansion in Russia."

20 Mathes, "The Origins of Confrontation Politics in Russian Universities," esp. 31–39. The most famous contemporary denunciation of the universities, directly referring to St. Peters-

burg University, is Pisarev, "Nasha universitetskaia nauka." See also the diary account of well-known censor, academician, and professor A. V. Nikitenko, dated 24 September 1861, quoted in Gessen, "Peterburgskii universitet osen'iu 1861 g.," 11.

21 The professoriate repeatedly failed to maintain a workable balance. This can be clearly seen in the 1899 Petersburg student disturbances, commonly pointed to as the dress rehearsal for the 1905 Revolution. See Kassow, *Students, Professors, and the State in Tsarist Russia*, esp. 5–6.

22 Mendeleev, "Dnevnik 1861 g.," 171.

23 Mendeleev, "Dnevnik 1861 g.," 178.

24 Sinel, *The Classroom and the Chancellery*, 30–31; and Kassow, "The University Statute of 1863," 249. When, in the mid–1870s, the Ministry of Popular Enlightenment tried to roll back some of the privileges granted by the law, chemist A. M. Butlerov responded: "At the present time I can express only the conviction that the university statute of 1863 guarantees the normal life and development of our universities and thus should not be subjected at the present time to changes which in any case are capable of destabilizing, to the harm of the enterprise, the surety and stability of order of things that exists in a given time." Butlerov response to Ministry survey, [1875], PFARAN f. 101, op. 1, d. 102, ll. 14–15, reproduced in Volkova, "Materialy k deiatel'nosti A. M. Butlerova v Peterburge," 8–9. Mendeleev's similar response, dated 12 December 1875, can be found in PFARAN f. 101, op. 1, d. 102, ll. 45–48.

25 Kassow, "The University Statute of 1863," 256; Mathes, "The Origins of Confrontation Politics in Russian Universities," 43; and Eimontova, *Russkie universiteti na grani dvukh epokh*, 322 and passim.

26 Tishchenko and Mladentsev, *Dmitrii Ivanovich Mendeleev, ego zhizn' i deiatel'nost'. Universitetskii period*, 75.

27 On Mendeleev's efforts to minimize his teaching requirements, see the letter from the director of the Technological Institute to the Department of Trade and Manufactures of the Ministry of Finances, 25 July 1866, RGIA f. 733, op. 158, d. 45, ll. 45–46ob. Note that he only began to rock the boat after his future at St. Petersburg University seemed more secure. On Mendeleev at the Institute, see Averbukh and Makarenia, *Mendeleev v Tekhnologicheskom Institute*.

28 The emphasis on the formulation of the periodic system in the process of textbook writing owes a great deal to the pioneering work of Soviet historian and philosopher of science B. M. Kedrov, especially: "Etapy razrabotki D. I. Mendeleevym periodicheskogo zakona"; "K istorii otkrytiia periodicheskogo zakona D. I. Mendeleevym"; "Razvitie D. I. Mendeleevym estestvennoi ('korotkoi') sistemy elementov"; "Nauchnyi metod D. I. Mendeleeva"; *Den' odnogo velikogo otkrytiia*; *Filosofskii analiz pervykh trudov D. I. Mendeleeva o periodicheskom zakone*; (with D. N. Trifonov), *Zakon periodichnosti i khimicheskie elementy*; *Mikroanatomiia velikogo otkrytiia*; and "Nad mendeleevskimi rukopisiami." For a complete bibliography, see "Bibliografiia osnovnykh nauchnykh trudov B. M. Kedrova." For English-language versions of this argument, see Kedrov, "Mendeleev, Dmitry Ivanovich"; and Graham, "Textbook Writing and Scientific Creativity." Although Kedrov's English article appeared substantially later than his Russian originals, the argument remained essentially unchanged.

Kedrov's basic goal was to perform a painstaking microhistorical investigation of Mendeleev's discovery of the periodic law, which he localized to the events of a single day, 17 February 1869. This date has a somewhat arbitrary quality, as Kedrov bases the investigation only on the four extant archival documents, and can be interpreted equally plausibly to discount the "one-day" theory of the discovery. See the compelling criticism in Trifonov, "Versiia-2"; and Dmitriev, "Nauchnoe otkrytie in statu nascendi."

However, in the process, Kedrov debunked a series of highly persistent myths about the origins of the periodic law, such as: The periodic system was found by experimental determination of atomic weights in sequence. (It was found during the course of writing a textbook.)

Mendeleev was sick and so could not present his paper on the periodic system to the March 1869 meeting of the Russian Chemical Society. (He was on a consulting trip for the Free Economic Society, investigating cheese-making cooperatives.) Mendeleev discovered the periodic table after waking from a dream. (It was clearly not.) For the false accounts, see, respectively, Morozov, *D. I. Mendeleev i znachenie ego periodicheskoi sistemy dlia khimii budushchago*, 41; Menshutkin, *Khimiia i puti ee razvitiia*, 229–230; and Mendeleev's friend A. A. Inostrantsev's report of Mendeleev's dream quoted at length in Lapshin, *Filosofiia izobreteniia i izobreteniia v filosofii*, 81; and also in Inostrantsev, *Vospominaniia*, 144.

29 Fisher, "Avogadro, the Chemists, and Historians of Chemistry," 222. On the history of the current system of atomic weights, see Kurinnoi, "Vozniknovenie sovremennoi sistemy atomnykh vesov." Mendeleev still felt a need to insist explicitly that atomic weights be determined in accordance with the Karlsruhe regime ten years later, when agreement was essentially universal. Mendeleev in protocol of the Russian Chemical Society, 12 October 1872, *MS*, II, 224–225.

30 On the dearth of adequate textbooks in Russia, see Gordin, "The Organic Roots of Mendeleev's Periodic Law"; and Parmenov, *Khimiia kak uchebnyi predmet v dorevoliutsionnoi i sovetskoi shkole.*

31 The classic accounts of the importance of textbooks in the formation of a basic set of concepts for the practicing scientist are Kuhn, *The Structure of Scientific Revolutions*; and Fleck, *The Genesis and Development of a Scientific Fact*. The literature on the historical function of chemical textbooks is growing rather rapidly. See, especially, Lundren and Bensaude-Vincent, *Communicating Chemistry*; Hannaway, *The Chemists and the Word*; Anderson, *Between the Library and the Laboratory*; Bensaude-Vincent, "A View of the Chemical Revolution through Contemporary Textbooks"; and Brock, *H. E. Armstrong and the Teaching of Science.*

32 On the publication history of the eight editions of *Principles* published by Mendeleev in his lifetime, see Kablukov, "Obzor izdanii 'Osnov khimii' D. I. Mendeleeva." Because of the language barrier, most histories of the periodic law in English are based on the translated (into French, English, and German) *fifth* edition of *Principles,* which was heavily revised after the discovery of Mendeleev's predicted elements. The first periodic system was formulated in the middle of writing the *first* edition, which has never been translated into a Western European language. Failure to take this into account has severely distorted Western interpretations. A useful but limited account based on Russian sources is Rawson, "The Process of Discovery."

33 *MS*, XIII, 60–61. Ellipses added.

34 The literature on the history of atomism is huge. Helpful studies of the revival of atomism in the nineteenth century include: Rocke, *Chemical Atomism in the Nineteenth Century*; Nye, "The Nineteenth Century Atomic Debates and the Dilemma of an 'Indifferent Hypothesis'"; idem, "Berthelot's Anti-Atomism"; Knight, *Atoms and Elements*; Alborn, "Negotiating Notation"; Brock, *The Atomic Debates*; Farrar, "Sir B. C. Brodie and His Calculus of Chemical Operation"; and Buchdahl, "Sources of Scepticism in Atomic Theory."

35 *MS*, I, 15.

36 From a published typescript (1864), ADIM II-A–17–9–5, reprinted in Mendeleev, *Izbrannye lektsii po khimii*, 25.

37 Quotation from the 7th volume of *Osnovy khimii* (1903), *MS*, II, 448.

38 The view of the periodic system as being merely reducible to these physical laws has been effectively disputed in Scerri, "The Periodic Table and the Electron."

39 On the crucial distinction between simple substances and elements, see Mendeleev, "Periodicheskaia zakonnost' khimicheskikh elementov (1871)," in Mendeleev, *Periodicheskii zakon. Klassiki nauki,* 102 (the first paragraph of his famous 1871 article). For further development of this distinction, see Bensaude-Vincent, "Mendeleev's Periodic System of Chemical Elements"; Makarenia, *D. I. Mendeleev i fiziko-khimicheskie nauki,* Chapter 4; and Dmitriev, "Nauchnoe otkrytie in statu nascendi."

40 Kedrov, *Den' odnogo velikogo otkrytiia*, 21.

41 See the transcripts of Mendeleev's 1864 and 1870/1 lectures in Mendeleev, *Izbrannye lektsii po khimii*.

42 Mendeleev, "Sootnoshenie svoistv s atomnym vesom elementov," *ZhRFKhO*, t. 1, no. 2–3 (1869): 60–77, reprinted in Mendeleev, *Periodicheskii Zakon. Klassiki nauki*, 16.

43 Kedrov, *Den' odnogo velikogo otkrytiia*, 24–26.

44 *MS*, XIV, 122.

45 For a criticism of B. M. Kedrov's "group" analysis, see Khomiakov, "K istorii otkrytiia periodicheskogo zakona D. I. Mendeleeva." I am skeptical of Kedrov's hypothesized method of construction, a game of "chemical solitaire." (Kedrov, *Den' odnogo velikogo otkrytiia*, Chapter 3.) Essentially, Kedrov claimed that Mendeleev wrote the symbols of all the elements on notecards with their atomic weights, and experimented with different elemental arrangements through simulations of solitaire. Kedrov even painstakingly "recreated" the way these various games "must" have taken place. There is shockingly little evidence for this conjecture, which has achieved such staying power in the popular consciousness that it was recently employed in a television special. There are only two sources of evidence for this account: Mendeleev's mention of notecards in the seventh edition of *Principles* (1903), *MS*, II, 453n, and the hearsay reminiscences of Mendeleev's youngest son Ivan, recorded twenty years after his father's death. Ivan's accounts generally represent bedtime fantasies Mendeleev embellished to interest his son. See the excerpts in Pogodin, "Otkrytie periodicheskogo zakona D. I. Mendeleevym i ego bor'ba za pervenstvo russkoi nauki." The best argument for the existence of the cards is that Mendeleev would have needed many rough drafts, and nothing near the requisite quantity exists. While this is true, the disturbing absence of the cards in Mendeleev's archive, or even any trace or recollection by a contemporary that they had existed, speaks just as strongly against the "chemical patience" account.

46 Mendeleev, "Sootnoshenie svoistv s atomnym vesom elementov (1869)," in Mendeleev, *Periodicheskii zakon. Klassiki nauki*, 18. I agree with Igor Dmitriev about the need to take Mendeleev's words at face value here. Dmitriev's reconstruction, based on the notion of "transitional elements," is articulated in Dmitriev, "Nauchnoe otkrytie in statu nascendi," 35.

47 Mendeleev, "Sootnoshenie svoistv s atomnym vesom elementov (1869)," in Mendeleev, *Periodicheskii zakon. Klassiki nauki*, 18.

48 For a detailed discussion of these elements, and the roots of the concept of "typicality" in Mendeleev's organic chemistry, see Gordin, "The Organic Roots of Mendeleev's Periodic Law."

49 Dmitriev, "Nauchnoe otkrytie in statu nascendi," 61–63. For an interpretation of the development of Mendeleev's law that differs from both Dmitriev's and mine, see Brooks, "Developing the Periodic Law." By one count, the original version of the periodic system made twenty-two mistakes, many of which Mendeleev had repaired by his 1871 German article, making it difficult to argue for a complete periodic law in early 1869. See Zamecki, "Mendeleev's First Periodic Table in Its Methodological Aspect," 122.

50 "Essai d'une ~~classification~~ système des éléments d'après leurs poids atomiques et fonctions chimiques." Mendeleev, *Nauchnyi arkhiv, t. 1*, 19 and 30.

51 Mendeleev, "Periodicheskaia zakonnost' khimicheskikh elementov (1871)," in Mendeleev, *Periodicheskii zakon. Klassiki nauki*, 131. There are similar statements in almost every subsequent publication on the periodic law.

52 See Chapter 7.

53 Mendeleev remained (in a perhaps conscious analogy to Newton) agnostic about the cause of periodicity: "I will not touch at all neither here, nor later, hypothetical notions which may be able to explain the essence of the law of periodicity, first, because the law is simple in and of itself; second, because the very subject is still too new, too poorly known in detail, (in order to be able to develop some kind of hypothesis); third, and this is the main point, because, in

my opinion, it is impossible, without the distortion of facts now known about the magnitude of atomic weights, to bring the law of periodicity into agreement with the doctrine of the atomic composition of bodies." Mendeleev, "Periodicheskaia zakonnost' khimicheskikh elementov (1871)," in Mendeleev, *Periodicheskii zakon. Klassiki nauki*, 124.

54 Mendeleev, "Sootnoshenie svoistv s atomnym vesom elementov (1869)," in Mendeleev, *Periodicheskii zakon. Klassiki nauki*, 10–14, quotation on 14.

55 Mendeleev, "Sootnoshenie svoistv s atomnym vesom elementov (1869)," in Mendeleev, *Periodicheskii zakon. Klassiki nauki*, 21–22. Emphasis in original. The two words *opyt* and *popytka* give the sense of a first try, a general outline. The fact that this is *not at all* a statement of the later periodic law has also been observed by Zamecki, "Mendeleev's First Periodic Table in Its Methodological Aspect," 124.

56 Mendeleev, "Sootnoshenie svoistv s atomnym vesom elementov (1869)," in Mendeleev, *Periodicheskii zakon. Klassiki nauki*, 14.

57 Mendeleev, "Ob atomnom ob"eme prostykh tel (1869)," in Mendeleev, *Periodicheskii zakon. Klassiki nauki*, 44 and 48.

58 Mendeleev, "O kolichestve kisloroda v solianykh okislakh (1869)," in Mendeleev, *Periodicheskii zakon. Klassiki nauki*, 54. Although we would now call this progression a phenomenon of valency (atomicity, in the parlance of the time), Mendeleev was hesitant to adopt this reading: "The doctrine of atomicity of the elements is a natural, although only formal, development of the type [theory] presentation and, like the latter, is appropriate primarily for the generalization only of substitution reactions" (p. 55).

59 "In a word, it seems to me that the quantity of elements, combined in a particle, is attributable to a few, still unformulated, but obviously simple and general laws which do not have anything in common with the understanding of the atomicity [i.e., valency] of elements." Mendeleev, "O kolichestve kisloroda v solianykh okislakh (1869)," in Mendeleev, *Periodicheskii zakon. Klassiki nauki*, 57. While the notion of law he employed in this article is stricter than before, he still did not have a sense of how to differentiate what would properly be called "laws" in his later parlance with what were mere "regularities." On the importance of the "oxygen limit" in Mendeleev's thinking, see Dmitriev, "Nauchnoe otkrytie in statu nascendi," 76–77.

60 Mendeleev, "Estestvennaia sistema elementov i primenenie ee k ukazaniiu svoistv neotkrytykh elementov (1870)," in Mendeleev, *Periodicheskii zakon. Klassiki nauki*, 74.

61 Mendeleev, "Estestvennaia sistema elementov i primenenie ee k ukazaniiu svoistv neotkrytykh elementov (1870)," in Mendeleev, *Periodicheskii zakon. Klassiki nauki*, 75. Emphasis in original.

62 Mendeleev, "Estestvennaia sistema elementov i primenenie ee k ukazaniiu svoistv neotkrytykh elementov (1870)," in Mendeleev, *Periodicheskii zakon. Klassiki nauki*, 89–90. In a draft of this article, dated 29 November 1870, one can see the process by which these thoughts congealed. In the following quotation, the words with lines through them were excised by Mendeleev and those in square brackets were added: "The application of the ~~law~~ [foundation] of periodicity for the seeking out of [undiscovered] elements and for the determination of their properties, in my opinion, comprises the sharpest form for the judgment of the practical applications for the scientific working out of chemical data [, of those results which are founded on a natural system of elements and on the totality of observations,] which we have about [already known] elements." Mendeleev, *Nauchnyi arkhiv, t. 1*, 181.

63 Mendelejew, "Die periodische Gesetzmässigkeit der chemischen Elemente." I quote from the Russian original from which Wreden made his German translation, as there are some discrepancies between the two.

64 Mendeleev, "Periodicheskaia zakonnost' khimicheskikh elementov (1871)," in Mendeleev, *Periodicheskii zakon. Klassiki nauki*, 107. Emphasis in original.

65 Mendeleev, "Periodicheskaia zakonnost' khimicheskikh elementov (1871)," in

Mendeleev, *Periodicheskii zakon. Klassiki nauki*, 123–125. Similarly, he concluded: "In the preceding I do not strive to establish a completed system and know that what I expressed demands corrections and supplements, but I propose that a reliable path of comparative study, such as I tried to hold to, will sooner bring us to that end which chemists have been trying to reach. Bold hypotheses often please our minds very much, often temporarily lead to success, but yet more often bring us to false results and they die of their own accord, especially if they are not based on laws, the searching out of which comprises the nearest task of scientific movement. In the preceding I tried to base myself on the laws of substitution, limits, and periodicity, and I think that these laws should lie at the basis of any generalization which touches the formation of various forms of compounds" (p. 174).

66 See Figurovskii, "Triumf periodicheskogo zakona D. I. Mendeleeva"; and Brush, "The Reception of Mendeleev's Periodic Law in America and Britain." Recently, Eric Scerri and John Worrall have persuasively argued—from historical and philosophical premises—that prediction was *not* especially important for the justification of the periodic law in the eyes of contemporary chemists. Scerri and Worrall, "Prediction and the Periodic Table." This does not change the fact that *Mendeleev* considered prediction to be crucial.

67 Mendeleev, *Nauchnyi arkhiv, t. 1*, 623.

68 Mendeleev, "Sootnoshenie svoistv s atomnym vesom elementov (1869)," in Mendeleev, *Periodicheskii zakon. Klassiki nauki*, 31. Emphasis in original.

69 Kedrov commentary in Mendeleev, *Nauchnyi arkhiv, t. 1*, 54.

70 Mendeleev, "Ob atomnom ob"eme prostykh tel (1869)," in Mendeleev, *Periodicheskii zakon. Klassiki nauki*, 42. Mendeleev suggested that perhaps indium could occupy the space under aluminum. On the problems of indium, see Dmitriev, "Problema razmeshcheniia indiia v periodicheskoi sisteme elementov."

71 Dmitriev, "Problema razmeshcheniia indiia v periodicheskoi sisteme elementov"; Trifonov, *Redkozemel'nye elementy*; idem, *Redkozemel'nye elementy i ikh mesto v periodicheskoi sisteme*; idem, *Problema redkikh zemel'*.

72 Mendeleev actually fluctuated between 44 and 45, but seemed more convinced of the value of 44. See the editor's comments in Mendeleev, *Periodicheskii zakon. Klassiki nauki*, 696.

73 Mendeleev, "Estestvennaia sistema elementov i primenenie ee k ukazaniiu svoistv neotkrytykh elementov (1870)," in Mendeleev, *Periodicheskii zakon. Klassiki nauki*, 90–91. Mendeleev's 1871 predictions are slightly more detailed but the same in essence: Mendeleev, "Periodicheskaia zakonnost' khimicheskikh elementov (1871)," in ibid., 150–152.

74 Mendeleev, "Estestvennaia sistema elementov i primenenie ee k ukazaniiu svoistv neotkrytykh elementov (1870)," in Mendeleev, *Periodicheskii zakon. Klassiki nauki*, 92–95.

75 Mendeleev, "Estestvennaia sistema elementov i primenenie ee k ukazaniiu svoistv neotkrytykh elementov (1870)," in Mendeleev, *Periodicheskii zakon. Klassiki nauki*, 95–98, quotation on 95.

76 Dobrotin, "K istorii otkrytiia germaniia (ekasilitsiia)"; Kedrov, "Die Vorhersage des Ekasiliziums und die Entdeckung des Germaniums"; and Dobrotin and Makarenia, "Prognozirovanie svoistv skandiia i germaniia v rabotakh D. I. Mendeleeva."

77 S. F. Savchenikov, "Otnosheniia mezhdu atomnymi vesami elementov," *Gornyi zhurnal*, 1871, reproduced in Mendeleev, *Nauchnyi arkhiv, t. 1*, 759–760.

78 Mendeleev, "Periodicheskaia zakonnost' khimicheskikh elementov (1871)," in Mendeleev, *Periodicheskii zakon. Klassiki nauki*, 153–156.

79 Mendeleev, "Estestvennaia sistema elementov i primenenie ee k ukazaniiu svoistv neotkrytykh elementov (1870)," in Mendeleev, *Periodicheskii zakon. Klassiki nauki*, 98. The confirmation of Mendeleev's atomic weight corrections, as in the case of yttrium, pleased him immensely, even if these did not have the impact his eka-elements would later. Mendeleev, "O

vese ittriia (December 1871)," in Mendeleev, *Periodicheskii zakon. Klassiki nauki,* 183.

80 This leaves aside what Mendeleev privately believed. It seems certain that *he* was convinced by the very fact he could make the predictions, although he certainly did not expect the rest of the chemical world to follow immediately. As he wrote in 1873, with reference to the rare earths: "Were the law not general and if it did not offer the key to the settling of the question of the elements, then I would have pushed at the obstacles and would have had to leave alone the exceptions (*corps à sérier*), which should not be allowed under a strict natural law. This sort of thing has not taken place. On the contrary, the law has found application with all elements, which produces a convincing proof of its correctness." Mendelejeff [Mendeleev], "Ueber die Anwendbarkeit des periodischen Gesetzes bei den Ceritmetallen," 45.

81 Mendeleev, "Estestvennaia sistema elementov i primenenie ee k ukazaniiu svoistv neotkrytykh elementov (1870)," in Mendeleev, *Periodicheskii zakon. Klassiki nauki,* 101. Ellipses added.

82 Upon the discoveries of the first two eka-elements, Mendeleev's 1871 German article was translated into both French and English as Mendeleef [Mendeleev], "La Loi Périodique des Éléments Chimiques (1879)"; and Mendeleef [Mendeleev], "The Periodic Law of the Chemical Elements (1879–1880)." The French translation appeared with a new preface by Mendeleev that emphasized the lawlike nature of his predictions.

83 For a detailed chronology of this discovery, see Kedrov, "Podrobnaia kommentirovannaia khronologiia otkrytiia galliia."

84 Urbain, "L'œuvre de Lecoq de Boisbaudran"; and Gramont, "Lecoq de Boisbaudran."

85 Lecoq de Boisbaudran, "Caractères chimiques et spectroscopiques d'un nouveau métal[. . .]"; idem, "Sur quelques propriétés du gallium"; idem, "Sur le spectre du gallium"; idem, "Nouvelles recherches sur le gallium"; idem, "Nouveau procédé d'extraction du gallium"; and idem, "Réactions chimiques du gallium."

86 Dmitriev, "Teoreticheskie issledovaniia P. E. Lekoka de Buabodrana." This point is also made in Urbain, "L'œuvre de Lecoq de Boisbaudran." On the correct atomic weight of gallium, see Lecoq de Boisbaudran, "Sur l'équivalent du gallium." For the empiricist caricature, see Kedrov, "Otkrytie galliia." Kedrov's position is Leninist in form: Lecoq de Boisbaudran was a spontaneous figure, while Mendeleev, as a vanguard, had true consciousness. This structure buttresses Kedrov's dialectical-materialist interpretation of Mendeleev.

87 As reported by A. Kuhlberg in *Berichte der Deutschen Chemischen Gesellschaft* 8 (1875–1876): 1680–1681.

88 Mendeleeff [Mendeleev], "Remarques à propos de la découverte du gallium (1875)," 970–971.

89 Lecoq de Boisbaudran, "Sur quelques propriétés du gallium," 1105.

90 Lecoq de Boisbaudran, "Nouvelles recherches sur le gallium." For the density experiments, see idem, "Sur les propriétés physiques du gallium."

91 Dobrotin and Makarenia, "Prognozirovanie svoistv skandiia i germaniia v rabotakh D. I. Mendeleeva," 56–60.

92 Nilson, "Om Scandium, en ny jordmetall."

93 Quoted in Dobrotin and Makarenia, "Prognozirovanie svoistv skandiia i germaniia v rabotakh D. I. Mendeleeva," 57.

94 Dobrotin and Makarenia, "Prognozirovanie svoistv skandiia i germaniia v rabotakh D. I. Mendeleeva," 58.

95 "As has already been remarked above, it is of rather special interest that from my identification the derived atomic weight of scandium gives exactly the same number as Mendeleev has conferred on the atom of his predicted basic substance *eka-boron,* that without a doubt is identical with scandium. Since now both the atomic weight and the properties of the element *eka-*

aluminum, also incidentally predicted by Mendeleev, coincide with those of the gallium discovered by Lecoq de Boisbaudran, so the speculations of the Russian chemist—which not only predicted the existence of the named substances but was also able to give the essential properties of them in advance—are thus confirmed in the most evident manner." Nilson, "Ueber das Atomgewicht und einige charakteristische Verbindungen des Scandiums," 1449–1450.

96 Walden, "Dmitri Iwanowitsch Mendelejeff," 4751; and Freund, *The Study of Chemical Composition,* 479–480.

97 Clève, "Sur le scandium," 421–422. This is a revision of Cleve, "Om Skandium."

98 Winkler, "Germanium, Ge, ein neues, nichtmetallisches Element." Some years later, Winkler defended his priority against T. Richter in "Zur Entdeckung des Germaniums."

99 Quoted in Lissner, "Sviazi D. I. Mendeleeva s gornoi akademiei vo Freiberge," 51.

100 Winkler, "Ueber die Entdeckung neuer Elemente im Verlaufe der letzten fünfundzwanzig Jahre," 15. In a draft letter citing Winkler for a Russian order of merit, Mendeleev suggested that the name symbolized the union of Russian and German science—a German element predicted by a Russian scientist. See ADIM I–A–59-1-26, quoted in Dobrotin, "K istorii otkrytiia germaniia (ekasilitsiia)," 58–59.

101 In the middle of 1871, Mendeleev wrote to his Heidelberg friend, chemist Emil Erlenmeyer: "My plan is long-term, but I am not afraid of length. I want first to go through every point with the rare elements, during which I will try to find affirmation of those changes in atomic weights which I proposed first for Ce, La, Di, then for Yt, Er. Then I will turn to Ti, Zr, Nb and Ta to study them and when I am the master of these chemical rarities I will try, studying the appropriate instances, to find among them one of my predicted eka-elements." Reproduced in Mendeleev, *Nauchnyi arkhiv, t. 1,* 707. See also the account by Kedrov, "Predvidenie i poiski D. I. Mendeleevym ekasilitsiia." The one element in Mendeleev's list here that cannot be found on modern periodic tables is "Di" for didymium, an element that was later understood to be a mixture of the two elements praseodymium and neodymium.

102 See the reproductions in Mendeleev, *Nauchnyi arkhiv, t. 1,* 186–187. On the Moscow job, see Mendeleev to A. Iu. Davydov, 28 September 1871, quoted in Kedrov, "Predvidenie i poiski D. I. Mendeleevym ekasilitsiia," 17.

103 The historical record contradicts the general assumption that Mendeleev "spent the rest of his life boldly examining the consequences [of periodicity] and defending its validity." Scerri, "The Evolution of the Periodic System," 81. Aside from the discoveries of his predicted elements, Mendeleev did follow the progress his system made abroad. He was especially interested in the works of British scientists Thomas Carnelley and William Crookes: [W. Crookes], "The Chemistry of the Future"; Carnelley, "Suggestions as to the Cause of the Periodic Law and the Nature of the Chemical Elements"; idem, "The Influence of Atomic Weight"; and idem, "Mendelejeff's periodisches Gesetz und die magnetischen Eigenschaften der Elemente." Both of these authors stressed the "lawlike" nature of the periodic system. On his personal copy of his German 1871 article, Mendeleev specifically cited "The Chemistry of the Future" as an excellent analysis.

104 These are mainly the systems of Mendeleev, Lothar Meyer, William Odling, and John Newlands. This is not the place for a detailed discussion of this dispute. For an introduction to the principal issues, see Van Spronsen, *The Periodic System of Chemical Elements*; and Cassebaum and Kauffman, "The Periodic System of the Chemical Elements."

105 Emphasis in original. The letter was written on 24 December 1869, ADIM I–A–44–1-6, reproduced in Figurovskii, *Dmitrii Ivanovich Mendeleev,* 97–100.

106 Wyrouboff, "On the Periodic Classification of the Elements," 31. Ellipses added.

CHAPTER 3

1 Levi, *The Periodic Table*, 60.

2 Letter of 29 April 1872, reproduced in Krätz, *Beilstein-Erlenmeyer*, 27. Ellipses added.

3 Mendeleev's working notebook, 27 November 1870-[1878], ADIM I-Zh–35–1–1, ll. 1–2.

4 ADIM I-Zh–35–1–1, l. 68. There are, of course, earlier tracks of the idea. See, for example, P. A. Il'enkov to Mendeleev, 14 November 1871, on ideal barometers. ADIM I-V–23–1–62. Mendeleev's gas work is typically neglected in the historiography of Russian science. The few exceptions are Kerova, Krotikov, and Dobrotin, "Issledovaniia D. I. Mendeleeva v oblasti fiziki gazov"; Kapustin, "O trudakh D. I. Mendeleeva po voprosam ob izmenenii ob"ema gazov i zhidkostei"; Kollerov, "Raboty D. I. Mendeleeva po issledovaniiu svoistv i sostava gazov"; and Gorbatsevich, "Raboty D. I. Mendeleeva v oblasti fizicheskikh konstant."

5 Daniel Todes has insightfully analyzed Pavlov's large-scale laboratory in *Pavlov's Physiology Factory*. Interestingly, Pavlov seems to have made no reference to the obvious predecessor of Mendeleev's laboratory, even though Mendeleev's views on factory organization (see Chapter 6) were influential in the design of the dog lab.

6 See, for example, Galison and Hevly, *Big Science*, especially Galison's introduction, "The Many Faces of Big Science"; and Galison, *Image and Logic*.

7 Trifonov, *Problema redkikh zemel'*, 47. See also Kerova et al., "Issledovaniia D. I. Mendeleeva v oblasti fiziki gazov," 73. A less sophisticated version of this same claim is offered by Kedrov, "Otkrytie periodicheskogo zakona D. I. Mendeleevym," 63.

8 Mendeleev, "Udel'nye ob"emy," *MS*, I, 183n1.

9 Mendeleev's personal copy of *Osnovy khimii*, 1st ed., ADIM Bib. 1010/1, table on inside cover. B. M. Kedrov, one of the most prominent scholars of Mendeleev's periodic law, misread Mendeleev's admittedly atrocious handwriting as "Ether is the lightest of all. Clarify with experiment." Kedrov, "Otkrytie periodicheskogo zakona D. I. Mendeleevym," 64.

10 See the valuable collection edited by Cantor and Hodge, *Conceptions of Ether*. On ether theories from Aristotle to the mid-eighteenth century, see Cantor and Hodge, "Introduction." For more on Newton's ether and his chemistry, see Thackray, *Atoms and Powers*. For a survey of ether theories from ancient Greece to the dawn of quantum theory, see Whittaker, *A History of the Theories of Aether and Electricity*; and Schaffner, *Nineteenth-Century Aether Theories*. Both of these latter books focus, however, almost exclusively on connections of the ether to electromagnetism, ignoring ether theories (like Mendeleev's) outside this purview.

11 Buchwald, "The Quantitative Ether in the First Half of the Nineteenth Century"; Cantor and Hodge, "Introduction," 49–50; and Whittaker, "The Aether."

12 Caneva, "Ampère, the Etherians, and the Oersted Connexion." See also Heimann, "Ether and Imponderables." Daniel Siegel has correctly cautioned that while the ether was used for unification, it should not be assumed that this "universalizing" function was understood the same way by all who claimed it. See his "Thomson, Maxwell, and the Universal Ether in Victorian Physics." The metaphor of political unification was especially important in the recently unified German Empire, where, as Norton Wise has pointed out, in physics, "[a] mechanical ether provided the only legitimate basis for unity." M. Norton Wise, "German Concepts of Force, Energy, and the Electromagnetic Ether," 275. The ether served a similar function in eighteenth-century Scotland, after the country had lost its political unity to London. See Christie, "Ether and the Science of Chemistry." On the persistence of ether modeling into the 1910s in England, see Goldberg, "In Defense of Ether"; and Stein, "'Subtler Forms of Matter' in the Pe-

riod Following Maxwell."

13 ADIM II-A–17–1–6; ADIM II-A–17–1–7. For more on planetary atmospheres and the ether, see also Chapter 8.

14 Mendeleev, "Udel'nye ob"emy (1856)," *MS*, I, 145, 152. We will see this atomic theory of the ether again in Chapter 8.

15 *MS*, I, 154. Mendeleev saw as one of the chief failures of Berzelius's electrochemical theory its inability to deal adequately with the devolution of heat and light under chemical action. Gerhardt provided a solution. See Mendeleev's master's summary: Mendeleev, *Polozheniia, izbrannyia dlia zashchishcheniia na stepen' magistra khimii*, 4. On Mendeleev's attachment to Gerhardt's chemistry, see Gordin, "The Organic Roots of Mendeleev's Periodic Law."

16 See, for example, the 1861 reference to Regnault's tables and formulas for gas pressure in *MS*, V, 60n1. On the border disputes between physics and chemistry in the nineteenth century, see Nye, *From Chemical Philosophy to Theoretical Chemistry*. Mendeleev's negotiation of the boundary between these two fields has generated some odd attempts at classifying Mendeleev as some sort of hybrid of the two, as in Veinberg, "Khimik ili fizik Mendeleev?"

17 Mendeleev treated density measures as tests for reaction products; that is, he was interested in gases as tools of research and not as research subjects themselves. *Analiticheskaia khimiia* (St. Petersburg: Obshchestvennaia Pol'za, 1866), reproduced in *MS*, VI, 7–8.

18 Mendeleev, "Zamechaniia po povodu raboty Endr'iusa nad szhimaemost'iu uglekisloty (1870)," *MS*, V, 110–111, quotation on 111.

19 Mendeleev, *Ob uprugosti gazov*, v. 1 (1875), *MS*, VI, 227. Mendeleev sided with P. S. Laplace and S. D. Poisson, who thought the transition from the earth's atmosphere to the ether should be sharp rather than gradual, although Mendeleev's position derived from chemical, not mathematical, argumentation. Mendeleev also wrote about such topics for popular audiences. See his encyclopedia article on air from 1875, *MS*, VI, 590–616.

20 Levitskii, "O sushchestvovanii soprotivliaiushcheisia sredy v nebesnom prostranstve"; von Asten, "Über die Existenz eines widerstehenden Mittels im Weltenraume." On the tremendous theological significance attributed to Encke's comet, see Smith and Wise, *Energy and Empire*, 91–94. Encke did not discover the comet, but was the first person to calculate accurately its period. See Batten, "Johann Franz Encke"; and Sekanina, "Encke, the Comet."

21 Mendeleev, minutes of the Russian Physical Society, 7 December 1876, *ZhRFKhO* 9, no. 1, otd. 1, ch. fiz. (1877): 2–3, quotation on 2.

22 On gas theories in this period, see Fox, *The Caloric Theory of Gases*. Gay-Lussac's research was considered a cornerstone of Napoleonic science. See Crosland, *The Society of Arcueil*, 251.

23 Fox, *The Caloric Theory of Gases*, Chapter 8, quotations on 298.

24 Duhem, *The Aim and Structure of Physical Theory*, 146–174, passim.

25 Gay-Lussac's law is generally known today as Charles's law. This is because Gay-Lussac, who formulated the relation in January 1802 after very detailed experimental studies, was also scrupulous about tracing intellectual genealogies. Charles had done similar experimental work fifteen years earlier. See Crosland, *Gay-Lussac*, ix. Mariotte's law is the common name throughout Europe for what is known in the Anglo-American tradition as Boyle's law. It was standard practice in the 1870s to use both interchangeably or hyphenated.

26 On the German chemical community, see Hufbauer, *The Formation of the German Chemical Community*; Homburg, "Two Factions, One Profession"; and Johnson, "Academic Chemistry in Imperial Germany." On England, see Bud and Roberts, *Science versus Practice*; and Roberts, "'A Plea for Pure Science.'" For France, see Crosland, "The Organisation of Chemistry in Nineteenth-Century France"; and Fell, "The Chemistry Profession in France." For an example of the hardships of organizing a chemical community in the provinces, see Cannizzaro's failed attempt in Italy, recounted in Cerruti and Torracca, "Development of Chemistry in Italy."

27 Kozlov and Lazarev, "Tri chetverti veka Russkogo khimicheskogo obshchestva," 124; Fig-

urovskii, "Khimiia v dopetrovskoi Rusi." On the Petrine transformation of natural philosophy in Russia, see Gordin, "The Importation of Being Earnest."

28 On these ventures, spearheaded by A. N. Engel'gardt and N. N. Sokolov, see Musabekov, "Pervyi russkii khimicheskii zhurnal i ego osnovateli"; and Brooks, "Russian Chemistry in the 1850s." Chemical laboratory space was especially scarce in St. Petersburg, as attested by noted chemist N. N. Beketov, "Istoriia khimicheskoi laboratorii pri Akademii nauk."

29 Bradley, "Voluntary Associations, Civic Culture, and *Obshchestvennost'* in Moscow."

30 Much of this account is from the standard source: Kozlov, *Vsesoiuznoe khimicheskoe obshchestvo imeni D. I. Mendeleeva.* For another interpretation, see Brooks, "The Formation of a Community of Chemists in Russia"; and idem, "The Evolution of Chemistry in Russia during the Eighteenth and Nineteenth Centuries." On the origin of the Society in informal Petersburg circles centered around Mendeleev, see Volkova, "Russkoe fiziko-khimicheskoe obshchestvo i Peterburgskii-Leningradskii universitet." On Russian culture in Heidelberg, see Birkenmaier, *Das russische Heidelberg.*

31 "Vnutrennie izvestiia" section of the 17 August 1861 issue of *Russkii invalid*, #17, p. 733, quoted in Kozlov, *Vsesoiuznoe khimicheskoe obshchestvo imeni D. I. Mendeleeva,* 13. Ellipses added. J. F. Fritzsche was irritated that he was excluded from the list of famous Petersburg chemists, and wrote a complaint to Mendeleev on 29 August 1861, reproduced in Tishchenko and Mladentsev, *Dmitrii Ivanovich Mendeleev, ego zhizn' i deiatel'nost'. Universitetskii period,* 195.

32 Mendeleev jotted down a list of those in favor of making the society official, including N. I. Lavrov, N. P. Fedorov, N. P. Nechaev, A. K. Krupskii, F. K. Beilstein, Mendeleev himself, J. F. Fritzsche, F. R. Wreden, N. A. Menshutkin, V. F. Petrushevskii, E. F. Radlov, G. A. Schmidt, V. V. Bek, P. P. Alekseev, F. N. Savchenkov, G. V. Struve, and L. N. Shishkov. Only four voted for keeping it informal: A. R. Shuliachenko, A. N. Engel'gardt, P. A. Lachinov, and G. G. Gustavson. Kozlov, *Vsesoiuznoe khimicheskoe obshchestvo imeni D. I. Mendeleeva,* 13–14. Engel'gardt in particular was loath to commit himself, as he partook of radical politics and was later exiled from Petersburg for his populist views. See Kozlov, "Nauchnaia i obshchestvennaia deiatel'nost' A. N. Engel'gardta"; and Wortman, *The Crisis of Russian Populism,* Chapter 5.

33 See the petition of 4 January 1868 quoted in Kozlov and Lazarev, "Tri chetverti veka Russkogo khimicheskogo obshchestva," 128.

34 Mendeleev's role behind the scenes was very prominent. He essentially wrote the first draft of the charter (ADIM II-A-5-5-14), a point unrecognized by Kozlov. See Krotikov, "K istorii organizatsii Russkogo khimicheskogo obshchestva," 83.

35 Owen, "The Russian Industrial Society and Tsarist Economic Policy"; Rieber, *Merchants and Entrepreneurs in Imperial Russia,* 200; idem, "Interest-Group Politics in the Era of the Great Reforms"; Zelnik, *Labor and Society in Tsarist Russia,* 284; and Hogan, *Forging Revolution,* 11.

36 Mendeleev claimed in *Ob uprugosti gazov* (1875) that Kochubei approached him in January 1872. Mendeleev wrote that Kochubei was eager for potential applications to steam engines, and thus gave hope that Mendeleev's research would be funded. *MS,* VI, 224. Some evidence from the notebooks, however, make this particular chronology suspect.

37 He emphasized that much of the practical benefit would come not from the actual results of the research, but from the process of laboratory investigation. *MS,* VI, 149n2, 487, and 490. For more on Mendeleev and gunpowder, see Gordin, "No Smoking Gun." The applications to gunpowder were not entirely cynical or opportunistic. In fact, in his first working notebook, Mendeleev began to look at the experimental implications of gunpowder explosions. ADIM I-Zh-35-1-1, l. 228. His first detailed publication on gases, "O szhimaemosti gazov (1872)," also indicated gunpowder as a potential application. *MS,* VI, 131n, 141.

38 The reason for the delay, he claimed, was the unforeseen need to redo most of Regnault's work, which he had originally planned on simply using for the realms where Regnault had taken data. Mendeleev, *Ob uprugosti gazov* (1875), Chapter 9, reproduced in *MS,* VI, esp. 583.

Mendeleev continued these arguments in a report to the Russian Technical Society, "Kratkii otchet o khode issledovanii nad uprugost'iu gazov, proizvodimykh D. I. Mendeleevym (1877)," *MS*, XXV, 321–324. Mendeleev was not entirely straightforward here. In his first publication on gases, "O szhimaemosti gazov (1872)," he structured the article on the a priori inadequacy of Regnault's data. *MS*, VI, 128–171. As recently as his 1869–1871 edition of *Principles of Chemistry*, Mendeleev considered Regnault's data beyond reproach. *MS*, XIII, 216n1.

39 Mendeleev to the Russian Technical Society, 28 March [1872], ADIM I-Zh-7–3–21, l. 1. On "control" in latter-day big science, see Galison, Hevly, and Lowen, "Controlling the Monster."

40 See Kochubei to Grand Prince Konstantin Nikolaevich, reproduced in V. E. Tishchenko and M. N. Mladentsev, *Dmitrii Ivanovich Mendeleev, ego zhizn' i deiatel'nost'. Universitetskii period*, 176–177.

41 Kochubei to Tolstoi, 11 April 1872, RGIA f. 733, op. 147, d. 1025, ll. 2–3ob. On the redesign of the laboratory, see *MS*, VI, 226.

42 See the letter from Mendeleev to his family, 10 May 1874, ADIM Alb. 1/53.

43 Petrushevskii was to conduct research on condensation of gases on solid surfaces and on the problems of measuring high temperatures; A. V. Gadolin was to work on artillery applications; V. L. Kirpichev was to examine the mechanical composition of glass for instruments; Gadolin and N. P. Petrov would also create various experimental mechanisms; and L. P. Semiachkin would work on naval technology. "Zhurnal pervogo zasedaniia komissii po issledovaniiu uprugosti parov i gazov pri raznykh davleniiakh," [21 March 1872], ADIM Alb. 2/622, 5.

44 "Zhurnal pervogo zasedaniia komissii po issledovaniiu uprugosti parov i gazov pri raznykh davleniiakh, predlozhennomu professorom D. I. Mendeleevym," [21 March 1872], ADIM Alb. 2/622, l. 2. The commission was chaired by academician and Inspector of Arsenals A. V. Gadolin, joined by V. L. Kirpichev, instructor at the Technological Institute in St. Petersburg, M. L. Kirpichev, his brother and an instructor at the Artillery Academy, L. P. Semiachkin, assistant to the chair of the fourth division of the Russian Technical Society, G. K. Brauer, Mendeleev's instrument maker and former mechanic of Pulkovo observatory, N. P. Petrov, professor at the Engineering Academy and the Technological Institute, and professors at St. Petersburg University F. F. Petrushevskii and D. I. Mendeleev.

45 Mendeleev did in fact report to the Technical Society, as recorded in their annual reports by Secretary L'vov: RGIA f. 90, op. 1, d. 1, l. 11 (1875), ll. 70–80ob. (1876), ll. 174–187ob. (1877). Mendeleev also submitted a summary on 23 April 1877 of his low-pressure results and predicted that he would soon be able to explore high pressures. RGIA f. 90, op. 1, d. 1, ll. 86–88ob.

46 Mendeleev, *Ob uprugosti gazov* (1875), *MS*, VI, 231. For reasons examined below, Mendeleev did few experiments after 1877, and his expenses were only 450 rubles above his 1875 total. "Spisok schetov predstavlennykh Imperatorskomu Russkomu Tekhnicheskomu Obshchestvu po raskhodam proizvedennym dlia proizvodstva opytov nad uprugost'iu gazov," 1872–1877 [January 1881], ADIM I-Zh-23–1–10. See also the 1871–1877 expense report in *MS*, XXV, 326–327.

47 Mendeleev, "Pul'siruiushchii nasos (1872)," *MS*, VI, 99–125, quotation on 106. Mendeleev's balance was highly praised abroad. See Salleron, "Sur la nouvelle balance de M. Mendeleef."

48 ADIM II-Zh–36–2–1. For a secondary, albeit cursory, discussion of these various instruments, see Dubravin, "O laboratornykh priborakh D. I. Mendeleeva dlia issledovaniia uprugosti gazov." After completing his gas research, Mendeleev returned the equipment to the Society with noticeable grace, as attested by the Society's thank-you note. P. A. Kochubei to Mendeleev, 18 December 1882, ADIM I-V–28–1–45; and L'vov to Mendeleev, undated, ADIM I-V–38–1–26.

49 On standardization in Russia, see Gordin, "Measure of All the Russias"; and Chapter 6 below. On the problems of metrology in general, see, for example, Kula, *Measures and Men*; O'Connell, "Metrology"; Schaffer, "Late Victorian Metrology and Its Instrumentation"; and the various essays in Wise, *The Values of Precision*.

50 Zapisnaia knizhka, 1874–1876, #20, entry of 21 June 1874, ADIM II-A–1–1–9, ll. 17–18, on the actual calibration experiments. The results of the calibration are presented in ADIM I-Zh–35–1–1, l. 294. The kilo calibration results appear in ADIM I-Zh–35–1–3, ll. 94–108.

51 Quoted from a personal letter by Iuzef Bogusskii to Paul Walden in the latter's "Dmitri Iwanowitsch Mendelejeff," 4764.

52 Iu. G. Bogusskii to Mendeleev, undated [Mendeleev noted that he answered it on 28 December 1881], ADIM I-V–3–2–44, ll. 2–3. Ellipses added.

53 This list is compiled from Kerova et al., "Issledovaniia D. I. Mendeleeva v oblasti fiziki gazov," 85. The erasure of assistants from the narratives of scientific investigation has been interestingly discussed in Shapin, *A Social History of Truth*, Chapter 8.

54 See Bogusskii's notebook, dated January 1876–May 1877, ADIM II-Zh–35–2–5, which is characterized by its careful, methodical system. Kaiander's notebooks were neatly written up and packed away in January 1874, and focused mostly on the preliminary calibration of Bogusskii's various high-pressure experiments. See his notebook, ADIM II-V–64–1–58. Gutkovskaia's 1875–1876 notebook is at ADIM II-Zh–35–2–7, Gemilian's at ADIM II-Zh–35–2–6.

55 See Mendeleev's notebooks, ADIM II-Zh–35–1–1, II-Zh–35–1–2, II-Zh–35–1–3, II-Zh–35–1–4, II-Zh–35–1–4a, II-Zh–35–1–5, II-Zh–35–1–5a. The exception was Gutkovskaia's notebook, which Mendeleev monitored for form and correctness of approach. ADIM II-Zh–53–2–7, dated 1875–1876. Perhaps she received special treatment because she was a woman, and Mendeleev doubted her ability in precision experimentation, although he gives no such hint here.

56 Mendeleev's citational practices bear this out. The gas articles mostly cite other work from the lab, and the finished products were often co-authored with the assistant responsible for that branch of work. See the rough draft for "O szhimaemosti gazov," ADIM II-Zh–36–1–8.

57 See M. L. Kirpichev's first three gas notebooks, ADIM II-Zh–35–2–1, II-Zh–35–2–2, and II-Zh–35–2–3. The latter two were dated 1876 by Mendeleev. This clearly refers to the date he managed to label them rather than when they were written, as Kirpichev had been dead over a year by this time. Much of the mathematical facility Oscar Sheynin attributes to Mendeleev was in fact due to Kirpichev or other amanuenses. Cf. Sheynin, "Mendeleev and the Mathematical Treatment of Observations in Natural Science."

58 Kirpichev's fourth notebook [1874], II-Zh–35–2–4.

59 The biographical information is from Mendeleev's eulogy to Kirpichev from the minutes of the Russian Chemical Society, 6 March 1875, *MS*, VI, 213–214. The quotation is from Mendeleev's only English article on gases: Mendeleeff [Mendeleev], "Researches on Mariotte's Law," 456. Throughout this article, Mendeleev makes ample references to his assistants.

60 For a summary of the procedures involved, see Kerova et al., "Issledovaniia D. I. Mendeleeva v oblasti fiziki gazov," 79–84.

61 *MS*, VI, 192.

62 Mendeleev's fourth working notebook, 1874–1877, ADIM II-Zh–35–1–4, ll. 20–21.

63 There is a counterpoint to these ideas in the construction of the first cloud chambers in Victorian England, which were intended to create not an analytic experiment (breaking a phenomenon of nature down into component parts), but a "mimetic experiment" (simulating the phenomena of meteorology in the laboratory). See Galison, *Image and Logic*, Chapter 2.

64 Mendeleev, "O temperature verkhnikh sloev atmosfery (May 1876)," *MS*, VII, 35.

65 Mendeleev repeatedly rejected mountain observations. For an example, see *MS*, VII, 36. Edward Frankland in England performed a successful series of meteorological observations from mountaintops. Russell, *Edward Frankland*, 429–433.

66 Mendeleev was given full discretion on setting the publication apparatus in motion. See his letters to Nikolai Avtonomovich, 14 May and 29 May 1874, ADIM II-Zh–36–2–4.

67 *MS*, VI, 223.

68 Mendeleev in the minutes of the Russian Physical Society, 7 October 1875, *ZhRFKhO* 1, no. 8, otd. 1 (1875): 260–265, on 264. V. Shatskii to Mendeleev, 31 May 1876, ADIM I-V–19–4–148, informed him that the book had sold eight copies already that month at 5.50 rubles apiece.

69 Kochubei to Mendeleev, 20 April 1881, ADIM Alb. 2/623.

70 Mendeleev, "Researches on Mariotte's Law (1877)," 498.

71 *MS*, VI, 172 and 174.

72 The pamphlet is reproduced in *MS*, VI, 181–186.

73 *MS*, VI, 186.

74 Golinski, "Barometers of Change."

75 Leicester, "Mendeleev's Visit to America." On the Philadelphia Exposition, see Rydell, *All the World's a Fair*; and the original catalog in McCabe, *The Illustrated History of the Centennial Exhibition*.

76 Mendeleev, *O barometricheskom nivelirovanii i o primenenii dlia nego vysotomera* (St. Petersburg: Tip. Dept. Udelov, 1876), reprinted in *MS*, VII, 53–193. On surveying and weather forecasting, see esp. 89n2, 59, and 73.

77 On meteorology in the "pre-scientific" stage, where the dominant theoretical influence remained book IV of Aristotle's *Meteorologia*, see Frisinger, *The History of Meteorology*. On the transformations within meteorology that allied it with the physical sciences, see Kutzbach, *The Thermal Theory of Cyclones*; Garber, "Thermodynamics and Meteorology"; Friedman, *Appropriating the Weather*; Khrgian, *Ocherki razvitiia meteorologii*; and Nebeker, *Calculating the Weather*. On the telegraph, see Fleming, *Meteorology in America*, Chapter 7; and Monmonier, "Telegraphy, Iconography, and the Weather Map," 15.

78 See Anderson, "The Weather Prophets"; and Tucker, "Voyages of Discovery on Oceans of Air." On Glaisher's newspaper reports, see Kutzbach, *The Thermal Theory of Cyclones*, 66.

79 Fleming, "Meteorological Observing Systems before 1870"; Fierro, *Histoire de la Météorologie*, 117–118; Davis, "Weather Forecasting and the Development of Meteorological Theory at the Paris Observatory"; Burton, "Robert FitzRoy and the Early History of the Meteorological Office"; Monmonier, "Telegraphy, Iconography, and the Weather Map," 15; and Khrgian, *Ocherki razvitiia meteorologii*, Chapter 8.

80 Kington, "The Development of Meteorology during the Enlightenment," 812; Cassidy, "Meteorology in Mannheim."

81 Meteorology in the provinces was interesting to Western Europe for several reasons, but it could serve cultural and political purposes at home as well. In Australia, for example, intercolonial cooperation on meteorology was one of the first zones where a new model for confederation was attempted before unification into the state of Australia in 1906. Home and Livingston, "Science and Technology in the Story of Australian Federation."

82 Fleming, "Meteorological Observing Systems before 1870," 251.

83 Khrgian, "Istoriia meteorologii v Rossii," 71–79; Pasetskii, *Adol'f Iakovlevich Kupfer*; Rykachev, *Istoricheskii ocherk Glavnoi fizicheskoi observatorii*; and Nezdiurov, *Ocherki razvitiia meteorologicheskikh nabliudenii v Rossii*, Chapters 1–3. On Wild in Switzerland, see Kutzbach, *The Thermal Theory of Cyclones*, 59. On the growth of the GFO under Wild, see Nezdiurov, *Ocherki razvitiia meteorologicheskikh nabliudenii v Rossii*, Chapter 3; and *Materialy dlia istorii akademicheskikh uchrezhdenii*, 319–559. Despite the dramatic growth in the quality and quan-

tity of observations, however, there was a lack of theoretical and experimental meteorology, the emphasis placed instead on Baconian collection. Typical of this approach was the work of the Permanent Secretary of the Academy of Sciences, which systematically gathered and synthesized two centuries of Russian observations: Veselovskii, *O klimate Rossii*.

84 Mendeleev, "O temperature verkhnikh sloev atmosfery (1876)," *MS*, VII, 36.

85 Mendeleev, "O temperaturakh atmosfernykh sloev (1876)," *MS*, VII, 252. On Gay-Lussac's ascents and meteorology, see Crosland, *Gay-Lussac*, 28–31; idem, *The Society of Arcueil*, 262–263; and Cawood, "Terrestrial Magnetism and the Development of International Collaboration." Glaisher's public ascents are discussed at length in Tucker, "Voyages of Discovery on Oceans of Air." For Mendeleev's criticisms of Glaisher, see Mendeleev, "O temperature verkhnikh sloev atmosfery (1876)," *MS*, VII, 37, 39. Mendeleev scoured the secondary literature on aerostats. See his notebook, ADIM II-Zh-35-1-5, l. 215.

86 Mendeleev, "O temperature verkhnikh sloev atmosfery (1876)," *MS*, VII, 48. Ellipses added.

87 Mendeleev, "O temperaturakh atmosfernykh sloev (1876)," *MS*, VII, 248. Ellipses added. Note that Mendeleev did not invoke the ether. For a contrast of "fictive" and realist attitudes toward the ether by British scientists, see Benson, "Facts and Fictions in Scientific Discourse."

88 Mohn, *Meteorologiia ili uchenie o pogode*, trs. N. Iordanskii and F. Kaputsin, ed. D. Mendeleev (1876).

89 This biographical information is taken from Kutzbach, *The Thermal Theory of Cyclones*, 240. Mendeleev actually met Mohn at the 400th anniversary of the University of Uppsala, as he related in September 1877. Mendeleev told him of the translation, and noted: "Professor Mohn was extraordinarily pleased that his meteorology was translated into Russian . . . [and] that here [in Russia] the matter of airflight for meteorological goals is undertaken." Mendeleev, "Po povodu 400-letiia upsal'skogo universiteta v Stokgol'me (1877)," *MS*, XV, 329. Ellipses added.

90 Mendeleev, preface in Mohn, *Meteorologiia* (1876), v.

91 Mendeleev, preface in Mohn, *Meteorologiia* (1876), viii.

92 Mendeleev, "Researches on Mariotte's Law (1877)," 499.

93 Minutes of Russian Chemical Society meeting, 2 March 1872, *MS*, VI, 126. Mendeleev persisted with this *theoretical* critique of Regnault's results even later in the decade. See Mendeleev, "Researches on Mariotte's Law (1877)," 455.

94 Navy Ministry to Prince A. M. Shirinskii-Shikhmazov, 9 July 1878, RGIA f. 733, op. 147, d. 1025, ll. 17–18. See also the correspondence in V. E. Tishchenko and M. N. Mladentsev, *Dmitrii Ivanovich Mendeleev, ego zhizn' i deiatel'nost'. Universitetskii period*, 228–229.

95 Mendeleev, *O soprotivlenii zhidkostei i o vozdukhoplavanii* (1880). This volume is also reprinted in *MS*, VII, 291–461. See also Vorob'ev, "D. I. Mendeleev i vozdukhoplavanie"; and idem, *Genezis russkoi vozdukhoplavatel'noi mysli v trudakh D. I. Mendeleeva*.

96 Mendeleev, *O soprotivlenii zhidkostei i o vozdukhoplavaniia* (1880), 7.

97 Mendeleev, *O soprotivlenii zhidkostei i o vozdukhoplavaniia* (1880), 74–77, quotation on 77. Mendeleev repeated his call for data-gathering at a general session of the Russian Physico-Chemical Society on 27 December 1879, *MS*, VII, 283–287.

98 Mendeleev, "Researches on Mariotte's Law (1877)," 455–456 and 499.

99 Mendeleev and M. L. Kirpichev, "Predvaritel'naia zametka ob uprugosti razriazhennogo vozdukha (1874)," *MS*, VI, 194. In his 1875 public lectures on chemical solutions, however, Mendeleev invoked billiard-ball models for gas behavior. *MS*, IV, 237–238.

100 On these priority claims, see: Kalifati and Sychev, "K 100-letiiu universal'noi gazovoi postoiannoi D. I. Mendeleeva"; Gloushkin, "Uravnenie sostoianiia ideal'nogo gaza D. I. Mendeleeva"; Kalifati, "Otkrytie D. I. Mendeleevym universal'noi gazovoi postoiannoi"; and Kireev, "Rabota D. I. Mendeleeva po uravneniiu sostoianiia ideal'nogo gaza." These claims have

been effectively dismissed by M. N. Kiseleva, who argues that while Mendeleev's ideal gas law was preceded by the 1870–1872 formulation by August Friedrich Horstmann. See her article, "K istorii otkrytiia uravneniia sostoianiia ideal'nogo gaza."

101 Kalifati, "Otkrytie D. I. Mendeleevym universal'noi gazovoi postoiannoi," 15; Kipins, "K istorii ustanovleniia uravneniia sostoianiia ideal'nogo gaza."

102 *MS*, VI, 229.

103 *V* is the volume of the gas in liters, *T* is temperature in Celsius, *P* is pressure in meters of mercury at 0°. Mendeleev proposed two slightly different versions of the law in mid-September 1874. I have presented the one he offered to the Russian Physical Society, since one would expect even less specific chemical information before that audience. He developed the empirical foundations of this "expression" (not a law), in Mendéléeff [Mendeleev], "Des écarts dans les lois relatives aux gaz."

104 See minutes of the Russian Physical Society, 17 September 1874, *MS*, VI, 212; and minutes of Russian Chemical Society, 12 September 1874, *MS*, VI, 211.

105 Mendeleev first began playing with the results for different gases (hydrogen, nitrogen, oxygen, carbon monoxide and dioxide) in a notebook while he was abroad in 1876. See Zapisnaia knizhka 1874–1876, #20, ADIM II-A-1–1–9, l. 7. See also his reports on Kaiander's results at the meetings of the Russian Physical Society, 4 November 1875 (*MS*, VI, 215–218) and 4 May 1876 (*MS*, VI, 631), and of the Russian Chemical Society, 6 November 1875 (*MS*, VI, 219). Mendeleev gathered these results in his small green gas notebook: ADIM II-Zh-35-1–4a.

106 Mendeleev to F. N. L'vov, 22 April 1878, reproduced in Tishchenko and Mladentsev, *Dmitrii Ivanovich Mendeleev, ego zhizn' i deiatel'nost'. Universitetskii period*, 185.

107 *MS*, VI, 667–669. Mendeleev's resignation proceedings were published as a report to the Russian Technical Society, *Ob opytakh nad uprugost'iu gazov* (St. Petersburg: Tip. brat. Panteleevykh, 1881), reproduced in *MS*, VI, 663–684.

108 *MS*, VI, 670.

109 *MS*, VI, 681.

110 See Mendeleev to Kochubei, 14 February 1881, *MS*, XXV, 328, thanking him for the honorary membership and again urging Gadolin as the logical successor. On Gadolin, see Larman, *Aksel' Vil'gel'movich Gadolin*. The one objection to Mendeleev leaving the project was from chemist K. I. Lisenko, who felt that Mendeleev had reneged on a moral obligation that the Technical Society had assumed by taking money from the state. Tishchenko and Mladentsev, *Dmitrii Ivanovich Mendeleev, ego zhizn' i deiatel'nost'. Universitetskii period*, 192–193. Mendeleev's last experiment was on the density of steam, published in the minutes of the Russian Physical Society, 6 January 1883, *MS*, VI, 685.

CHAPTER 4

1 Boborykin, "Ni vzad—ni vpered [16 March 1876]," 1. Ellipses added. See also Neznakomets, "Nedel'nye ocherki i kartinki [29 December 1875]." A similar mood had gripped the United States two decades earlier, as described in Moore, "Spiritualism and Science," 475. I will often refer to local Petersburg events as "Russian" Spiritualism, which is a reasonable oversimplification given the concentration of the Russian movement in this city.

2 Spiritualism falls into the category of "fringe sciences"—such as Mesmerism, phrenology, and physiognomy—that were defined as outside the purview of "established" science as a consequence of recent nineteenth-century professionalization. See, for example, Winter, *Mesmerized*; and Cooter, *The Cultural Meaning of Popular Science*. Since Spiritualism is now widely considered a "pseudoscience," it is tempting to relate its rise to gullible individuals resisted by a few

valiant scientific truth-seekers. This, however, obscures the flow of history, since whether Mendeleev was "correct" in debunking mediums was not a foregone conclusion, but the very issue under debate. As a result, I will not attempt to evaluate who was "right" in this chapter. While it may be unsatisfying to walk away without a final verdict, the documents do not allow us an unambiguous answer to such understandable worries.

3 The Russian term for the movement, *spiritizm*, is perhaps more felicitously translated "Spiritism," but I have opted throughout to render it as "Spiritualism." The Russian term comes from the French *spiritisme*, which primarily refers to the doctrines of the school of French mystic Allan Kardec. The Russian movement, however, was much more heavily influenced by the Anglo-American Spiritualism movement, which emphasized physical effects and was more likely to entertain a scientific agnosticism. I shall also keep the term capitalized to differentiate it from a generalized hostility to materialism. The secondary literature on Russian Spiritualism is sparse. See Carlson, "Fashionable Occultism"; Berry, *Spiritualism in Tsarist Society and Literature*; and idem, "Mediums and Spiritualism in Russian Literature during the Reign of Alexander II."

4 There were other Spiritualist commissions in the late nineteenth century, but Mendeleev's was the first to consist exclusively of scientists and declare as its mission the *investigation*, not the debunking, of the phenomena. On other attempts, see Doyle, *The History of Spiritualism*, 306; Podmore, *Mediums of the 19th Century*, II, 151; and Putnam, *Agassiz and Spiritualism*.

5 Mendeleev himself was the prime perpetrator of this narrative. In 1898, Mendeleev wrote of the 1875–1876 Spiritualism episode: "When A. M. Butlerov and N. P. Vagner began to proselytize strongly for Spiritualism, I decided to battle against superstition, for which a commission was formed at the Physical Society. I did a lot there, and they gathered at my apartment. My view is well expressed in the public lectures of 15 December 1875 and 24 and 25 April 1876, especially in the latter. Professors have to act against professorial authority. The result was as it should be—they tossed Spiritualism. I don't regret that I worked so hard." Mendeleev, "Spisok moikh sochinenii (1899)," in Shchukarev and Valk, *Arkhiv D. I. Mendeleeva*, 74–75. This narrative has been used to argue several false contentions: that Mendeleev was a dialectical materialist, that his actions exemplified the scientific method, or that he articulated a coherent positivist worldview. The Soviet tradition of claiming Mendeleev as a dialectical materialist is discussed in Chapter 9. Such writers portray the battle against Spiritualism as perhaps the best example of his materialist outlook. See, for example, Figurovskii, *Dmitrii Ivanovich Mendeleev*, 111; Ionidi, *Mirovozzrenie D. I. Mendeleeva*, 146; and Belov, *Filosofiia vydaiushchikhsia russkikh estestvoispytatelei*, 248. On Mendeleev as exemplar of the scientific method, see: Rice, "Mendeleev's Public Opposition to Spiritualism"; Volgin and Rabinovich, "Dostoevskii i Mendeleev," translated as "Dostoevsky and Mendeleev"; and Makarenia and Nutrikhin, *Mendeleev v Peterburge*, 148–153. On anti-Spiritualism as positivism, see Vucinich, *Science in Russian Culture*, 160. Don C. Rawson, in the best secondary piece on Mendeleev's Commission, errs by importing Mendeleev's metaphysical framework from his 1903 ether pamphlet (see Chapter 8) as an explanation for his behavior in 1875. See his "Mendeleev and the Scientific Claims of Spiritualism."

6 On American Spiritualism, see Moore, *In Search of White Crows*; Carroll, *Spiritualism in Antebellum America*; Douglas, *The Feminization of American Culture*, Chapter 6; and Kerr, *Mediums, and Spirit-Rappers, and Roaring Radicals*. Many Spiritualists placed the phenomena of rappings and table-motion in the metaphysical framework offered by the reputed clairvoyant and healer Andrew Jackson Davis, the Poughkeepsie Seer, who argued that the spirits of Galen and Swedish mystical theologian Emanuel Swedenborg told him of the dawn of the new faith. Carroll, *Spiritualism in Antebellum America*, Chapter 2. On Swedenborgianism, which was heavily promoted in the 1840s as a religion by such groups as the Fourierists, see Goldfarb and Goldfarb, *Spiritualism and Nineteenth-Century Letters*, 29; and Swedenborg, *Concerning Heaven and Its Wonders*. On Davis, see his *The Magic Staff*; as well as Brown, *The Heyday of Spiritualism*,

Chapter 2; and Podmore, *Mediums of the 19th Century*, I, Chapter 11.

7 On Mesmerism in England, see Winter, *Mesmerized*. On Mesmerism's controversial history in *ancien régime* France and its frequent clashes with natural philosophers, see Darnton, *Mesmerism and the End of the Enlightenment in France*; and Schaffer, "Self Evidence."

8 This is the thesis of the classic work on Spiritualism and science: Oppenheim, *The Other World*. See also Gauld, *The Founders of Psychical Research*, 75; Nelson, *Spiritualism and Society*; and Turner, *Between Science and Religion*. For the general history of English Spiritualism, see Barrow, *Independent Spirits*. On worries about American commercialism, see Brown, *The Heyday of Spiritualism*, 247.

9 As one historian has commented: "D. D. Home was in many ways *the* outstanding Victorian medium. Virtually alone he survived with his reputation intact." Palfreman, "Between Scepticism and Credulity," 206. Emphasis in original. Home was very eager to allow experimentation and even exposed other mediums as frauds: Home, *Lights and Shadows of Spiritualism*; Dunraven, *Experiences in Spiritualism with D. D. Home*. For biographical information, see his autobiography, *Incidents in My Life*, and the rather enthusiastic biographies: Burton, *Heyday of a Wizard*; and Jenkins, *The Shadow and the Light*.

10 Crookes's published works on psychical phenomena can be found in Medhurst, *Crookes and the Spirit World*. See also Oppenheim, *The Other World*, 338–354; Palfreman, "William Crookes"; and Doyle, *The History of Spiritualism*, I, Chapter 11. For biographical information, see Brock, "A British Career in Chemistry."

11 The complete letter is reprinted in Fournier d'Albe, *The Life of Sir William Crookes*, 196–198.

12 On Home's visit to Russia, see Goldfarb and Goldfarb, *Spiritualism and Nineteenth-Century Letters*, 149; and Burton, *Heyday of a Wizard*, 243–244.

13 On Lodge, see Jolly, *Sir Oliver Lodge*; Wilson, "The Thought of Late Victorian Physicists"; and Root, "Science, Religion, and Psychical Research." On Varley, see Noakes, "Telegraphy Is an Occult Art." Alfred Russel Wallace is a particularly interesting case, since his conversion to Spiritualism was part of the reason he changed his mind on the adequacy of natural selection to account for human intelligence, a major point of disagreement with Charles Darwin. See Wallace, *Miracles and Modern Spiritualism*; Turner, *Between Science and Religion*, Chapter 4; and Kottler, "Alfred Russel Wallace, the Origin of Man, and Spiritualism."

14 He ridiculed Wallace and Crookes, arguing that Spiritualism was evidence that empiricism failed without proper dialectical-materialist method. Engels, *Dialectics of Nature*.

15 Quoted in Britten, *Nineteenth Century Miracles*, 352. Camille Flammarion, a well-known French Spiritualist scientist, considered Aksakov's authority "very great" as to the authenticity of mediums. See his *Mysterious Psychic Forces*, 66.

16 This text appeared in 1890 in German and was translated into French and Russian: Aksakof, *Animisme et Spiritisme*; and Aksákow, *Animismus und Spiritismus*. See also his historical account, Aksákoff, *Vorläufer des Spiritismus*. Hess notes that Aksakov's writings are still part of the Spiritualist canon, along with Crookes's and Lodge's (*Spirits and Scientists*, 183–184).

17 Aksakov, *Spiritualizm i nauka*, quotation on i. Mendeleev's copy is kept at ADIM Bib. 42/1.

18 Bykov, *Aleksandr Mikhailovich Butlerov*, 160.

19 Reproduced in Krätz, *Beilstein-Erlenmeyer*, 19–20.

20 Often, Butlerov is presented as if he were the dominant Spiritualist and Aksakov followed his lead, instead of the other way around, e.g., in Inglis, *Natural and Supernatural*.

21 Brooks, "Alexander Butlerov and the Professionalization of Science in Russia"; Rocke, "Kekulé, Butlerov, and the Historiography of Chemical Structure."

22 For example, in his survey history of science in Imperial Russia, Alexander Vucinich declares: "During this period, Butlerov the scientist was completely separated from Butlerov the

spiritualist." Butlerov only succumbed to Spiritualism when "influenced" by the upper classes of Petersburg society (Vucinich denies any corresponding "influence" upon his scientific work), and Vucinich erroneously claims that Butlerov made "no serious effort" to reconcile science with Spiritualism. Vucinich, *Science in Russian Culture*, 145. Most accounts of Butlerov prefer to ignore Spiritualism completely, such as Butlerov, *Nauchnaia i pedagogicheskaia deiatel'nost'*; Gumilevskii, *Aleksandr Mikhailovich Butlerov*; and the essays in *A. M. Butlerov*. The counterexample and by far the best biography remains Bykov, *Aleksandr Mikhailovich Butlerov*.

23 Markovnikov, "Vospominaniia i cherty iz zhizni i deiatel'nosti A. M. Butlerova," 94–95.

24 PFARAN f. 22, op. 2, d. 157, quoted in Bykov, "O nauchnom metode A. M. Butlerova," 112n. Regrettably, very few substantial archival sources on Butlerov's Spiritualism have survived. Mikhail, Butlerov's son, related that on 5 April 1928 his mentally ill wife burned a great deal of Butlerov's correspondence to Aksakov, which Mikhail had kept out of the Academy of Sciences archive for sentimental reasons. These documents doubtless contained information on Spiritualism. Musabekov, "Novye materialy ob A. M. Butlerove," 230.

25 Editor's footnote in Vagner, "Pis'mo k redaktoru [April 1875]," 855n. Mendeleev heavily annotated his copy (ADIM Bib. 42/2). The *Messenger of Europe* was a central organ in the cultural politics of the day, as described in Pogorelskin, "'The Messenger of Europe.'"

26 Vagner, "Pis'mo k redaktoru," 859.

27 The article came adorned with an editorial footnote similar to that of Stasiulevich: "The name of this article's author, holding such distinguished renown and authority in the scientific world, professor of Chemistry at St. Petersburg University A. M. Butlerov, involuntarily turns particular attention to it and in any case justifies its appearance in a journal." Editor's footnote to Butlerov, "Mediumicheskiia iavleniia [November 1875]," 300n.

28 Butlerov, "Mediumicheskiia iavleniia," 303.

29 Butlerov, "Mediumicheskiia iavleniia," 347.

30 Vagner, "Mediumizm [October 1875]," 869.

31 The criticisms, on the other hand, were published in a broader array of journals. For example, a translation of physiologist William Carpenter's scathing attack on Crookes was published to display proper scientific procedures. Karpenter, "Fiziologicheskoe ob"iasnenie nekotorykh iavlenii spiritizma [May 1875]."

32 Markov, "Magiia pod krylom nauki [3 June 1875]," 1–2. For Vagner's response, see his "Eshche po voprosu o mediumizme [9 June 1875]." A similar connection between Spiritualism and the Great Reforms is made in E. K., "Spiritizm i nasha intelligentsia [25 March 1875]."

33 Rachinskii, "Po povodu spiriticheskikh soobshchenii g. Vagnera [May 1875]," 381 and 399.

34 Shkliarevskii, "Chto dumat' o spiritizme? [June and July 1875]."

35 This account is taken from the papers of the Spiritualist Commission: Mendeleev, *Materialy dlia suzhdeniia o spiritizme*, 3–4.

36 Eval'd, who originally served as chair of the Commission, resigned early on the grounds that he was a committed anti-Spiritualist and did not want to waste his time: "I don't have the least [desire] either to study so-called Spiritualist phenomena, or to investigate their causes, or to convince jealous Spiritualists in the erroneous significance of Spiritualism. . . ." Eval'd to Mendeleev, 28 October 1875, ADIM Alb. 4/85. Ellipses added. Kovalevskii also only went to the first two meetings. At the third meeting D. K. Bobylev and D. A. Lachinov joined.

37 Mendeleev, *Materialy dlia suzhdeniia o spiritizme*, 5–8.

38 Zapisnaia knizhka, 1874–1876, #20, ADIM II-A–1–1–9, ll. 67–69.

39 Vagner to Mendeleev, 4 June 1875, ADIM Alb. 4/74.

40 Butlerov went with Aksakov to England on this recruiting trip. He wrote back an excited letter to V. V. Markovnikov on 9 October 1875: "Having been in London, I saw things which would have been enough to convince [me] of the reality and the authenticity of the facts, if I

had not been convinced earlier. After my trip Aksakov saw still greater wonders in Newcastle and hopes to bring here mediums for our scientific commission. I will soon publish an article on this subject." Butlerov, *Nauchnaia i pedagogicheskaia deiatel'nost'*, 306.

41 Mendeleev, *Materialy dlia suzhdeniia o spiritizme*, 11.

42 Mendeleev, *Materialy dlia suzhdeniia o spiritizme*, 14–18.

43 Vagner, "[Letter to the editor] [3 December 1875]," 3. Ellipses added.

44 Mendeleev, *Materialy dlia suzhdeniia o spiritizme*, 20n.

45 Mendeleev, *Materialy dlia suzhdeniia o spiritizme*, 27. Mendeleev added in a footnote that Butlerov did not attend this and the following seances but sat in the back room with the others.

46 The minutes report of Mendeleev: "Sitting in the semicircle, he at first clearly distinguished the white handkerchiefs on the mediums in normal positions, and then it seemed to him that the position of the elder medium's handkerchief had significantly changed. Having checked this impression several times, he lit a match for clarification and managed to notice that the elder medium was sitting, leaning towards the curtain, half-turning his face back to the right." Mendeleev, *Materialy dlia suzhdeniia o spiritizme*, 29.

47 Mendeleev, *Materialy dlia suzhdeniia o spiritizme*, 28. Interestingly, as related in an appendix, before the session Kraevich had asked whether the mediums possessed anything sharp on their persons and Aksakov suggested that he search them, but Kraevich demurred (p. 69).

48 For her account, see Trirogova-Mendeleeva, *Mendeleev i ego sem'ia*, 28. Ol'ga could not have heard this story from her father the next day, as she was staying with her mother at their cottage outside Moscow under the strained marital circumstances described in the next chapter. Notice also that these memoirs were written substantially after Mendeleev's death in 1907, diminishing the memory of his daughter, who was anyway only about ten at the time. For Mendeleev's account of the tear, see Mendeleev, *Materialy dlia suzhdeniia o spiritizme*, 30n.

49 Aksakov, *Pamiatnik nauchnogo predubezhdeniia*, 6.

50 Mendeleev, *Materialy dlia suzhdeniia o spiritizme*, 30–31.

51 Aksakov, "[Letter to the editor] [7 December 1875]." Mendeleev kept a copy at ADIM Alb. 4/80a.

52 Mendeleev to Aksakov, 8 December 1875, ADIM Alb. 4/153.

53 Aksakov to Mendeleev, 11 December 1875, ADIM Alb. 4/169. Ellipses added. Publicly, Aksakov complained that "fairness demands that the witnesses (in this case Mr. Butlerov and I) should also be given the floor at the end of the lecture. . . . This is how it is done, at least in England, in public lectures on Spiritualism." Aksakov, "[Letter to the editor] [7 December 1875]."

54 Mendeleev asked the St. Petersburg branch of the Slavic Friendship Committee, in particular its subcommittee to aid the victims of the uprising in Bosnia and Herzegovina, to arrange for the venue and facilitate approval of the public lecture by the Ministry of Internal Affairs. Letter of 4 December 1875, ADIM I-V–19–4–64. The Committee later wanted publication rights, but Mendeleev refused. See I. Iankumov to Mendeleev, 17 December [1875], ADIM I-V–19–4–63.

55 Mendeleev, *Materialy dlia suzhdeniia o spiritizme*, 308.

56 Mendeleev, *Materialy dlia suzhdeniia o spiritizme*, 314–315.

57 Boborykin, "Voskresnyi fel'eton [21 December 1875]," 1. Mendeleev's speaking style has been compared repeatedly to "grating rocks" for its wild fluctuations in pitch and speed. Even the most laudatory praises of the chemist could only say that the content made up for defects in eloquence. See Tishchenko, "Vospominaniia o D. I. Mendeleeve," 127; Veinberg, *Iz vospominanii o Dmitrii Ivanoviche Mendeleeve kak lektor*, 2; Kablukov, "Dmitrii Ivanovich Mendeleev [1907]," 96; and idem, "Dmitrii Ivanovich Mendeleev [1927]," 102.

58 Shkliarevskii, "Kritiki togo berega [January 1876]," 494. The title here is a reference to a novel by Russian thinker A. I. Herzen. See also [Tsvet], *Spiritizm i spirity*. This fiery piece was published early in 1876, and Mendeleev heavily annotated his copy (ADIM Bib. 42/4), which

may have served as the rhetorical model for his April lectures.

59 Vagner to Mendeleev, 1 January 1875 [*sic*: 1876], ADIM Alb. 4/52. Vagner maintained relations with the Commission and was particularly interested in the experimental investigations proposed for January. He turned to Mendeleev, he said, because he was the only one he considered a true friend to Butlerov and himself. See Vagner to Mendeleev, 6 January 1875 [*sic*: 1876], ADIM Alb. 4/53.

60 Mendeleev, *Materialy dlia suzhdeniia o spiritizme*, 32–34.

61 Mendeleev, *Materialy dlia suzhdeniia o spiritizme*, 37.

62 Mendeleev, *Materialy dlia suzhdeniia o spiritizme*, 68. At the thirteenth session, on 25 January 1876, Aksakov pleaded that mediums were often nervously ill and hence there had to be some procedure for cancellation without penalizing the Spiritualists. An advance warning of three days was decreed (p. 40).

63 Mendeleev, *Materialy dlia suzhdeniia o spiritizme*, 38. There remained other procedural issues. For example, there were extended debates over whether seances should continue in Mendeleev's apartment or should be moved to Aksakov's, where Spiritualist phenomena had already been reported. F. F. Petrushevskii went to inspect the apartment in January and had no objections, but wanted Mendeleev to clear the decision. Petrushevskii to Mendeleev, 22 January 1876, ADIM I-V–19–4–85. One can see from this exchange that Mendeleev's decision on such matters carried final authority, even though officially he was an equal member to the rest.

64 Aksakov to Mendeleev, 19 February 1876, ADIM Alb. 4/178.

65 Mendeleev, *Materialy dlia suzhdeniia o spiritizme*, 46 and 50.

66 Claire's exit sparked some interest among the newspapers, whose appetite had been whetted by the December lecture but had not been able to glean any information from the Commission or the Spiritualists since. Aksakov, irritated that newspapers were claiming that Claire left because she was exposed as a fraud, wrote to the major dailies that she stayed in Petersburg as long as scheduled and that *he* had removed her from the seances. Aksakov, "Pis'mo v redaktsiiu [12 March 1876]," 2. Claire performed successful seances at Aksakov's house after withdrawing from the Commission. See "Spiriticheskiia podvigi [1 March 1876]."

67 It was for just this reason, Mendeleev noted later, that the Commission on 9 May 1875 had rejected female mediums as undesirable. Since this stipulation remained unwritten, Aksakov was able to bring one in with impunity, Mendeleev charged, to upset the investigators. Mendeleev, *Materialy dlia suzhdeniia o spiritizme*, 94n.

68 Mendeleev, *Materialy dlia suzhdeniia o spiritizme*, 79–80, 83.

69 Bobylev said on 29 January that he saw Claire's hands moving actively (pp. 88–89), and others saw her cheating with her legs (pp. 117–118). Mendeleev defended his aggressiveness as showing the need for instruments to supplement the senses. See his statement to the minutes of 11 January 1876, ADIM Alb. 4/150.

70 For the English case, see Owen, *The Darkened Room*. On American female mediums, see Braude, *Radical Spirits*; and Moore, *In Search of White Crows*, Chapter 4. For an alternative interpretation, see Douglas, *The Feminization of American Culture*.

71 Mendeleev later offered an interest-theory explanation for why Spiritualism attached itself to the women's movement. According to him, women used Spiritualism to argue that they could perform the same labor as men by using spirits to assist them, thus minimizing the wage differential between the two sexes. Mendeleev, *Materialy dlia suzhdeniia o spiritizme*, 327.

72 Mendeleev to Aksakov, 18 February 1876, ADIM Alb. 4/177; and Aksakov to Mendeleev, 19 February 1876, ADIM Alb. 4/178.

73 Mendeleev to anonymous, 19 February 1876, ADIM Alb. 4/176.

74 Mendeleev, *Materialy dlia suzhdeniia o spiritizme*, 169.

75 Shapin, *A Social History of Truth*; and Shapin and Schaffer, *Leviathan and the Air Pump*.

76 For a contemporary example of this reasoning, see Johnson, "Spiritualism Tested by Sci-

ence [May 1858]," 28.

77 On the decline of noble status after Emancipation, see Manning, *The Crisis of the Old Order in Russia*, 9 and 39; Hamburg, *Politics of the Russian Nobility*; Wirtschafter, *Social Identity in Imperial Russia*, 27–28; Rieber, *Merchants and Entrepreneurs in Imperial Russia*, 149; and idem, "The Sedimentary Society," 357. My argument depends only on the *perception* of an economic decline; the evidence for a real decline is more equivocal. See Becker, *Nobility and Privilege in Late Imperial Russia*; and Munting, "Economic Change and the Russian Gentry."

78 Mendeleev, *Materialy dlia suzhdeniia o spiritizme*, 45 (see also p. 151). Of course, following Shapin, one can easily argue that Mendeleev was merely replacing trust in social status with a different kind of trust: trust in Mendeleev's interpretation of the scientific method and its embodiment in institutions and instruments.

79 His objection to personal statements is particularly interesting: "Such personal statements based on subjective impressions either have no significance for research and then should not be admitted, or they are significant and then the Commission itself, which was formed, one would think, in order that personal subjective impressions be replaced with impersonal observations, has no significance." Aksakov, "[Letter to the Commission] [11 March 1876]," 2–3.

80 Butlerov's was prefaced by a comment from the newspaper editors: "We publish every word with the greatest pleasure, since we sincerely would very much like this affair to be explicated before the public openly and clearly, without giving cause for any reproaches directed against the Physical Society." Butlerov echoed Aksakov almost exactly on the question of instruments. Butlerov, "[Two letters to the editor] [8 March 1876]," 2.

81 Mendeleev continued to taunt Aksakov after the Commission. For example, picking up on a statement that Aksakov had once made that, although a person could fake the mediumistic phenomena of a table tilting towards himself, humans were incapable of tilting tables away with palms placed flat on the surface. Mendeleev said that he could do this, and he would do so in public if Aksakov would agree to "give your word that after this you will write and say nothing in favor of mediumism in any of the European languages." If Mendeleev failed, he would drop his opposition. Mendeleev to Aksakov, 18 April 1875 [*sic*: 1876], ADIM Alb. 4/180. He never sent the letter, but he published a variant of this wager later in the *Materialy*.

82 Mendeleev, *Materialy dlia suzhdeniia o spiritizme*, 51–52.

83 Mendeleev, *Materialy dlia suzhdeniia o spiritizme*, 59.

84 Mendeleev, *Materialy dlia suzhdeniia o spiritizme*, 60. Emphasis in original.

85 "Ot Komisii dlia izsledovaniia mediumicheskikh iavlenii [25 March 1876]."

86 Butlerov and Aksakov, "[Letter to the editor] [29 March 1876]." Emphasis in original. Ellipses added.

87 Vagner, "Za i protiv [12 April 1876]," 5. Some journalists echoed his claims, arguing that the Commission was "purely police-judicial" characterized by "police-style observation, police penetratingness, craftiness, [and] slovenliness." "Vnutrennee obozrenie [April 1876]," 255–257.

88 Mendeleev, *Materialy dlia suzhdeniia o spiritizme*, 91–92n. Ellipses added. Vagner recognized Mendeleev's good intentions in their private correspondence: "When you read my first article on mediumism in the *Messenger of Europe*, you had the idea of investigating these phenomena. You said 'here is something interesting.' At the same time there arose in you the desire to correct me and Butlerov back onto the true path. You reproached us that we had not turned with this matter to a scientific society, that we did not go by the well-known and correct path of all scientific researches and discoveries. . . . Good, honest intentions, wonderful convictions of a true scientist and a good colleague!" Vagner to Mendeleev, 19 February 1876, ADIM Alb. 4/56. Ellipses added.

89 For Mendeleev's public lectures on chemistry, see Mendeleev, *Izbrannye lektsii po khimii*. On chemists' public lectures generally, see Brooks, "Public Lectures in Chemistry in Russia." Mendeleev had planned to speak immediately after the official statement's release, but the police

took three weeks to approve it. V. Gaevskii to Mendeleev, 21 March 1876, ADIM Alb. 4/163.

90 See the advertisement in *Novoe Vremia*, 24 April 1876, #54: 1. This time Mendeleev donated the proceeds to a commission of the Russian Technical Society that aided needy scholars.

91 Pavel Ivanovich Nikitin to Mendeleev, [26 April 1876], ADIM I-V–23–1–107; General Petersburg news, *Golos*, 28 April (10 May) 1876, #117: 2.

92 Mendeleev, *Materialy dlia suzhdeniia o spiritizme*, 329–330. On the earth revolving around the sun, see p. 322.

93 Mendeleev, *Materialy dlia suzhdeniia o spiritizme*, 357. For example, as he responded to attacks on his liquid-expansion work in 1884: "The correct understanding of a scientific dispute instills respect not only to the truth itself, but also to the people who dedicate their time in searching out a piece of it, however small. Without an understanding of the conditions of a scientific dispute, it is not only useless, but even harmful to the flow of thoughts directed to a conscious unfolding of truth." Mendeleev, "Eshche o rasshirenii zhidkostei (otvet professoru Avenariusu," *ZhRFKhO* 16, fiz. otd. (1884): 475–492, in *MS*, V, 164.

94 Mendeleev, *Materialy dlia suzhdeniia o spiritizme*, 381. Ellipses added. Mendeleev would repeatedly insist that open publication was the best way to defeat false doctrines. As he said in 1878 (not in reference to Spiritualism): "If people want that there will be those who battle against false doctrines, it is necessary to give the opportunity for each doctrine to express itself. How is one supposed to believe a voiced unilateral accusation, not having heard the other side? Everyone, even if they don't say it, will think: perhaps if they are banning it and accusing it without discussion, and if it isn't entirely true, then maybe there is a part of a new truth [in it]?" Letter of 21 August 1878, reproduced in Tishchenko and Mladentsev, *Dmitrii Ivanovich Mendeleev, ego zhizn' i deiatel'nost'. Universitetskii period*, 83.

95 Butlerov to Stasiulevich, 1 May 1875, PD f. 293, op. 1, d. 287, ll. 1–1ob. This letter was ignored, provoking a harsher response from the usually mild-mannered chemist: Butlerov to Stasiulevich, 8 May 1875, PD f. 293, op. 1, d. 287, ll. 2–2ob. Butlerov was again denied a hearing in 1879 in the same journal, even though it had published an explicit attack on him. Butlerov to Stasiulevich, 4 January 1879, PD f. 293, op. 1, d. 287, ll. 3–3ob.

96 Butlerov to Stasiulevich, undated, PD, f. 293, op. 1, d. 116, ll. 1–2, on 1ob.

97 Foreign Spiritualists tended to exaggerate the importance of Church censorship of Spiritualism in Russia, as in Britten, *Nineteenth Century Miracles*, 355.

98 Quoted in Bykov, *Aleksandr Mikhailovich Butlerov*, 162. Ellipses added. Oddly, Vagner concurred, writing after Mendeleev's December lecture: "I can also answer your lecture with lectures, but I don't want to do that. This lecture could create a scandal and unrest in the public—which I of course do not want. But I am fully convinced that this lecture would destroy the entire effect of your lecture and would clearly show the public the goal and the form of activity of the Commission." Vagner to Mendeleev, 22 December [1875], ADIM Alb. 4/60.

99 General Petersburg news, *Golos*, 27 April (9 May) 1876, #116: 3. See also Letanin, "Pervaia lektsiia g. Mendeleeva o spiritizme [26 April 1876]."

100 Boborykin, "Ni vzad—ni vpered," pt. 2, 1.

101 The Russian Physical Society granted Mendeleev full authority over the publication of the minutes in a 30–0 vote (Petrushevskii to Mendeleev, 24 March [1876], ADIM Alb. 4/161). On the rest of the publication decisions, see ADIM II-A–8–2–1 and ADIM II-A–8–2–2. In the Soviet edition of Mendeleev's collected works, only the first preface and the two lectures were included. Most egregiously, Mendeleev's footnotes were not reproduced.

102 These footnotes perform an entirely different function from those in his *Principles of Chemistry*. The first edition of *Principles* placed technical material in smaller print, while in later editions, this material was moved into footnotes. Mendeleev also incorporated revisions concerning advances in chemistry into footnotes. In *Principles*, footnotes solved the problem of revision and also served to divide advanced readers from beginners. In the *Materialy*, however,

Mendeleev employed footnotes as *necessary* companions to the text.

103 Mendeleev, *Materialy dlia suzhdeniia o spiritizme*, 1.

104 These articles consist not only of the two public lectures, but also of E. G. Beketova's translation of Lavoisier's attack on Mesmerism, laboratory investigations of supposed mediumistic phenomena, and other "instructive" examples to show both historically and methodologically different attacks on Spiritualism. These juxtapositions of scientific and historical articles with the supposedly unadorned (but heavily footnoted) minutes of the Commission further persuaded the reader of the seriousness of the issues involved. For example, the article by S. K. Kvitka, a young student at the Mining Academy, was a scientific study of table motions that concluded that the phenomena were real but caused by magnetic forces. Mendeleev included this piece as a juxtaposition because it showed that a mere student could properly investigate Spiritualism with nothing beyond the introductory physical techniques, even though he disagreed with the conclusions. Mendeleev, *Materialy dlia suzhdeniia o spiritizme*, 163. Kraevich did not want to include the Kvitka piece on the grounds that it did not reveal anything new, missing Mendeleev's point about how its banality served to dismiss Spiritualism. [Kraevich to Mendeleev], [March or April 1876], ADIM Alb. 4/164. Kvitka, for his part, was delighted by the inclusion, and by Mendeleev's praise that he exhibited strong "common sense." See V. M. Garshin to E. S. Garshina, 2 April 1876, in Garshin, *Polnoe sobranie sochinenii*, III, 78. I would like to thank L. B. Bondarenko for this reference.

105 Mendeleev, *Materialy dlia suzhdeniia o spiritizme*, ix. Mendeleev in his personal notebook reported that 1374 copies of the *Materials* were printed with 14 more for the censors, to a total of 1388. This was markedly less than the 1787 copies of Mohn's meteorology that he published for similar aerostatic ends. Zapisnaia knizhka, 1874–1876, #20, ADIM II-A–1–1–9, l. 105.

106 Mendeleev, *Materialy dlia suzhdeniia o spiritizme*, x.

107 As seen in Chapter 3, meteorology at precisely this time was publicly transforming from amateur predictions to respectable science. On the "taming" of meteorology in Britain, and on efforts to disentangle it from the occult, see Anderson, "The Weather Prophets." In the late eighteenth century, meteorology was often cited to argue for the plausibility of Mesmerism. Sutton, "Electric Medicine and Mesmerism," 377.

108 See, for example, Akatov, *O pozitivnykh osnovakh noveishago spiritualizma*, 67, 88. Critics of Spiritualism would also cite the ether's properties to deny the possibility of spirit materialization, as in Shkliarevskii, "Kritiki togo berega," 476–478. For secondary accounts of this linkage, see Wynne, "Physics and Psychics"; Oppenheim, *The Other World*, 218; Wilson, "The Thought of Late Victorian Physicists"; Benson, "Facts and Fictions in Scientific Discourse," 834; and Cantor, "The Theological Significance of Ethers," 146. In other cases, Spiritualists cited the ether as a model hypothesis in order to justify their own hypothesis of spiritual forces. Butlerov and Vagner were among such claimants. See Butlerov, "Koe-chto o mediumizme"; and Vagner, "Peregorodichnaia filosofiia i nauka [13 and 20 July 1883]."

109 Aksakov, *Razoblacheniia*, xvi; emphasis in original. Aksakov reserved special scorn for the official statement of the Commission, which he attacked line-by-line, contradicting each statement of the Commission from logical, empirical, and psychological standpoints. He eventually published this excerpt separately as *A Monument of Scientific Prejudice* (*Pamiatnik nauchnogo predubezhdeniia*). Spiritualists considered this a definitive refutation of the Commission's "pathetic report." Akatov, *O pozitivnykh osnovakh noveishago spiritualizma*, 4.

110 See Gordin, "Loose and Baggy Spirits," for a detailed discussion of Dostoevsky's position.

111 Aksakov maintained his defense of Spiritualism until his death. In his response to von Hartmann's *Philosophy of the Unconscious*, Aksakov declared: "In the decline of life I ask myself sometimes, 'Have I in truth done well to have devoted so much time and toil and money to the

study and the publication of facts in this domain? Have I not wasted my existence, with no result to justify all my pains?' Yet always I seem to hear the same reply: 'A life on earth can have no higher aspiration than to demonstrate the transcendental nature of man's being; to prove that he is called to a destiny loftier than the existence which he knows.' I cannot then regret that I have devoted my whole life to the pursuit of this aim; although it be by methods which science shuns or spurns—methods which I hold far trustier than any other which science has to show. And if it be in the end my lot to have laid one stone of that temple of the spirit, built up from century to century by men true of heart—this will be the highest, and the only, recompense which I ever strove to gain." Quoted in Inglis, *Natural and Supernatural,* 450.

112 For his spirit photography, see his "Sine ira et studio [1894]"; and Pribytkov, "O fotografiiakh [1894]." Vagner's seances with Eusapia took place in Naples. See Vagner to A. D. Butovskii, 2 December (N.S.) 1883, PD f.1, op. 2, d. 174, ll. 3–4ob.; and 6 February (N.S.) 1884, PD f. 1, op. 2, d. 174, ll. 6–7ob. On Eusapia, see Münsterberg, "My Friends, the Spiritualists."

113 Strakhov began his attacks on Spiritualism in 1876, but his dispute with Butlerov only heated up in the 1880s. See "Tri pis'ma o spiritizme. Pis'mo pervoe [15 November 1876]"; "Tri pis'ma o spiritizme. Pis'mo II [22 November 1876]"; "Tri pis'ma o spiritizme. Pis'mo III [29 November 1876]"; "Eshche pis'mo o spiritizme [1 February 1884]"; "Fizicheskaia teoriia spiritizma [26 February 1885]"; and "Zakonomernost' stikhii i poniatii [11 and 26 November 1885]." For an informative biography of Strakhov, see Gerstein, *Nikolai Strakhov.* Strakhov was on casual terms with Mendeleev, whom he liked to needle over metaphysics. Strakhov to Mendeleev, 26 April 1889, ADIM Alb. 2/248. For Butlerov's side of the dispute, see especially: "Umstvovanie i opyt [7 February 1884]"; "O 'vozmozhnom' i 'nevozmozhnom' v nauke [1883]"; *Empirizm i dogmatizm v oblasti mediumizma*; and "Chetvertoe izmerenie prostranstva i mediumizm [February 1878]." Vagner also joined in: "Peregorodichnaia filosofiia i nauka"; and "Razdvoennaia filosofiia [3 April 1884]."

114 Carlson, "Fashionable Occultism," 138; idem, *"No Religion Higher Than Truth,"* 4–5. Spiritualist publication soared, including a translation of Podmore's history of Spiritualism and a bibliography of the occult: Podmore, *Spiritizm*; and Antonoshevskii, *Bibliografiia okkul'tizma.* For present-day interest, see Stephens, "The Occult in Russia Today."

115 Mendeleev received some limited correspondence about the Commission. Wilhelm K. Döllen, an astronomer from Pulkovo, wrote to Mendeleev about astrophysicist Carl Friedrich Zöllner's theories about the fourth dimension and Spiritualism, and asked him to debunk Henry Slade should he ever come to Petersburg. (Slade did, but Mendeleev kept out of it.) Döllen to Mendeleev, 26 January 1878, ADIM I-V–11–1–80. On a more dramatic note, Mendeleev received letters from a woman who signed herself as "Mother of the Family," who in her first missive condemned Mendeleev for his denial of spirits as "a criminal in the eyes of God and the entire Christian world." Amid her accusations of materialism and corruption, she sent Mendeleev to Hell, a declaration Mendeleev interpreted, rather hastily, as a death threat. Two weeks later, she apologized for her severe tone. ADIM Alb. 4/170 and 4/171.

116 General Petersburg news, *Golos,* 11 (23) March 1876, #71: 2; and *Novoe Vremia,* 11 April 1876, #41: 3. The results of these later investigations were equivocal. To the delight of Pribytkov, the editor of *Rebus,* some doctors found no evidence of trickery in exploratory seances. Pribytkov, *Mediumicheskie iavleniia pered sudom vrachei.* The Church, on the other hand, continued to oppose Spiritualism as "a diseased phenomenon in the area of contemporary thought." Verzhbolovich, *Spiritizm pred sudom nauki i khristianstva,* 53.

117 *S.-Peterburgskie Vedomosti,* 4 May 1876, quoted in Aksakov, *Razoblacheniia,* 260.

118 Mendeleev, *Neftianaia promyshlennost' v Severo-Amerikanskom shtate Pensil'vanii i na Kavkaze* (1877), in *MS,* X, 82; and Zapisnaia knizhka, 1874–1876, #20, ADIM II-A–1–1–9, ll.

80–83.

119 Pribytkov, "Professor Mendeleev priznaet mediumicheskiia iavleniia [1894]"; and Aksakov, "Po povodu odnogo iz 'pshikov' professora Mendeleeva [1894]." Historian Maria Carlson has even taken this "admission" as a change of heart. See her "Fashionable Occultism," 138; and *"No Religion Higher Than Truth,"* 25.

120 Mendeleev, "Spiriticheskie uzly [18 May 1904]," 3.

CHAPTER 5

1 Flaubert, *Madame Bovary*, 341.

2 On the early Academy as bastion of manners, see Gordin, "The Importation of Being Earnest."

3 One of these concerned the term "Russian," which historian G.-F. Müller traced to Swedish, while Lomonosov insisted on a Slavic etymology, as chronicled in Black, *G.-F. Müller and the Imperial Russian Academy*, Chapter 5. On increasing Russian representation, see Schulze, "The Russification of the St. Petersburg Academy of Sciences in the Eighteenth Century."

4 Vucinich, *Empire of Knowledge*, 34.

5 Soboleva, *Organizatsiia nauki v poreformennoi Rossii*, 100. Chemistry was a partial exception. See Pogodin, "Akademiia nauk i puti razvitiia khimii v dorevoliutsionnoi Rossii."

6 Soboleva, *Bor'ba za reorganizatsiiu peterburgskoi Akademii nauk*, 189–196.

7 Tishchenko, "D. I. Mendeleev i Russkoe khimicheskoe obshchestvo"; Kozlov, *Vsesoiuznoe khimicheskoe obshchestvo imeni D. I. Mendeleeva*, 48.

8 V. V. Markovnikov, "[Jubilee Speech]," 63–64.

9 Markovnikov to Butlerov, 11 February 1870, reproduced in Plate, "Novye materialy k biografii V. V. Markovnikova," 78. See also F. Beilstein to E. Erlenmeyer, 29 April (11 May) 1872, reproduced in Krätz, *Beilstein-Erlenmeyer*, 26.

10 Reproduced in Kniazev, "D. I. Mendeleev i tsarskaia Akademiia nauk," 307.

11 Kniazev, "D. I. Mendeleev i tsarskaia Akademiia nauk," 309.

12 Kniazev, "D. I. Mendeleev i tsarskaia Akademiia nauk," 310–311. After Mendeleev's death, the Academy cited this election as corresponding member to excuse its failure to elect him as a full member, claiming they had recognized Mendeleev already in 1876, long before other societies did. Beketov, "Dmitrii Ivanovich Mendeleev."

13 For more on Zinin, see Figurovskii and Solov'ev, *Nikolai Nikolaevich Zinin*.

14 Beketov strongly advocated the periodic law. See Beketov, "Znachenie periodicheskoi sistemy D. I. Mendeleeva"; and Ulanovskaia, "N. N. Beketov o periodicheskom zakone."

15 Historians have traditionally bypassed this issue by ignoring Spiritualism. Given the paucity of surviving Butlerov personal correspondence, all I can offer is a dual conjecture. First, Butlerov considered the issue of "Russifying" the Academy so important that he needed to shelve his personal hostility. Second, Butlerov, raised according to a noble code of ethics by his grandfather, perhaps considered it more "gentlemanly" to bury the hatchet.

16 Volkova, "Materialy k deiatel'nosti A. M. Butlerova v Akademii nauk," 1300–1301. L. N. Shishkov, a Petersburg chemist, wrote to Butlerov as early as 25 May 1863 (PFARAN f. 22, op. 2, d. 255) that he had just heard from Mendeleev that there were no jobs under the new charter that could bring Butlerov to Petersburg, as quoted in Figurovskii and Musabekov, "Vydaiushchiisia russkii khimik L. N. Shishkov," 64. Mendeleev's own rapid rise in the capital shows this to be somewhat disingenuous. Mendeleev, as chairman of the physico-mathematical faculty at the University (since 1867), eventually generated a loophole in the 1863 statute to allow for Butlerov's hire. See Krotikov and Filimonova, "Ocherk pedagogicheskoi deiatel'nosti D. I. Mendeleeva v Peterburgskom universitete (1867–1881 gg.)," 144–145. On Mendeleev's

stalling, see Beilstein to Butlerov in Bykow and Bekassowa, "II. F. Beilsteins Briefe an A. M. Butlerow"; and Markovnikov to Butlerov, 31 October [1867], reproduced in Bykov, *Pis'ma russkikh khimikov k A. M. Butlerovu*, 246.

17 *MS*, XV, 296; and Bykov, "Dva otzyva D. I. Mendeleeva ob A. M. Butlerove," 116. Emphasis in original.

18 See the survey of Butlerov's career at the University in Volkova, "A. M. Butlerov i Peterburgskii universitet." Also, in February 1869, Butlerov unsuccessfully tried to find a German translator for the first edition of Mendeleev's *Principles of Chemistry*. See Bykov, "Materialy k istorii trekh pervykh izdanii," 283n. *Principles* was translated into German in its fifth edition in the 1880s.

19 Glinka, "Aleksandr Mikhailovich Butlerov v chastnoi i domashnei zhizni," 191 and 197. Later historians repeatedly compared the two most prominent chemists of the Imperial period. See, for example, the rather strained effort in 1928 by their former student V. E. Tishchenko, "A. M. Butlerov i D. I. Mendeleev v ikh vzaimnoi kharakteristike"; and the somewhat better attempt by Bykov, "Svet i teni v nauchnoi biografii."

20 On 13 April 1879, Butlerov almost single-handedly led a protest in the general meeting of the Academy against the nomination of Leopold Schröder as adjunct in Sanskrit to the third division (historical-philological) of the Academy, on the grounds of being too junior. Butlerov, *Nauchnaia i pedagogicheskaia deiatel'nost'*, 225. The Schröder case attracted the attention of the Academy's critics. See Lamanskii, "Eshche plemiannik i sanskritolog."

21 Butlerov, *Nauchnaia i pedagogicheskaia deiatel'nost'*, 381, 226. Beketov may have been Butlerov's first choice. He initially accepted the nomination, and then retracted when he felt that Butlerov really favored Mendeleev. N. N. Beketov to A. M. Butlerov, 28 March 1880 and 14 October 188[0], reproduced in Bykov, *Pis'ma russkikh khimikov k A. M. Butlerovu*, 56–57, and 59. On 29 November Beketov wrote to Butlerov explaining how glad he was he was not Mendeleev's competitor, given Butlerov's lovely citation to the Academy on his behalf (p. 61). For an interesting analysis of Beketov's decision, see Dmitriev, "Skuchnaia istoriia."

22 Minutes reproduced in Kniazev, "D. I. Mendeleev i tsarskaia Akademiia nauk," 324. Emphasis in original. For the voting margins required for election, see the 1836 statute, reproduced in *Ustavy Akademii nauk SSSR*, 92–119.

23 See the facsimile in Volkova, "Materialy k deiatel'nosti A. M. Butlerova v Akademii nauk," 1302.

24 PFARAN f. 24, op. 1, d. 135, l. 1. Also included are newspaper clippings on the incident.

25 Lamanskii, "Otkrytoe pis'mo [. . .] g. Baklundu [23 November 1880]"; and idem, "Otkrytoe pis'mo [. . .] O. V. Struve [23 November 1880]."

26 "Novye vybory v Akademii [26 November 1880]."

27 Butlerov, *Nauchnaia i pedagogicheskaia deiatel'nost'*, 234–236; K. S. Veselovskii to A. M. Butlerov, 17 December 1880, PFARAN f. 22, op. 2, d. 48, ll. 4–4ob.

28 Shmulevich and Musabekov, *Fedor Fedorovich Beil'shtein*, 7–8, 33–35; Hjelt, "Friedrich Konrad Beilstein"; and Richter, "K. F. Beilstein, sein Werk und seine Zeit."

29 Beilstein to Butlerov, 6 (18) February 1866: "I am Mendeleev's successor at the *Technological Institute* and am busying myself dealing with my imposed duties. That is no small affair, when I tell you that my predecessor—who, as you know, is not really a practical chemist—never bothered with the work of *Praktikanten* and went at most for a few minutes into the laboratory every 1/4 of the year." Reproduced in Bykow and Bekassowa, "II. F. Beilsteins Briefe an A. M. Butlerow," 278. Emphasis in original.

30 Butlerov, *Nauchnaia i pedagogicheskaia deiatel'nost'*, 242–258, quotation on 243. Butlerov wrote a hostile letter to French chemist Adolphe Wurtz, demanding that he retract his recommendation of Beilstein given the way the Academy had treated Mendeleev. See Volkova, "Materialy k deiatel'nosti A. M. Butlerova v Akademii Nauk," 1303–1304. For the full texts of the

debates over Beilstein, see "Predlozhenie i balotirovanie professora F. F. Beil'shteina."

31 PFARAN f. 24, op. 1,, d. 135, l. 14.

32 The date was 7 October 1886. Shmulevich and Musabekov, *Fedor Fedorovich Beil'shtein*, 58.

33 Quoted in Tishchenko, "D. I. Mendeleev i Russkoe khimicheskoe obshchestvo," 1528. Ellipses added.

34 Letter of 17 November 1880, quoted in Tishchenko, "D. I. Mendeleev i Russkoe khimicheskoe obshchestvo," 1528. Emphasis in original.

35 PFARAN f. 22, op. 1, d. 38, quoted in Shmulevich and Musabekov, *Fedor Fedorovich Beil'shtein*, 52–53. Emphasis in original.

36 "Delo Mendeleeva [21–25 November 1880]." On Mendeleev's economic work, see Chapter 6.

37 "[Adres D. I. Mendeleevu po povodu izbraniia ego v pochetnye chleny Russkogo fiziko-khimicheskogo obshchestva]," in Butlerov, *Sochineniia*, III, 166.

38 Quoted in Plate, "Novye materialy k biografii V. V. Markovnikova," 79. See also Volkova, "Materialy k deiatel'nosti A. M. Butlerova v Peterburge (1869–1886)," 12–15.

39 See, for example, Markovnikov's funeral oration for Butlerov, "Moskovskaia rech' o But-lerove," 174; and letters sympathizing with Butlerov's *Rus'* article, such as A. L. Potylitsyn to Butlerov, 4 February 1883, in Bykov, *Pis'ma russkikh khimikov k A. M. Butlerovu*, 360.

40 PFARAN f. 22, op. 1, d. 48, ll. 1–3.

41 PFARAN f. 22, op. 1, d. 44, ll. 2–44.

42 I. S. Aksakov to A. M. Butlerov, 21 January 1882, PFARAN f. 22, op. 2, d. 1, l. 1. For Ivan's biography, see Lukashevich, *Ivan Aksakov*.

43 Ivan Aksakov, "[Editor's introduction] [13 February 1882]," 3. Ellipses in original. The original quotation from which Aksakov drew Butlerov's title (printed in German in the Russian article) is: "die Akademie ist doch keine Russische, sondern eine Kaiserliche Akademie!"

44 Despite their disagreements, their relations remained cordial until 1886, the year they both died. See, for example, Butlerov to I. S. Aksakov, 20 April 1885, PD f. 3, op. 4, d. 82, ll. 1–2ob.

45 Butlerov, *Sochineniia*, III, 118.

46 Butlerov, *Sochineniia*, III, 119.

47 Butlerov, *Sochineniia*, III, 123.

48 Butlerov, *Sochineniia*, III, 125.

49 Butlerov, *Sochineniia*, III, 126.

50 Butlerov, *Sochineniia*, III, 129–130.

51 This is the central argument of Lincoln, *The Great Reforms*.

52 *Strana*, 27 November 1880, #93: 7. See also *S.-Peterburgskie Vedomosti*, 21 December 1880 (2 January 1881), #352: 2; *Golos*, 20 November (2 December) 1880, #321: 3; *Golos*, 21 November (3 December) 1880, #322: 2; *Golos*, 23 November (5 December) 1880, #324: 2–3; *Golos*, 24 November (6 December) 1880, #325: 3; *Golos*, 25 November (7 December) 1880, #326: 3.

53 Menshutkin, *Zhizn' i deiatel'nost' Nikolaia Aleksandrovicha Menshutkina*, 236.

54 *Golos*, 7 (19) December 1880, #338: 3; "Mendeleevskii obed [8 December 1880]."

55 As Butlerov wrote to Markovnikov on 10 November 1882: "Then they recalled your atti-tude to the Mendeleev election, and I had to guarantee that you both, in case of election, would accept the calling with gratitude. I hope that I was not mistaken and that you won't leave me in a very uncomfortable position with a refusal or a statement, like Mendeleev did, that 'I never wanted to be elected.'" Quoted in Butlerov, *Nauchnaia i pedagogicheskaia deiatel'nost'*, 268.

56 McReynolds, *The News under Russia's Old Regime*, Chapter 4; Ambler, *Russian Journalism and Politics*, 34.

57 McReynolds, "Imperial Russia's Newspaper Reporters"; and idem, "V. M. Doroshev.
58 Grossman, "Rise and Decline of the 'Literary' Journal, 1880–1917."
59 Dostoevsky advocated the creation of a voluntary Free Russian Academy of Sciences independent of Imperial control. Dostoevskii, *Polnoe sobranie sochinenii*, XXVII, 54. Saltykov-Shchedrin, in his *Diary of a Provincial in Petersburg*, blamed Germans for problems in the Academy of Sciences. Saltykov-Shchedrin, *Polnoe sobranie sochinenii*, X, 347–348.
60 Modestov, "Russkaia nauka i obshchestvo [4 December 1880]," 1.
61 The secondary literature on the Academy affair mimics these two strands exactly. For "bureaucracy" accounts, see: Kniazev, "D. I. Mendeleev i tsarskaia Akademiia nauk"; idem, "D. I. Mendeleev i imperatorskaia akademiia"; Tishchenko and Mladentsev, *Dmitrii Ivanovich Mendeleev, ego zhizn' i deiatel'nost'. Universitetskii period*, 195–218; and Figurovskii, *Dmitrii Ivanovich Mendeleev*, 192. For nationalist or "German"-centered accounts, see: Vucinich, *Empire of Knowledge*, 56–57; Kiseleva, "K voprosu o neizbranii D. I. Mendeleeva v Rossiiskuiu Akademiiu nauk"; and Romanovskii, *Nauka pod gnetom rossiiskoi istorii*, 84–88. For a combination of the two, charging that the Baltic Germans were agents of bureaucratic reaction, see Leicester, "Mendeleev and the Russian Academy of Sciences."
62 *Journal de St. Pétersbourg*, 13 January 1881, #11, kept in PFARAN f. 24, op. 1, d. 135, l. 12. Emphasis in original.
63 "Bezzakonniki v Akademii nauk [7 December 1880]," 3.
64 See, for example, "[Nasha Akademiia nauk i trebovaniia ee ustava] [17 December 1880]"; and "K voprosu o peresmotre ustava Akademii nauk [12 December 1880]."
65 "Voskresnye nabroski [16 November 1880]," 2. In fact, hopeful reformers of the Academy in late Imperial Russia frequently referred to Mendeleev's case to ply their cause, and the Soviet Academy used him to signal Soviet Russia's new direction: "In the first place, our Soviet Academy should lead the study and wide propagandizing of Mendeleev's works, remembering its deepest difference from the old Imperial Academy, which demonstratively slammed the door before this titan of scientific thought." Kedrov, "Ob otnoshenii k mendeleevskomu nasledstvu," 207. See also Tolz, *Russian Academicians and the Revolution*, 46.
66 I exclude the important exception of Jews. The general process of expansion by assimilation is discussed in LeDonne, *The Russian Empire and the World*. On the connection with industrialization, see Geyer, *Russian Imperialism*.
67 *Novoe Vremia*, 26 November (8 December) 1880, #1706: 1. The Polish silence was due to the Russian occupation of Poland and the contemporaneous forced Russification.
68 *Pokrok* was quoted in *S.-Peterburgskie Vedomosti*, 28 November (10 December) 1880, #328: 2. On Pan-Slavism "from below," see Hunczak, "Pan-Slavism or Pan-Russianism."
69 *Strana*, 13 November 1880, #89: 1. See also Petrushevskii, "Postupok Akademii nauk [31 December 1880]," 1. Newspapers were seminal in constructing a national consciousness during the Great Reforms. See McReynolds, *The News under Russia's Old Regime*, 44–46.
70 V. Zh., "Novyi podvig Akademii nauk [15 November 1880]," 1.
71 I found only one example: "Truth is universal, and thus science, as the highest expression of such truth, is the only honest cosmopolitanism. From this point of view the affront, which our Academy arranged all by itself, having voted against D. I. Mendeleev, resounds throughout Europe, and thus the very question of the present scandal deserves, even outside of all national ideas, some attention." V. P., "D. I. Mendeleev i Akademiia [19 November 1880]," 1.
72 Whittaker, *The Origins of Modern Russian Education*, 188; and Vucinich, *Empire of Knowledge*, 44.
73 By 1917, the vast majority were Great Russian, peppered with six Ukrainians, four Germans, one Croat, one Belorussian, and one Georgian. Tolz, *Russian Academicians and the Revolution*, 7, 19; Vucinich, *Empire of Knowledge*, 46.
74 Morozov, "Iz istorii nashei Akademii nauk [4 December 1880]," 3, for the case of N. Ia.

Ozeretskovskii (1775–1827), who tried to spearhead a commission to boost Russian member-ship in the Academy; Black, *G.-F. Müller and the Imperial Russian Academy*, 92 and 96, on hos-tility to Müller's German history of Siberia; and Kablukov, "Iz vospominanii o khimii v Moskovskom universitete," 729, for Markovnikov's resentment of non-Russian studies of Rus-sia.

75 See Gordin, "The Importation of Being Earnest," 30–31, for a refutation of this com-monly held view of Peter's intentions.

76 Bulgakov, "Nemetskaia partiia v Russkoi akademii [1881]," 430–431. This is one of the only "thick journal" pieces on the affair.

77 Some journalists would claim that by "Russians" they just meant loyal subjects of the Empire, who spoke Russiakh and had a Russian University degree, be they Russian, Polish, Ger-man, Armenian, Tatar, or whatever: *Novoe Vremia*, 30 December 1880 (11 January 1881): #1738: 1.

78 This vilification of Baltic Germans was so common that *The Voice* had already received two warnings in 1867 for their attacks on the group. The accumulation of twenty warnings for various reasons shut down this paper in 1883. Balmuth, *Censorship in Russia*, 34 and 101.

79 See Thaden, *Russification in the Baltic Provinces and Finland*; Weeks, *Nation and State in Late Imperial Russia*; Haltzel, "The Reaction of the Baltic Germans to Russification." On rela-tions before Russification, see Thaden with Thaden, *Russia's Western Borderlands*.

80 The censorship law of 1865 freed Moscow and Petersburg papers from pre-publication censorship, and persecuted German newspapers in the provinces saw this as an opportunity. See Henriksson, "The *St. Petersburger Zeitung*," 366–367; and idem, "Nationalism, Assimilation, and Identity," 352.

81 Especially in "Zur Nichtwahl Mendelejew's [24 December 1880]." For attacks on the *Zeitung*, see *Novoe Vremia*, 28 November (10 December) 1880, #1708: 1; "Novaia vylazka aka-demicheskikh nemtsev [28 December 1880]," 1.

82 Antonovich, "Predislovie [January 1881]," 241.

83 When attention finally shifted from Mendeleev, however, this journal that published the above complaint did not have much time to take advantage of it. For the very issue that dis-placed Mendeleev, the assassination of the Tsar, led to a clampdown on dissenting media. On the media reaction to the assassination of Alexander II, see McReynolds, *The News under Russia's Old Regime*, 92–95.

84 This case is argued in Dmitriev, "Skuchnaia istoriia." The negative case against Mendeleev is presented in "Zur Nichtwahl Mendelejew's [24 December 1880]."

85 M. D. L'vov to Butlerov, 20 August 1880, reproduced in Bykov, *Pis'ma russkikh khimikov k A. M. Butlerovu*, 206.

86 Quoted in Tishchenko and Mladentsev, *Dmitrii Ivanovich Mendeleev, ego zhizn' i deiatel'nost'. Universitetskii period*, 216.

87 Quoted in "Derptskie otgoloski akademicheskogo skandala [15 December 1880]," 1.

88 Kiseleva, "K voprosu o neizbranii D. I. Mendeleeva v Rossiiskuiu akademiiu nauk."

89 Veselovskii to Litke, 21 December 1880, PFARAN f. 24, op. 1, d. 135, ll. 10–11ob. As late as 1884, Veselovskii was still upset enough about Butlerov to send a packet of thirteen let-ters to conservative journalist K. D. Kavelin, one of which was forwarded to Butlerov (PD 20,398).

90 PFARAN f. 24, op. 2, d. 3, reproduced in Kniazev, "D. I. Mendeleev i tsarskaia Akademiia nauk," 324–325.

91 See the account in Ol'khovskii, "Tainyi arest akademika Famintsyna."

92 "Doklady gen.-ad. A. R. Drentel'na Aleksandru II," 165–166.

93 S. F. Glinka's account, reproduced in Makarenia, Filimonova, and Karpilo, *D. I. Mendeleev v vospominaniiakh sovremennikov*, 95. In 1886, A. S. Famintsyn led a brief and incon-

sequential effort to get Mendeleev elected as Butlerov's successor in the chemistry chair. In his personal notes, Famintsyn wrote: "As far as I know, the chief cause of the unacceptability of the above-mentioned candidate for an academic chair was his supposedly unpleasant character (*nrav*)." Quoted in Kniazev, "D. I. Mendeleev i tsarskaia Akademiia nauk," 327.

94 Kapustina-Gubkina, *Semeinaia khronika*, 213–217; Mendeleeva, *Mendeleev v zhizni*, 1–21; and Ozarovskaia, *D. I. Mendeleev v vospominaniiakh O. E. Ozarovskoi*, 135. Even Beilstein smirked about the matter in a letter to Germany: "Mendeleev appears, since the recent, epoch-making discoveries, to have lost the desire to compress gases anymore & and busies himself at the moment with: 'plaiser d'amour'—also not such a bad thing." Beilstein to Erlenmeyer, 22 February (6 March) 1878, reproduced in Krätz, *Beilstein-Erlenmeyer*, 60.

95 "Biograficheskie zametki o D. I. Mendeleeve (1906)," in Shchukarev and Valk, *Arkhiv D. I. Mendeleeva, t. 1*, 19.

96 Mendeleeva, *Mendeleev v zhizni*, 23–27.

97 Mendeleev had unsuccessfully tried to initiate divorce proceedings in a letter to Feozva on 16 January 1878, reproduced in Tishchenko and Mladentsev, *Dmitrii Ivanovich Mendeleev, ego zhizn' i deiatel'nost'. Universitetskii period*, 336. Divorce in Imperial Russia straddled the uneasy boundary between religious affairs and family law. During the Great Reforms, formal marital separations increased, and in 1881 began to skyrocket to about 1,500 civil separations a year. For the crime of infidelity, however, the injured party could either sue for divorce under the Orthodox Church, forfeiting the right to press criminal charges, or remain married but prosecute under criminal law. For a detailed analysis of family law, see Wagner, *Marriage, Property, and Law in Late Imperial Russia*. On the boundary between criminal and clerical law in divorce, see Engelstein, *The Keys to Happiness*, 51–52.

98 The appeal for divorce, on Mendeleev's admissions of infidelity, is contained in the Report to the Holy Synod by Isidor, Metropolitan of St. Petersburg and Novgorod, 26 January 1882, RGIA f. 796, op. 163, d. 1805, ll. 1–4. The official dissolution was granted in ibid., ll. 5–5ob.

99 Mendeleeva, *Mendeleev v zhizni*, 41.

100 Ozarovskaia, *D. I. Mendeleev v vospominaniiakh O. E. Ozarovskoi*, 139.

101 As one commentator put it, the 11 November rejection was "one of the reasons which greatly assisted in the creation of [Mendeleev's] popularity." B. N. Menshutkin in Mendeleev, *Periodicheskii zakon* (1926), 173.

102 Quoted in Raikov, "Iz vospominanii zoologa Aleksandra Mikhailovicha Nikol'skogo," 82–83.

103 Mendeleev's domestic honors also spike on his 70th birthday in 1904. For a complete list, see Skvortsov, "Uchenyi titul D. I. Mendeleeva."

104 Mendeleev's statement, roughly translated, was: "If you want to hear the applause, then you also have to hear the heckling." Quoted in Dobrotin et al., *Letopis' zhizni i deiatel'nosti D. I. Mendeleeva*, 209. Popular literature was littered with laudatory portrayals of virtuous bandits fighting for justice. See Brooks, *When Russia Learned to Read*, Chapter 5.

105 Mendeleev to A. I. Popova, 21 January 1881, ADIM Alb. 1/430, quoted in I. S. Dmitriev, "Skuchnaia istoriia."

106 Mendeleev also encouraged subscription for the Zinin and Voskresenskii prizes, shifting some attention from himself to the Chemical Society. *S.-Peterburgskie Vedomosti*, 1 (13) December 1880, #331: 2; and *Golos*, 16 November 1880, #317: 2.

107 Quotations from Mendeleev to K. M. Feofilaktov, director of Kiev University, reprinted in *Golos* 7 (19) December 1880, #338: 3; and *Novoe Vremia*, 5 (17) December 1880, #1715: 3–4.

108 Reproduced in Georgievskii, "Literaturnoe nasledstvo D. I. Mendeleeva," 39. See also Mendeleev's statement to the Russian Chemical Society upon being elected a permanent mem-

ber: "Rejected *there*, I am elected *here*, and I speak honestly that for me this election by an autonomous institution is much more pleasant." *Golos*, 20 December 1880 (1 January 1881), #351: 3. Emphasis in original.

109 *MS*, XXV, 676.

110 Kniazev, "D. I. Mendeleev i tsarskaia Akademiia nauk," 329; and Rykachev to Chief Bureau of Weights and Measures, 19 January 1898, RGIA f. 28, op. 1, d. 218, l. 5. For more on the poor relations between the Academy and the Chief Bureau under Mendeleev's tenure, see the recollections by M. N. Mladentsev published in Tishchenko and Mladentsev, *Dmitrii Ivanovich Mendeleev, ego zhizn' i deiatel'nost'. Universitetskii period*, 381–382.

111 Witte to Academy of Sciences, 4 December 1899, PFARAN f. 6, op. 1, d. 16, ll. 48–49. Witte also noted later: "It is to the discredit of the Academy of Sciences that when it came to choose a chemist to fill a vacancy in its ranks it would not elect [Mendeleev] because of his contentious personality and the fact that he had been critical of its work; instead it chose a chemist of little reputation." Witte, *The Memoirs of Count Witte*, 168. Konstantin Konstantinovich, president of the Academy, responded that since German influence had tapered somewhat, he could now put Mendeleev forth. Letter dated 19 December 1899, reproduced in Tishchenko and Mladentsev, *Dmitrii Ivanovich Mendeleev, ego zhizn' i deiatel'nost'. Universitetskii period*, 217. One can only presume that Mendeleev declined. For an example of Western views that Mendeleev was indeed a member, see Jaubert, "Mendeléeff," 97.

CHAPTER 6

1 Turgenev to K. S. Aksakov, 16 (28) January 1853, in Turgenev, *Pis'ma*, v. 2, 186.

2 ADIM II-A–10–2–13, published as Mendeleev, "Kakaia zhe Akademiia nuzhna v Rossii." On its publication history, see Meilakh, "Posleslovie."

3 Mendeleev, "Kakaia zhe Akademiia nuzhna v Rossii," 179.

4 Mendeleev, "Kakaia zhe Akademiia nuzhna v Rossii," 180–182.

5 Mendeleev, "Kakaia zhe Akademiia nuzhna v Rossii," 189.

6 Poirier, *Lavoisier*, Chapters 14–15; Brock, *Justus von Liebig*, Chapter 5.

7 His earlier faith in industrialization, some have argued, was drawn from his childhood in Tobol'sk, Siberia, where his mother reopened her inherited glass factory in nearby Aremzianka after the family was financially crippled by his father's sudden blindness. Supposedly, surrounded by industry in his youth, Mendeleev attempted to recreate those years on a larger scale. See Almgren, "Mendeleev," Chapter 1; Figurovskii, *Dmitrii Ivanovich Mendeleev*, Chapter 2; Almgren, "D. I. Mendeleev and Siberia." On Mendeleev's early biography, see Kapustina-Gubkina, *Semeinaia khronika*; and Tishchenko, "Dmitrii Ivanovich Mendeleev." Mendeleev maintained an interest in glass manufacturing both for sentimental reasons and because glass offered a prime example of what he called "indeterminate chemical compounds," i.e., solutions. See *MS*, XVII, 21–34 and especially 47–401 (Mendeleev's exhaustive 1864 compilation of information on the glass industry). See also Barzakovskii and Dobrotin, *Trudy D. I. Mendeleeva v oblasti khimii silikatov*. I contend that there is little continuity between such youthful musings and his later economic work. Such a continuity requires a representation of Mendeleev's mature views that allows for far greater democratic participation than Mendeleev *ever* advocated, even in the early 1860s at his most liberal.

8 For a useful survey, see Vucinich, *Social Thought in Tsarist Russia*.

9 Zapisnaia knizhka, 1874–1876, #20, ADIM II-A–1–1–9, l. 106.

10 Mendeleev, "Ob organizatsii sel'skokhoziaistvennykh opytov (1866)," *MS*, XVI, 28. Ellipses added. The protocol for observations is reproduced in *MS*, XVI, 37–55. On Mendeleev and agriculture, see Mendeleev, *Raboty po sel'skomu khoziaistvu i lesovodstvu*; Makarenia, "Vo-

ploshchenie mechty"; Kerova and Gasanova, *Boblovo*; and Ivanov, "D. I. Mendeleev i pishchevaia promyshlennost'."

11 Mendeleev, "Ob obshchestve dlia sodeistviia sel'skokhoziaistvennomu trudu (1870)," *MS*, XVI, 285. Ellipses added. Mendeleev would later call these "my first economic thoughts." *MS*, XVI, 260.

12 "Spisok moikh sochinenii (1899)," in Shchukarev and Valk, *Arkhiv D. I. Mendeleeva*, 57, 59. See Almgren, "Mendeleev," Chapter 4.

13 See, especially, Ulam, *Prophets and Conspirators in Prerevolutionary Russia*; Daly, *Autocracy under Siege*, 31 and passim; and Venturi, *Roots of Revolution*, 633–720.

14 Quoted in Zaionchkovsky, *The Russian Autocracy in Crisis*, 199–200.

15 Zaionchkovsky, *The Russian Autocracy in Crisis*. On the countryside in this period, see Pearson, *Russian Officialdom in Crisis*, 119. It is important not to overstate the discontinuity across the events of the assassination, as the transition from Reform to Counter-Reform was often gradual, as stressed in Freeze, *The Parish Clergy in Nineteenth-Century Russia*.

16 Taranovski, "Alexander III and His Bureaucracy," 217. See also Wortman, *Scenarios of Power. Volume 2*, 256–263 and 341–344; and Whelan, *Alexander III and the State Council*.

17 This is true even though under Nicholas II Mendeleev achieved rank IV on the Table of Ranks (the highest of any chemist), and the Order of Aleksandr Nevskii and the Order of the Knights of the White Eagle: Director of the Chancellery of the Ministry of Finances to Mendeleev, 8 June 1905, RGIA f. 28, op. 1, d. 1076, l. 32; and Witte to Mendeleev, 1 April 1902, ADIM Alb. 1/492. These honors served a complicated function of micromanaging status. See Bennett, "*Chiny, Ordena,* and Officialdom"; and idem, "Evolution of the Meanings of Chin."

18 On ministerial politics in the Great Reforms and after, see Rieber, "Interest-Group Politics in the Era of the Great Reforms"; idem, "The Sedimentary Society"; Lincoln, "Reform and Reaction in Russia"; and Orlovsky, *The Limits of Reform*.

19 For an official general history of the Ministry of Finances, see *Ministerstvo Finansov*.

20 The scope of these efforts is staggering; he offered his services in attracting foreign capital to Russia, shipbuilding, metallurgy, and energy development through coal mining, to name just a few. Shipbuilding: *MS*, XXV, 631–638; and *MS*, XXI, 53–63. International finance: Mendeleev to Ivan Pavlovich Shipov, 29 October 1905, RGIA f. 28, op. 1, d. 1076, ll. 38–38ob. Metallurgy: *MS*, XV, 323 (dated 1876); Mendeleev, ed., *Ural'skaia zheleznaia promyshlennost' v 1899 g.* (St. Petersburg: 1900), reprinted in *MS*, XII; Bardin and Rikman, "Raboty D. I. Mendeleeva v oblasti metallurgii." Coal: Kudriavtseva and Shekhter, *D. I. Mendeleev i ugol'naia promyshlennost' Rossii*.

21 See the series of works by Thomas C. Owen: "Impediments to a Bourgeois Consciousness in Russia"; *Russian Corporate Capitalism from Peter the Great to Perestroika*, 8, 125, and 153; *Capitalism and Politics in Russia*; and "Entrepreneurship and the Structure of Enterprise in Russia." See also Rieber, *Merchants and Entrepreneurs in Imperial Russia*, 27, 79; Rimlinger, "Autocracy and the Factory Order in Early Russian Industrialization"; Gorlin, "Problems of Tax Reform in Imperial Russia"; Blackwell, "The Russian Entrepreneur in the Tsarist Period"; and Bowman, "Russia's First Income Taxes."

22 Gregory, *Russian National Income*, 192.

23 Gerschenkron, *Economic Backwardness in Historical Perspective*, 123, and Chapter 6 more generally; and idem, *Europe in the Russian Mirror*. Olga Crisp in general accords with Gerschenkron's state-led development approach. See her *Studies in the Russian Economy before 1914*, especially Chapter 1. See also Gregory and Sailors, "Russian Monetary Policy and Industrialization." A vocal opposition has argued that any growth resulted directly from Emancipation and the contemporary demographic boom, and that in concrete instances, like the passing of the 1891 tariff, state measures actually caused inflation, bankruptcy, and unemployment: Kahan,

Russian Economic History, 65–68; Owen, *Russian Corporate Capitalism from Peter the Great to Perestroika*, 16; idem, "The Russian Industrial Society and Tsarist Economic Policy," 604–605; and Gatrell, "The Meaning of the Great Reforms."

24 The state-led aspects of Mendeleev's economics made him a useful political tool for the Soviets, who republished his works repeatedly (although with serious editorial cuts to blunt his heavily capitalist inclinations). See Mendeleev, *Problemy ekonomicheskogo razvitiia Rossii*; and idem, *Granits poznaniiu predvidet' nevozmozhno*.

25 See especially Walicki, *The Controversy over Capitalism*; and Szporluk, *Communism and Nationalism*, Chapter 13.

26 The still influential text that helped create the notion of a Witte system, often motivated by Gerschenkronian assumptions, is Von Laue, *Sergei Witte and the Industrialization of Russia*.

27 Von Laue, "Factory Inspection under the Witte System." This factory inspection system collapsed after the summer 1899 crisis in European financial markets cut credit to Russian government and industry. Interest rates rose and investors dumped securities; the Ministry of Internal Affairs besieged the inspection system. See Hogan, *Forging Revolution*, 47–48.

28 Foreign investment made up 50% of all new capital formation in Russia from 1893 to 1914. See, especially, McKay, *Pioneers for Profit*; Crisp, *Studies in the Russian Economy before 1914*, Chapter 7; Carstensen, "Foreign Participation in Russian Economic Life"; and idem, *American Enterprise in Foreign Markets*. The two American companies discussed by Carstensen were in fact the largest commercial industrial firms in Russia on the eve of World War I. All of these writers cite economics or corporate strategies rather than state inducements as the primary motivations for investing in the Russian economy.

29 Gregory and Sailors, "Russian Monetary Policy under Industrialization."

30 Witte, *The Memoirs of Count Witte*, 66 and 322.

31 Witte, *The Memoirs of Count Witte*, 168.

32 On the Baku oil industry in this period, see: McKay, "Baku Oil and Transcaucasian Pipelines"; idem, "Entrepreneurship and the Emergence of the Russian Petroleum Industry"; idem, "Restructuring the Russian Petroleum Industry in the 1890s"; Tolf, *The Russian Rockefellers*; Stackenwalt, "Dmitrii Ivanovich Mendeleev and the Emergence of the Modern Russian Petroleum Industry"; and Tishchenko and Mladentsev, *Dmitrii Ivanovich Mendeleev, ego zhizn' i deiatel'nost'. Universitetskii period*, 296–327.

33 The trip to the United States was important for crystallizing Mendeleev's views on the failure of liberal democracy to exploit resources sufficiently. He was dismayed at the 1876 World's Fair in Philadelphia that American scientists displayed no interest in oil. He advocated, however, the American approach of cutting oil-company subsidies and taxes. His complete views, along with his disparaging account of the United States, were published in *Neftianaia promyshlennost' v severo-amerikanskom shtate Pensil'vanii i na Kavkaze* (St. Petersburg: 1877), reprinted in *MS*, X, 17–244. Here, the Great Reforms were still his ideal: "The peaceful, just, and desirable further resolution of social tasks by way of gradual and strictly conducted internal reforms—this is expected from the Slavic peoples and first of all from Russia. . . . [T]hey see that Russia, moving thus upon her freedom to expand and grow, will achieve the highest position among other nations, and they hope that then Slavdom will introduce in European civilization such transformations, which they can't even contemplate now" (152–153). Mendeleev rarely expressed views in such concordance with the Slavophiles.

34 Mendeleev, "Proiskhozhdenie nefti [1877]"; protocol of the Russian Chemical Society, 15 January 1876, *MS*, X, 14; Mendeleev, *Neftianaia promyshlennost'* (1877), Chapter 4. See also Kharichkova, "Zaslugi Dmitriia Ivanovicha Mendeleeva v oblasti izucheniia nefti i neftianoi tekhniki"; Nametkin, "Trudy D. I. Mendeleev v oblasti izucheniia nefti i neftianoi promyshlennosti."

35 The few Mendeleev scholars who have attempted economic studies have used the oil

work as a lens for his evolving views: Stackenwalt, "The Economic Thought and Work of Dmitrii Ivanovich Mendeleev"; Almgren, "Mendeleev," Chapter 5.

36 Parkhomenko, *D. I. Mendeleev i russkoe neftianoe delo.*

37 Kovalevskii to Mendeleev, 11 May 1901, ADIM Alb. 1/445. Emphasis in original.

38 Kovalevskii to Mendeleev, 12 and 13 May 1901, ADIM Alb. 1/446–447.

39 For a statistical assessment of this claim, see Wheatcroft, "The 1891–92 Famine in Russia."

40 Anonymous, "Rabochii vopros v Rossii," undated, ADIM Alb. 1/448.

41 Mendeleev, "Rabochii vopros v Rossii," ADIM Alb. 1/449.

42 Notes on the back of Kovalevskii's original letter, ADIM Alb. 1/445.

43 Stackenwalt, "The Economic Thought and Work of D. I. Mendeleev," especially Chapter 2. Stackenwalt claims Mendeleev based his economic thought on positivism, ignoring Mendeleev's explicitly anti-positivist metaphysics (see Chapter 8 below).

44 Mendeleev, "Osnovy fabrichno-zavodskoi promyshlennosti. Toplivo (1897)," in Mendeleev, *S dumoiu o blage rossiiskom*, 28.

45 Mendeleev, "O vozbuzhdenii promyshlennogo razvitiia v Rossii," ADIM Bib. 1022/11b, in *MS*, XX, 80. See also Mendeleev, "Osnovy fabrichno-zavodskoi promyshlennosti. Toplivo (1897)," 40. Mendeleev visited Leiden University on its 300th anniversary as a representative to the institution that Peter the Great used to train so many of his foreign specialists. There he praised Peter as a modernizer. Zapisnaia knizhka, 1874–1876, #20, ADIM II-A–1–1–9, ll. 50–52.

46 Mendeleev, "Uchenie o promyshlennosti. Vstuplenie v biblioteku promyshlennykh znanii (1900)," in Mendeleev, *S dumoiu o blage rossiiskom*, 67. See also Mendeleev, "Osnovy fabrichno-zavodskoi promyshlennosti. Toplivo (1897)," 129n3.

47 Mendeleev, "Pis'ma o zavodakh: Pis'mo pervoe (1885)," *MS*, XX, 108.

48 Mendeleev, "Uchenie o promyshlennosti (1900)," 100. Adam Smith's thought had been well known in Russia even before his *Wealth of Nations* was translated in 1802–1806. See Taylor, "Adam Smith's First Russian Disciple."

49 Quoted in Gindin, "D. I. Mendeleev o razvitii promyshlennosti v Rossii," 212.

50 Mendeleev, "Uchenie o promyshlennosti (1900)," 167n1.

51 Mendeleev, "Ob usloviiakh razvitiia zavodskago dela v Rossii (1882)," *MS*, XX, 30.

52 Mendeleev, "Ob usloviiakh razvitiia zavodskago dela v Rossii (1882)," *MS*, XX, 33.

53 Mendeleev, *Zavetnye mysli*. The version in *MS*, XXIV has so many erasures by the Soviet censorship as to be virtually unreadable. Reviewers greeted the original rather favorably; one in particular pointed to the strong evolutionary framework: *Novoe Vremia*, 11 (24) June 1904, #10156: 2. On public relations more generally, see Mendeleev to Kovalevskii, 11 April 1893, *MS*, XXI, 103; Fedor Kalinich Tranilin to Mendeleev, 5 November 1889, ADIM Alb. 2/292–293; Owen, *Russian Corporate Capitalism from Peter the Great to Perestroika*, 5; Pogodin, "Bor'ba D. I. Mendeleeva za razvitie otechestvennoi khimicheskoi promyshlennosti"; Musabekov, "D. I. Mendeleev i konstantinovskii zavod v Iaroslavle."

54 Mendeleev, *Zavetnye mysli*, 20n4.

55 Mendeleev, *Zavetnye mysli*, 26.

56 Mendeleev, *Zavetnye mysli*, 136–137. On the contemporary Russian urban crisis see Hamm, "The Breakdown of Urban Modernization."

57 Mendeleev, *Zavetnye mysli*, 180.

58 Agriculture, for example, was only possible part of the year and its labor force could not expand elastically. Mendeleev, *Zavetnye mysli*, 96.

59 Minister of Finances I. A. Vyshnegradskii asked for Mendeleev's assistance in reviewing the tariff in September 1889. Mendeleev always noted that his involvement was solicited. Mendeleev, "[Letter to the editor] [23 June 1891]." Mendeleev wrote this letter in response to

an article claiming he worked on the tariff "on his own initiative." The guilty article described the merits of targeted protectionism fairly. See "Novyi tamozhennyi tarif [21 June 1891]."

60 All figures are in terms of kopecks/*pud*, where a *pud* is 36 pounds. Crisp, *Studies in the Russian Economy before 1914*, 23–29.

61 See the avowals by contemporary observer Fenin, *Coal and Politics in Late Imperial Russia*, 175. Mendeleev collected pamphlets attacking both him and the 1891 tariff: Neyman-Spallart, *Proteksionizm i vsemirnoe khoziaistvo*; Sergeev, *Voprosy russkoi promyshlennosti*, 52–73; and Krestovnikov, *"Tolkovyi tarif" D. I. Mendeleeva*. Marginal annotations show that he read these texts very carefully. Historian Richard Wortman erroneously but understandably interprets the popular designation of the "Mendeleev tariff" as a result of Mendeleev's calls for protectionism, rather than the other way around. Wortman, *Scenarios of Power. Volume 2*, 266. Witte also turned to S. M. Propper and the daily *Birzhevye Vedomosti* as a crucial forum for defending the Witte system. See McReynolds, *The News under Russia's Old Regime*, 128–129.

62 Mendeleev, "Opravdanie protektsionizma [11 July 1897]," 2.

63 Mendeleev, "Opravdanie protektsionizma [11 July 1897]," 3.

64 Mendeleev, *Tolkovyi tarif* (1892), 46.

65 Its only certain benefit was to leverage Germany in a two-year commercial dispute. Kahan, *Russian Economic History*, 99–102; Rieber, *Merchants and Entrepreneurs in Imperial Russia*, 117; Owen, *Capitalism and Politics in Russia*, 47, 136.

66 Nemchinov, "Ekonomicheskie vzgliady D. I. Mendeleeva."

67 Mendeleev, "O sovremennom razvitii nekotorykh khimicheskikh proizvodstv, v primenenii k Rossii i po povodu vsemirnoi vystavki 1867 goda (1867)," *MS*, XVIII, 75. On List's protectionist system, see Szporluk, *Communism and Nationalism*.

68 Reproduced in Shchukarev and Valk, *Arkhiv D. I. Mendeleeva*, 36. Ellipses added.

69 Mendeleev, *Materialy dlia peresmotra obshchago tamozhennago tarifa Rossiiskoi Imperii po evropeiskoi torgovle* (1889), *MS*, XVIII, 277.

70 Mendeleev, *Zavetnye mysli*, 320.

71 Mendeleev, *Tolkovyi tarif*, iv and 56.

72 Mendeleev in 1889, quoted in Veinberg, *Iz vospominanii o Dmitrii Ivanoviche Mendeleeve kak lektor*, i.

73 Mendeleev to N. F. Zdekauer, 1889, reproduced in Volkova, "Perepiska D. I. Mendeleeva s inostrannymi uchenymi," 739. Ellipses added.

74 Zapisnaia knizhka, 1874–1876, #20, ADIM II-A–1–1–9, l. 106.

75 "Conservatism is a great and inevitable affair, but there is no need at all to worry about it in the specific matter of education, because [education] consists first of all in the transfer of science, and [science] is the collection of past and generally accepted wisdom, [and] because people imbued with science are essentially and unavoidably conservative to a certain degree and one must not let the crowd teach them from the straining of a conservative society, but leave [the task to] wise men who themselves seek the highest principles by uniting them with the crowd. . . ." Mendeleev, *Zavetnye mysli*, 278.

76 Zapisnaia knizhka, 1878–1879, ADIM II-A–1–1–20, ll. 50a–51, quoted in Zabrodskii, *Mirovozzrenie D. I. Mendeleeva*, 140. This clearly opposes Stalin-era claims that Mendeleev subconsciously believed in Bolshevik doctrines. See, for example, Fritsman, "D. I. Mendeleev," 115. For more on such interpretations, see Chapter 9.

77 Mendeleev, *Tolkovyi tarif*, 3. This chemical analogy had a different valence earlier in his career. For example, in the draft of a November 1871 letter to Erlenmeyer on organic chemistry, Mendeleev wrote: "[Hermann] Kolbe finds an emperor among princes, and you don't want an emperor, but nonetheless recognize princes. He is a final and consequential absolutist and you, forgive me for this, are a separatist. If only you would allow for atoms a commune, equality, and fraternity. Why the need to translate human relations to our so tiny atoms? If there is inequality

among them, this does not force us, in my opinion, to create for them great and small princedoms. The mastery is of another order. . . . On the other hand I, God save me, am not a communist, but I unwillingly stand on the view of chemical communism, considering all of you who revolve around Kolbe chemical monarchists." Mendeleev, *Nauchnyi arkhiv, t. I: Periodicheskii zakon*, 707. Ellipses added. On Kolbe and politics, see Rocke, *The Quiet Revolution*.

78 Mendeleev, *Zavetnye mysli*, 83.

79 "It is impossible to think that the matter is tending to the formation of a united general empire (or republic), but I think that it is moving toward the near elimination of small powers, or to their assimilation, as well as of [certain] large powers. . . ." Mendeleev, *Dopolneniia k poznaniiu Rossii* (1907), 11. Ellipses added. In Mendeleev's metaphysics, addressed in Chapter 8, this propensity against universal assimilation was identified with individual "spirit" or "soul."

80 Mendeleev, "Uchenie o promyshlennosti (1900)," 171n.

81 ". . . I consider the affairs of state not only theoretical but also purely practical, which thus needs its own special preparation, and the correct motion forward is for me thinkable not by decision of the majority, [which] is composed of all sorts of types of people, but only under the influence of leading individual people, the choice of whom under any form of direction remains accidental in a certain sense, [and] is most probable under the rule of monarchism, [which is] not interested in particulars, but determined only by national interests." Mendeleev, *Zavetnye mysli*, 148.

82 Departament Torgovli i Manufaktur Ministerstva Finansov, *Fabrichno-zavodskaia promyshlennost' i torgovlia Rossii* (1893), 3.

83 Mendeleev, *Zavetnye mysli*, Chapter 2; Mendeleev, *K poznaniiu Rossii*, 6th ed. (1907), 14n4. On Russian hostility to Malthusianism, see Todes, *Darwin without Malthus*.

84 Mendeleev, *Dopolneniia k poznaniiu Rossii*, 5.

85 Mendeleev, *Zavetnye mysli*, 364.

86 Mendeleev, *Dopolneniia k poznaniiu Rossii*, 17.

87 Mendeleev, *Zavetnye mysli*, 89 (quotation) and also 328.

88 Mendeleev, "O vozbuzhdenii promyshlennogo razvitiia v Rossii," *MS*, XX, 87. See also Mendeleev, "Pervaia nadobnost' russkoi promyshlennosti (1888)," *MS*, XXI, 27–28; Mendeleev to Witte, April 1902, *MS*, XVI, 330. A Ministry of Trade and Industry was in fact created in the early twentieth century.

89 See Ken Alder's fine account, "A Revolution to Measure"; and Crosland, "'Nature' and Measurement in Eighteenth-Century France." On the social implications of metrological standardization, see Porter, "Objectivity as Standardization"; and Schaffer, "Late Victorian Metrology and Its Instrumentation." On unification of measures and empire building, see Kula, *Measures and Men*. An interesting Latourian reading of standardization as the circulation of particulars ("immutable mobiles") is given by O'Connell, "Metrology."

90 I translate the Russian "measures and weights" as "weights and measures," which is more euphonious to the English reader. The secondary literature on Mendeleev's metrology is vast. For some of the better sources, see Mladentsev, "Uchrezhdenie Glavnoi Palaty mer i vesov i eia deiatel'nost'"; *D. I. Mendeleev. Ego nauchnoe tvorchestvo i raboty v Glavnoi Palate mer i vesov*; Egorov, "Dmitrii Ivanovich Mendeleev (nekrolog)"; Shost'in, *D. I. Mendeleev i problemy izmereniia*; Azernikov and Vasil'kova, *Mendeleev—metrolog*; Boitsov, *D. I. Mendeleev*; Mendeleev, *Trudy po metrologii*; and Brooks, "Mendeleev and Metrology."

91 I deal with this progression at greater length in "Measure of All the Russias." See also Kamentseva and Ustiugov, *Russkaia metrologiia*. Mendeleev's draft of the 1899 law, "Proekt, sostavlennyi Podkommisieiu v Glavnoi Palate mer i vesov 18–21 fevralia 1897 g.," with his comments, is kept at ADIM Bib. 1038/6.

92 *MS*, XXII, 25–26. Interestingly, almost all of the domestic advocates of the metric system in Russia were chemists. This was a more general phenomenon, as noted in another context:

"To engineers such relations [of length to mass and specific gravity] are of small moment, and consequently among English-speaking engineers the metric system is making no progress, while, on the other hand, the chemists have eagerly adopted it." Harkness, *On the Progress of Science as Exemplified in the Art of Weighing and Measuring*, 64–65.

93 *MS*, XIII, 16n2, 18–20.

94 Mendeleev's first venture into this field was an entry in a contest to measure alcohol content. Mendeleev and E. Radlov, 30 July 1863, RGIA f. 18, op. 8, d. 253. Mendeleev's 1864 doctoral dissertation is reproduced in *MS*, IV, 1–152. Mendeleev also modified the "Alcoholometry" chapter in his translation of Wagner's *Technology* (1862), *MS*, XV, 230–288. For a survey of alcoholometry and Mendeleev's involvement in the reform of the vodka farm, see Bondarenko, "Iz istorii russkoi spirtometrii"; Ivanov, "D. I. Mendeleev"; and Christian, *"Living Water"*, 364. On the reform in the context of the Great Reforms, see idem, "A Neglected Great Reform."

95 For the legend of Mendeleev's "creation" of the 80-proof standard, see Pokhlebkin, "Mendeleev i vodka." His view is effectively refuted by Dmitriev, "Natsional'naia legenda." Dmitriev's argument is briefly summarized in Gordin, "The Science of Vodka."

96 The Mendeleev hire was approved directly by Witte on 1 May 1893: RGIA f. 40, op. 1, d. 1, ll. 66–66ob.

97 There were, however, bureaucratic requests to accelerate the introduction of the metric system as it became more necessary for international trade. Mendeleev to V. I. Mikhnevich, 19 March 1894, RGIA, f. 28, op. 1, d. 195, ll. 2–3ob.; and Mendeleev to V. M. Verkhovskii, 20 February 1898, RGIA f. 28, op. 1, d. 220, ll. 2–3ob.

98 "Otchet za 1876 god o deiatel'nosti Imperatorskago Russkago Tekhnicheskago Obshchestva . . . ," 23 April 1877, RGIA f. 90, op. 1, d. 1, ll. 70–80ob., on 76ob.

99 The final vote was 3 to 11. See the text of Butlerov's objection in Volkova, "Materialy k deiatel'nosti A. M. Butlerova v Peterburge," 15.

100 On the role of international agreements in inducing Russian participation in the metric system, and on the role of Russian scientists in establishing such agreements, see Radovskii, "K uchastiiu russkikh uchenykh v mezhdunarodnykh soglasheniiakh o edinstve mer i vesov"; and Guillaume, *La Convention du Mètre*. On Russian metrological politics, see Kamentseva and Ustiugov, *Russkaia metrologiia*, 181–186; and Hallock and Wade, *Outlines of the Evolution of Weights and Measures*, 71, 94.

101 Mendeleev was proposed for nomination by G. Tresca of the French Conservatoire, and the two kept in close touch for several decades. Zapisnaia knizhka, 1874–1876, #20, ADIM II-A–1–1–9, entry of 20 June 1874, l. 12; and Mendeleev to Tresca, 5 May 1894, ADIM I-A–25–1–11. For a meticulous chronicle of Mendeleev's work on the commission, see Kamenogradskaia, *Deiatel'nost' D. I. Mendeleeva v S.-Peterburgskom universitete*, 148–156.

102 *Metricheskaia reforma v SSSR*; Arutiunov, *50 let metricheskoi reformy v SSSR*; Vlasov, "Istoricheskaia spravka po vvedeniiu metricheskoi sistemy v SSSR."

103 Witte, *The Memoirs of Count Witte*, 168.

104 The Ministry of Finances's proposal for the transformation of the Depot is "O preobrazovanii Depo obraztsovykh mer i vesov," 26 April 1893, ADIM Bib. 1044/78.

105 Mendeleev also undertook serious efforts to standardize electrical units in Russia. In this case, there was substantial unanimity from the Russian Technical Society, the Academy of Sciences, and the Chief Bureau that the metric units (ohm, volt, ampere) should be adopted, and Mendeleev—just as Glukhov had a decade earlier—pressed for adherence to international metric electrical protocols. Likewise, he established a long overdue electrical laboratory at the Chief Bureau. See Mendeleev to V. I. Kovalevskii, 8 January 1895, *MS*, XXII, 750–751; Mendeleev to Kovalevskii, 16 March 1894, *MS*, XXV, 538; Mendeleev to Department of Trade and Manufactures, 23 October 1893, RGIA f. 28, op. 1, d. 184, ll. 2–2ob.; and Glukhov to the Department,

23 April 1886, RGIA f. 28, op. 1, d. 149, ll. 2–2ob. On regulations for electrical measurement in Russia, see: Egorov, "O pravitel'stvennoi vyverke elektricheskikh izmeritel'nykh priborov"; and "Vremennyia pravila dlia ispytaniia i poverki elektricheskikh izmeritel'nykh priborov." On the electrical laboratory, see Lebedev, "Elektricheskoe otdelenie Glavnoi Palaty mer i vesov."

106 These inevitability statements are quite frequent; see, for example, *MS*, XXII, 26; and XXV, 560. On the use of decimal accounting in other systems, see XXII, 325.

107 Mendeleev to Kovalevskii, 21 December 1892, *MS*, XXII, 29.

108 *MS*, XXII, 30 and 47.

109 This highly respected firm prepared many national standards in this period. Mendeleev had even consulted with them during his gas work. Zapisnaia knizhka, 1874–1876, #20, ADIM II-A-1-1-9, ll. 3–4. On the difficulties of producing reliable standards, see Matthey, "The Preparation in a State of Purity of the Group of Metals Known as the Platinum Series." For Mendeleev's progress reports, see: Mendeleev to V. I. Kovalevskii, 30 October 1893, *MS*, XXII, 727–30; *MS*, XXII, 175–213; Mendeleev to Witte, [1895?], *MS*, XXII, 752; and the final report sent to Witte, "Vozobnovlenie prototipov ili osnovnykh obraztsov russkikh mer vesa i dliny v 1893–1898 gg. (1898)," *MS*, XXII, 393–721.

110 *MS*, XXII, 44, 731, 746; and XXV, 552. Mendeleev minutely outlined the requisite weighing procedures in XXII, 215–223. The final table of conversions, also sent to the International Metric Commission in Paris, is reproduced in XXII, 763–769.

111 Throughout, I have translated the word *palatka* as bureau (lowercase), to show the symmetry between these smaller units and the Chief Bureau (*Palata*) in Petersburg.

112 Glukhov, "Zapiska o sostoianii Depo obraztsovykh mer i vesov i voobshche o merakh i vesakh v Rossii," 28 December 1889, RGIA f. 28, op. 1, d. 167, ll. 1–7, on 1ob.; and Glukhov to Department of Trade and Manufactures, 22 March 1886, RGIA f. 28, op. 1, d. 148, ll. 1–2ob.

113 The model Mendeleev most closely approximated was the German Physikalisch-Technische Reichsanstalt, described in Cahan, *An Institute for an Empire*.

114 The cities and regions visited were: Nizhnii Novgorod, Kazan, Saratov, Riazan, Moscow, Orel, Kursk, Tomsk, Krasnoiarsk, Irkutsk, St. Petersburg (both port and land customs houses), Verzhbolovskii, Odessa, Warsaw, Grantskii, Riga, Lodz, Lublin, Tver, Maloiaroslavets, Kaluga, Tula, Vladimir, Kostroma, Iaroslavl', Rybinsk, Syzran', Penza, Samara, Ufa, Zlatoust', Cheliabinsk, Ekaterinburg, Perm, Irbit, Smolensk, Chernigov, Kiev, Berlin, Munich, Vienna, and Paris. The reports were published as: Lamanskii, "Iz otcheta, predstavlennago i. o. inspektora Glavnoi Palaty mer i vesov S. I. Lamanskii"; Blumbakh, "Dannyia o vyverke mer i vesov v Sibiri"; Skinder and Lamanskii, "Materialy dlia sostavleniia instruktsii o vyverke torgovykh mer i vesov"; Dobrokhotov, "Otchet o komandirovke v Tver', Moskvu, Maloiaroslavets . . ."; Egorov, "Otchet po komandirovke v goroda Varshavu, Lodz' i Liublin"; idem, "Otchet o komandirovke v goroda: Riazan' . . ."; and idem, "Otchet o komandirovke v goroda: Smolensk. . . ."

115 Blumbakh, "Dannyia o vyverke mer i vesov v Sibiri," 129.

116 ". . . [A] wider application of the metric system cannot be considered until local verification establishments are built and well organized." Mendeleev to Verkhovskii, 20 February 1898, RGIA f. 28, op. 1, d. 220, ll. 2ob.–3.

117 *MS*, XXV, 549. For a few of the statements on the importance of these local bureaus for the metric reform, see *MS*, XXII, 327, 328, 800, and 838. Its relation to the renewal of prototypes is described in a letter to V. M. Verkhovskii on 20 February 1898, in *MS*, XXII, 770.

118 *MS*, XXII, 794. Mendeleev's assistants also used this rhetoric, at one point claiming that "such control is possible only upon the existence of an entire net of special local verification establishments. . . ." Egorov, "Otchet po komandirovke v goroda Varshavu, Lodz' i Liublin," 74.

119 In favor of using status quo institutions was the Director of Moscow Assaying District

to Department of Trade and Manufactures, 11 April 1899, RGIA f. 28, op. 1, d. 482, ll. 3–4ob.

120 Mendeleev in Ministerstvo Finansov, Departament Torgovli i Manufaktur, *Zhurnaly zasedanii kommissii po peresmotru deistvuiushchikh o merakh i vesakh uzakonenii* (St. Petersburg: V. Kirshbaum, 1897), ADIM Bib. 1034/6, ll. 10, 17, and 22.

121 *MS*, XXII, 792–797.

122 *MS*, XXII, 792, 795, 847. For the actual regulations on stamping and verification, see "Vremennaia (1898 g.) instruktsiia No. 1"; and "Vremennaia (1898 g.) instruktsiia No. 2."

123 They all had the same training protocols, the same notions of measurement, and, thanks to the renewal of prototypes, the "same" standards in hand to carry to their separate zones. See Mendeleev, "Programma dlia ispytaniia v znanii metrologicheskikh priemov dlia lits. . . ." Women could even serve as competent verifiers, but only at a maximum ratio of one woman for each five men, and no more than two women per bureau. See Mendeleev to Kovalevskii, 21 October 1902, *MS*, XXII, 825–826. Mendeleev approved of female workers purely for financial reasons, as women were cheaper to hire than men. This clearly parts from the feminist appraisal of Mendeleev by one of his female employees: Ozarovskaia, *D. I. Mendeleev po vospominaniiam O. E. Ozarovskoi*. Mendeleev also felt that "persons of the female gender" (as he invariably called them) should be hired because tedious precision suited women well. This is especially interesting since in Prussia precision was gendered male as something that required great patience and intellect. See Olesko, "The Meaning of Precision," 126.

124 On the need for reserve verifiers, see *MS*, XXII, 793, 838. Mendeleev explained that verifiers needed to circulate more than police and tax-collectors and thus could perform some of their functions. On wagons, see *MS*, XXII, 839.

125 Egorov and Dobrokhotov, "Reviziia vesov i gir' v Gosudarstvennom Banke"; and Egorov, Dobrokhotov, and Müller, "Reviziia vesov i gir' v Pochtamte i pochtovykh otdeleniiakh g. S. Peterburga." Regulations for the conduct of these inspections were spelled out in "Instruktsiia dlia proizvodstva vnezapnykh revizii."

126 On shorter terms: *MS*, XXII, 840. On self-sufficiency: *MS*, XXII, 837. On statistics: *MS*, XXII, 842, 847–848.

127 Statement from the "Note from the protocol of the fourth [all-Russian trade-industrial] conference," 9 August 1896, *MS*, XXII, 329–330. As he wrote to V. M. Verkhovskii even earlier (1898): "[The metric system's] immediate introduction as obligatory would be directly deleterious to the success of the matter. *It will come in its own time, and one must think it will be soon.*" Mendeleev to Verkhovskii, 20 February 1898, RGIA f. 28, op. 1, d. 220, l. 3. Emphasis in original. Mendeleev had advocated optional implementation to Verkhovskii as far back as 21 December 1892, *MS*, XXII, 32. On the French origin of the metric system and on the difficulty of introducing it in practice, see Kennelly, *Vestiges of Pre-Metric Weights and Measures*; and Favre, *Les Origines du Système Métrique*. On the contemporary status of the metric system in Europe and America, see Eastburn, *The Metric System*.

128 One of Mendeleev's final metrological manifestoes was a 1906 plea arguing for the full implementation of the metric reform by the establishment of the entire array of local bureaus (*MS*, XXV, 609). In fact, the battle was all but lost by 1906: All twenty existing bureaus were completed by the end of 1904. The Ministry of Finances determined, in consultation with the Senate, that although the bureaus that existed displayed extraordinary profitability, they had already saturated the highly industrialized zones, and any further construction would yield diminishing returns. This, in addition to the fact that the Russo-Japanese War had sapped the Imperial Treasury, doomed any further budget increases for the Central Bureau. [D. I. Mendeleev], "Predstavlenie Ministerstva Finansov v Gosudarstvennyi Sovet o tom zhe [dalneishem ustroistve mestnykh poverochnykh uchrezhdenii v Imperii i o potrebnykh dlia sego kreditakh, a ravno o nekotorykh izmeneniiakh v deistvuiushchem zakone o merakh i vesakh i v shtate Glavnoi Palate mer i vesov]," 24 May 1907, ADIM Bib. 1052/5.

129 Wirtschafter, *Social Identity in Imperial Russia*, 8.

130 On the historical growth and permeation of such systems thinking in Western, and especially American, thought and culture, see Hughes and Hughes, *Systems, Experts, and Computers*.

131 Mendeleev to Procurator of the Holy Synod K. P. Pobedonostsev, 21 October 1898, *MS*, XXV, 588–589; Mendeleev, "Zaiavlenie o reforme kalendaria (1898/9)," *MS*, XXII, 774–777; and Mendeleev, "Kalendarnoe ob"edinenie (1900)," *MS*, XXII, 360.

CHAPTER 7

1 Lomonosov, *Polnoe sobranie sochinenii*, VIII, 206.

2 *MS*, XXIII, 97–98. Ellipses added.

3 Mendeleev, "Spisok moikh sochinenii (1899)," in Shchukarev and Valk, *Arkhiv D. I. Mendeleeva*, 82–83.

4 *MS*, XXIII, 234 (see 74 and 228 on examinations; 83 on the "Academy of Instructors").

5 *MS*, XXIII, 252.

6 Mendeleev to Witte, 15 October 1895, RGIA f. 560, op. 26, d. 29, ll. 21–44ob., on l. 21ob. Ellipses added. In the late 1870s, Mendeleev sent a similar proposal to Minister of the Interior M. T. Loris-Melikov. See the document in Tishchenko and Mladentsev, *Dmitrii Ivanovich Mendeleev, ego zhizn' i deiatel'nost'. Universitetskii period*, 94–96.

7 Mendeleev to Witte, 15 October 1895, RGIA f. 560, op. 26, d. 29, l. 25ob.

8 This according to a genealogy by Mendeleev's brother Pavel in 1880: "Iz rodoslovnoi, sostavlennoi v 1880 g. bratom Pavlom Ivanovichem (sluzhil togda v Novgorode)," in Shchukarev and Valk, *Arkhiv D. I. Mendeleeva*, 11. On Mendeleev's youth, see Mladentsev and Tishchenko, *Dmitrii Ivanovich Mendeleev*, v. 1; and Kapustina-Gubkina, *Semeinaia khronika*.

9 According to Mendeleev's autobiographical notes of 1906, his father was born on 18 February 1783 and his mother on 16 January 1793, although he suspected, given his memory of their appearance, that they were actually older than that. Mendeleev, "Biograficheskie zametki o D. I. Mendeleeve," in Shchukarev and Valk, *Arkhiv D. I. Mendeleeva*, 13–15.

10 Mendeleev, *K poznaniiu Rossii*, 6, 132–141.

11 D. Mendeleev, ed., *Ural'skaia zheleznaia promyshlennost' v 1899 g.* (St. Petersburg: Demakov, 1899), reprinted in *MS*, XII, 561–562 and 581; Dobrotin et al., *Letopis' zhizni i deiatel'nosti D. I. Mendeleeva*, 416.

12 On the Siberia myth, see Bassin, "Inventing Siberia"; and idem, "Turner, Solov'ev, and the 'Frontier Hypothesis.'" On Lomonosov's importance in creating this myth, see Boele, *The North in Russian Romantic Literature*.

13 Mendeleev, *Issledovanie vodnykh rastvorov po udel'nomu vesu* (1887), in Mendeleev, *Rastvory*, 379.

14 Mendeleev's writings on solutions are collected in *MS*, III and IV; Mendeleev, *Nauchnyi arkhiv: Rastvory*; and Mendeleev, *Rastvory*. On Mendeleev's solutions work, see Storonkin and Dobrotin, "Ob osnovnom soderzhanii ucheniia D. I. Mendeleeva o rastvorakh"; idem, "Kratkii ocherk ucheniia D. I. Mendeleeva o rastvorakh"; Shchukarev, "Uchenie ob opredelennykh i neopredelennykh soedineniiakh v trudakh russkikh uchenykh"; and Val'den [Walden], "O trudakh D. I. Mendeleeva po voprosu o rastvorakh."

15 Dolby, "Debates over the Theory of Solution," 327–331. See also Hiebert, "Developments in Physical Chemistry at the Turn of the Century"; idem, "The Energetics Controversy and the New Thermodynamics"; and Servos, *Physical Chemistry from Ostwald to Pauling*.

16 Walden, "Dmitri Iwanowitsch Mendelejeff," 4777. On the glass factory, see p. 4772.

17 Walden, "Dmitri Iwanowitsch Mendelejeff," 4779–4880.

18 Sacks, *Uncle Tungsten*.

19 Reproduced in Tishchenko and Mladentsev, *Dmitrii Ivanovich Mendeleev, ego zhizn' i deiatel'nost'. Universitetskii period*, 350.

20 Recollection by M. N. Mladentsev, published in Tishchenko and Mladentsev, *Dmitrii Ivanovich Mendeleev, ego zhizn' i deiatel'nost'. Universitetskii period*, 382.

21 Bischoff, Review of Mendelejeff's *Grundlagen der Chemie*, 264. Emphasis in original.

22 This point is nicely articulated in Mendeleev's German obituary: "Then it must be stressed that this book was of decisive significance for the *origin* and *further development* of his 'periodic system of elements': it originated first during the working out of the 'Principles' and was then first employed precisely in this textbook, and each further edition of the book noted and commented on each new success of the 'System,' and in each edition the author himself always introduced new thoughts and enlargements to the system." Walden, "Dmitri Iwanowitsch Mendelejeff," 4736. Emphasis in original.

23 *MS*, II, 258. Mendeleev admitted in the preface to the third edition (*MS*, XXIV, 4) that he had not realized in 1869 how widely applied his principle could be.

24 *MS*, XXIV, 13. By the terms of the divorce, Mendeleev's first wife received his University salary, so he had to support his new family with consulting and royalties from *Principles*.

25 *MS*, II, 328n13.

26 Mendeleev, *Izbrannye lektsii po khimii*, 156.

27 Review of Mendeléeff's *Principles of Chemistry*, 3d. English edition, *The Nation* 80, no. 2083 (1905): 438.

28 *MS*, XXIV, 41.

29 Mendeléeff [Mendeleev], "Comment j'ai trouvé le système périodique des éléments," 543.

30 This historical approach first appeared in the fifth edition, and was highlighted in the eighth. *MS*, XXIV, 27. The ten portraits of chemical titans in the fifth edition were Lavoisier, Dalton, Berthollet, Gay-Lussac, Davy, Gerhardt, Graham, Dumas, Kirchhoff, and Wöhler. There was also a ninth posthumous edition (1927), spearheaded by Mendeleev's widow. Mendeleev, *Osnovy khimii*, 9th ed. For the publication history of *Principles*, see Volkova, "Osnovy khimii i periodicheskii zakon."

31 Mendeleev, "Comment j'ai trouvé le système périodique des éléments," 533 (quotation) and 546. On Mendeleev's and others' efforts to articulate quantitative periodicity, see Trifonov, "K istorii voprosa ob analiticheskom vyrazhenii periodicheskogo zakona"; idem, "O matematicheskom modelirovanii periodicheskoi sistemy elementov"; idem, *O kolichestvennoi interpretatsii periodichnosti*; and Trifonov and Dmitriev, "O kolichestvennoi interpretatsii periodicheskoi sistemy." For earlier attempts, see *MS*, II, 432 and 508n; and Tchitchérine, "Le système des éléments chimiques."

32 Diary entry of 19 July 1905, in Shchukarev and Valk, *Arkhiv D. I. Mendeleeva*, 34–35. Ellipses added.

33 Mendeleev, "Predislovie (1905)" 54.

34 For his lecture notes, see "Biografii N'iutona, Zherara i Gei-Liussaka, Louvuaz'e i dr.," ADIM II-A–17–1–5. On Newton's third law and valency, see Mendeleev, "Periodicheskaia zakonnost' khimicheskikh elementov (1871)," in Mendeleev, *Periodicheskii zakon. Klassiki nauki*, 74; Mendeleev in protocol of Russian Chemical Society, 2 December 1882, *ZhRFKhO* 15, no. 1, otd. 1 (1883): 3; and Tishchenko, "Vospominaniia o D. I. Mendeleeve," 130. A cheeky correspondent perceptively critiqued this view: G. B. Nefedov to Mendeleev, 26 March 1905, ADIM II-V–24-N (Nefedov).

35 Letter dated 26 January 1883 (N.S.), ADIM I-A–56–1–17, quoted in Dobrotin et al., *Letopis' zhizni i deiatel'nosti D. I. Mendeleeva*, 220.

36 Mendeleev, "Popytka prilozheniia k khimii odnogo iz nachal estestvennoi filosofii N'iu-

tona (1889)," in Mendeleev, *Periodicheskii zakon. Klassiki nauki*, 537. For a helpful analysis of Newton's somewhat ambiguous third law, see Home, "The Third Law in Newton's Mechanics."

37 Mendeleev, "Popytka prilozheniia k khimii odnogo iz nachal estestvennoi filosofii N'iutona (1889)," 532.

38 Mendeleev, "Popytka prilozheniia k khimii odnogo iz nachal estestvennoi filosofii N'iutona (1889)," 554. Ellipses added.

39 Mendeleev, "The Periodic Law of Chemical Elements (1889)," 636.

40 Mendeleev, "The Periodic Law of Chemical Elements (1889)," 649–650. Ellipses added. The utopian speculations Mendeleev refers to will be discussed in the following chapter.

41 Mendeleev, "O priemakh tochnykh, ili metrologicheskikh vzveshivanii (1896)," *MS*, XXII, 217. Mendeleev made similar arguments in his important work on the vibrations of weights (*MS*, VII, 555–599).

42 He further commented in *On the Elasticity of Gases* (1875) that if one knew the weight of one liter of air, one would have a good measure of the local *g*. *MS*, VI, 324; and Zapisnaia knizhka, 1874–1876, #20, ADIM II-A-1-1-9, ll. 8–9 [undated but probably Spring 1874].

43 Blumbakh telegram to Mendeleev, 26 May 1897, ADIM Alb. 1/458; Blumbakh, "Geograficheskoe polozhenie Glavnoi Palaty mer i vesov." On the pendulum experiments, see Gorbatsevich, "Raboty D. I. Mendeleeva v oblasti fizicheskikh konstant," 43; and Veinberg, "Khimik ili fizik Mendeleev?" 77. The French were particularly interested in this work as a continuation of Léon Foucault's. See Mendeleev, "La Balance de Précision."

44 Mendeleev, "K izucheniiu napriazheniia tiazhesti po pomoshchi nesvobodnogo padeniia tel (1905)," *MS*, XXII, 387; and Mendeleev, "Podgotovka k opredeleniiu absoliutnago napriazheniia tiazhesti v Glavnoi Palate mer i vesov pri pomoshchi dlinnago maiatnika s zolotym sharom."

45 My thanks to Loren Graham for his insistence on the Lavoisier comparison.

46 Bensaude-Vincent, "Between History and Memory."

47 Amusingly, Mendeleev once accompanied his geologist friend Aleksandr Inostrantsev on an expedition, and was so tired that he slept through the entire journey and then the entire day and a half of the excursion. Inostrantsev, *Vospominaniia*, 142–143.

48 Shelley, *Frankenstein*, 7.

49 Mendeleev, review of E. Hoffman, *Severnyi Ural i beregovoi krebet Pai-Khoi* (St. Petersburg: 1856), reprinted in *MS*, XV, 128–148, on 128. Ellipses added.

50 Quoted in Ozarovskaia, *D. I. Mendeleev po vospominaniiam O. E. Ozarovskoi*, 129–130. Mendeleev disagreed with L. N. Tolstoy over the value of the scientific method and the primacy of industrialization over agriculture. The two never met. See Dobrotin and Karpilo, "D. I. Mendeleev o L. N. Tolstom." While at Heidelberg, Mendeleev had been strongly affected by A. I. Goncharov's *Oblomov*. See Almgren, "Mendeleev," 99. For an effort to portray Mendeleev as a cultured humanist, see Dobrotin and Karpilo, *Biblioteka D. I. Mendeleeva*, Chapter 4. They neglect to mention his adventure reading. Mendeleev's son Ivan also portrayed his father as a scholar of high literature. Given that Ivan was quite young when Mendeleev died, however, it is hard to see where he acquired this impression. See Tishchenko and Mladentsev, *Dmitrii Ivanovich Mendeleev, ego zhizn' i deiatel'nost'. Universitetskii period*, 355 and 366–367. On Mendeleev's antagonism to Dostoevsky, see Gordin, "Loose and Baggy Spirits."

51 Quoted in Kapustina-Gubkina, *Semeinaia khronika*, 187.

52 For Kuindzhi's biography, see Manin, *Kuindzhi*. On the Wanderers (*peredvizhniki*), see Gray, *The Russian Experiment in Art*, Chapter 1; and Valkenier, *Russian Realist Art*.

53 Mendeleeva, *Mendeleev v zhizni*, 55.

54 Vagner, "Eshche dva slova o kartine Kuindzhi [13 November 1880]."

55 Mendeleev, "Pred kartinoiu A. I. Kuindzhi [13 November 1880]," 2. Ellipses added. Mendeleev's argument here sounds eerily similar to the epistemes of Michel Foucault in his in-

fluential book, *The Order of Things.*

56 On Mendeleev as host of these gatherings, see Inostrantsev, *Vospominaniia*, 196.

57 Mendeleeva, *Mendeleev v zhizni*, 125 and 159; Trirogova-Mendeleeva, *Mendeleev i ego sem'ia*, 32; and Volkova, "D. I. Mendeleev i kniga," 102–103.

58 Editorial commentary in Mendeleev, *Nauchnyi arkhiv: Osvoenie krainego severa*, 15–16. On the Trans-Siberian Railroad, see Marks, *Road to Power.*

59 Sergei Witte claimed credit for himself: "On my initiative, construction was begun on the icebreaker *Ermak* toward the end of 1898. My immediate goal was to use the ship for keeping Petersburg and major Baltic ports open the year-round; my long-range goal was to use it to determine if it would be possible to sail the northern route to the Far East." Witte, *The Memoirs of Count Witte*, 288.

60 Mendeleev to Witte, "Ob issledovanii Severnogo Poliarnogo Okeana," 14 November 1901, reprinted in Mendeleev, *Nauchnyi arkhiv: Osvoenie krainego severa*, 272–280, on 273.

61 Quoted in Shpitser, "D. I. Mendeleev po vospominaniiam V. I. Kovalevskogo," 104. On a 22 January 1902 letter to S. N. Evreinov, requesting an audience with the Prince, Mendeleev noted: "The Grand Prince refused." ADIM II-A–14–2–5, quoted in Dobrotin et al., *Letopis' zhizni i deiatel'nosti D. I. Mendeleeva*, 437.

62 Witte, *The Memoirs of Count Witte*, 289. They supposedly quarreled because the latter refused to take Mendeleev along as a passenger. Mendeleev claimed he quit because he did not feel right conducting a scientific investigation under the control of a ship's captain. Figurovskii, *Dmitrii Ivanovich Mendeleev*, 252; and Fritsman, "D. I. Mendeleev i problema Arktiki."

63 Mendeleev "Spisok moikh sochinenii (1899)," in Shchukarev and Valk, *Arkhiv D. I. Mendeleeva*, 77. This balloon journey even made it into Mendeleev's terrifically incomplete autobiographical notes: "Biograficheskie zametki o D. I. Mendeleeve," in ibid., 21–22.

64 Mendeleev still engaged in work on aviation. He was solicited to write prefaces for Russian texts on flight, and he collected monographs on the topic. Mendeleev, unpublished preface to V. V. Kotov's brochure, "Samolety-aeroplany, pariashchie v vozdukhe," *MS*, VII, 544–552; Drzewiecki, *Les Ouiseaux considérés comme des Aéroplanes Animés*, ADIM Bib. 145/2. See also Vorob'ev, *Genezis russkoi vozdukhoplavatel'noi mysli v trudakh D. I. Mendeleeva*, 4–5.

65 Menning, *Bayonets before Bullets*, 233; Vorob'ev, "D. I. Mendeleev i vozdukhoplavanie," 132.

66 Rykachev, *Podniatie na vozdushnom share*, 5–6.

67 Rykachev, *Podniatie na vozdushnom share*, 30.

68 Mendeleev, "Vozdushnyi polet iz Klina vo vremia zatmeniia (1887)." This was reprinted in *MS*, VII, 471–546. On the *Northern Herald*, see Rabinowitz, "'Northern Herald.'"

69 Mendeleev, "Vozdushnyi polet iz Klina vo vremia zatmeniia (1887)," 93. For the Technical Society's involvement, see pp. 89–90.

70 Mendeleev, "Vozdushnyi polet iz Klina vo vremia zatmeniia (1887)," 88. Mendeleev reported on the corona in the minutes of the Russian Physical Society, 29 September 1887, *ZhRFKhO* 19, otd. II, no. 7 (1887): 336–337. Gay-Lussac and Biot's original 1809 scientific balloon flight that inspired Mendeleev had investigated magnetic field strength variance with altitude, not meteorology. Cawood, "Terrestrial Magnetism and the Development of International Collaboration in the Early Nineteenth Century," 562.

71 Quoted in Vorob'ev, "D. I. Mendeleev i vozdukhoplavanie," 129.

72 Mendeleev, "Vozdushnyi polet iz Klina vo vremia zatmeniia (1887)," 97–98 and 115. On the Minister's regulation see p. 98. The stipulation was warranted. In England, aeronauts like Coxwell were crucial to the success of ascents sponsored by the British Association for the Advancement of Science. See Tucker, "Voyages of Discovery on Oceans of Air," 153.

73 See Pang, "The Social Event of the Season." On the Montgolfiers' original balloon flight as an Enlightenment pastime and the spread of such mixed scientific and popular expeditions,

see Reynaud, *Les Freres Montgolfier et leurs étonnantes machines*; and Crouch, *The Eagle Aloft*.

74 Quoted in Tishchenko, "Dmitrii Ivanovich Mendeleev," 28. Emphasis in original. Mendeleev repeated this point in the *Northern Herald:* "Then the thought occurred to me that the preparations were all made and this was known everywhere, and if the balloon did not fly, then this would lead to a very bad impression, not only with respect to the moment, but to the entire fate of aerostatic ascents in our country." Mendeleev, "Vozdushnyi polet iz Klina vo vremia zatmeniia (1887)," 123.

75 Quoted in Makarenia, Filimonova, and Karpilo, *D. I. Mendeleev v vospominaniiakh sovremennikov*, 224. Ellipses in original. See also Winkler to Mendeleev, 4 August 1887 (N.S.), reproduced in Volkova, "Ukrepiteli periodicheskogo zakona," 324.

76 Walden, "Dmitri Iwanowitsch Mendelejeff," 4783. Ellipses added.

77 Quoted in Walden, "Dmitri Iwanowitsch Mendelejeff," 4783, 4783n.

78 Tucker, "Voyages of Discovery on Oceans of Air," 171.

79 Makarenia and Filimonova, "D. I. Mendeleev na s"ezde Britanskoi assotsiatsii v 1887 godu."

80 Despite substantial work by recent historians on the cultural construction of femininity, there is markedly little discussion of the concomitant construction of its complement, masculinity. Important exceptions include: Nye, *Masculinity and Male Codes of Honor in Modern France*; Bederman, *Manliness & Civilization*; Mangan and Walvin, *Manliness and Morality*; and Roper and Tosh, *Manful Assertions*.

81 See, respectively, Engelstein, *The Keys to Happiness*, 217; and Noble, *A World without Women*. In another context, nineteenth-century American lawyers constructed their profession as a masculine domain, which necessarily traded on this masculinity to preserve a republic of laws. Grossberg, "Institutionalizing Masculinity."

82 Dawson, "The Blond Bedouin"; MacKenzie, "The Imperial Pioneer and Hunter and the British Masculine Stereotype in Late Victorian and Edwardian Times."

83 Mendeleeva, *Mendeleev v zhizni*, 76–77.

84 Ozarovskaia, *D. I. Mendeleev po vospominaniiam O. E. Ozarovskoi*, 32.

85 Johanson, *Women's Struggle for Higher Education in Russia*; Koblitz, "Science, Women, and the Russian Intelligentsia"; Stites, *The Women's Liberation Movement in Russia*, 30–77; Ruane, *Gender, Class, and the Professionalization of Russian City Teachers*, Chapter 3.

86 Tishkin, "Peterburgskii universitet i nachalo vysshego zhenskogo obrazovaniia v Rossii," 32. In Mendeleev's library, he heavily commented upon the women's education sections of Fribes, *Po voprosam o vospitanii detei [. . .]*, ADIM Bib. 42/11.

87 Krotikov and Filimonova, "Ocherk pedagogicheskoi deiatel'nosti D. I. Mendeleeva v Peterburgskom Universitete (1881–1890 gg.)," 112–113.

88 Mendeleev to Menshutkin, 21 February 1885, PD f. 160, d. 4, ll. 186–192ob., on ll. 189–189ob. Mendeleev retained similar views after leaving University service. See Mendeleev to Witte, 15 October 1895, RGIA f. 560, op. 26, d. 29, ll. 21–44ob., on 31ob.

89 D. A. Tolstoi is often painted by historians as a *bête noire* of reaction, which, in some ways, he was. This portrayal has been particularly evident in studies on Mendeleev because of his personal antipathy toward Tolstoi, whom he blamed for his rejection by the Academy of Sciences. Tolstoi was, however, also a very capable administrator for Alexander III. See Freeze, *The Parish Clergy in Nineteenth-Century Russia*, Chapter 7; Whelan, *Alexander III and the State Council*, 64–71. On his role in the 1884 statute, see Sinel, *The Classroom and the Chancellery*; Pushkin, "*Raznochintsy* in the University," 38; and Brower, "Social Stratification in Russian Higher Education."

90 Kassow, *Students, Professors, and the State in Tsarist Russia*, 28 and 40; Daly, *Autocracy under Siege*, 108; Sinel, *The Classroom and the Chancellery*, 129; Naimark, *Terrorists and Social Democrats*, 132; Morrissey, *Heralds of Revolution*; and Mathes, "University Courts in Imperial Russia."

91 Student petition of 11 December 1887, ADIM Alb. 2/155, quoted in Makarenia and Filimonova, *D. I. Mendeleev i Peterburgskii universitet*, 41.

92 Assistant to the Minister of Internal Affairs to I. D. Delianov, 18 December 1887, RGIA f. 733, op. 150, d. 272, ll. 206–206ob.

93 Alexander Vucinich correctly notes but overemphasizes Mendeleev's hostility to nihilism in his *Science in Russian Culture*, 162; and "Mendeleev's Views on Science and Society." On student nihilism at St. Petersburg University, see Brower, *Training the Nihilists*; Volk, "Revoliutsionnye izdaniia narodovol'cheskikh kruzhkov Peterburgskogo universiteta v 1882 g."; and Zhukova, "Revoliutsionnye studencheskie kruzhki S.-Peterburgskogo universiteta."

94 Unsigned report, 20 March 1890, RGIA f. 1405, op. 91, d. 10717, ll. 4–7, quotation on l. 4ob.

95 Quoted in Figurovskii, *Dmitrii Ivanovich Mendeleev*, 198–199.

96 The text of the response was: "By order of the Minister of Popular Enlightenment, the attached paper is returned to Actual State Counselor Professor Mendeleev, since no Minister and none in the service of *His Imperial Excellency* has the right to accept such papers." Quoted in Tishchenko, "Dmitrii Ivanovich Mendeleev," 25. Emphasis in original.

97 Menshutkin, *Zhizn' i deiatel'nost' Nikolaia Aleksandrovicha Menshutkina*, 52.

98 Quoted in Veinberg, *Iz vospominanii o Dmitrii Ivanoviche Mendeleeve kak lektor*, 37.

99 Entry of 19 July 1905, in Shchukarev and Valk, *Arkhiv D. I. Mendeleeva*, 35. Ellipses added. See also Tishchenko and Mladentsev, *Dmitrii Ivanovich Mendeleev, ego zhizn' i deiatel'nost'. Universitetskii period*, 120–121.

100 Mendeleev, "Spisok moikh sochinenii (1899)," in Shchukarev and Valk, *Arkhiv D. I. Mendeleeva*, 77. Mendeleev, *Issledovanie vodnykh rastvorov* (1887), in Mendeleev, *Rastvory*, 396.

101 Mendeleev quoted in Tishchenko and Mladentsev, *Dmitrii Ivanovich Mendeleev, ego zhizn' i deiatel'nost'. Universitetskii period*, 119; Witte, *The Memoirs of Count Witte*, 168. The faculty at the University made some efforts to persuade him to rescind his resignation, but he preferred to go into retirement. Tishchenko and Mladentsev, *Dmitrii Ivanovich Mendeleev, ego zhizn' i deiatel'nost'. Universitetskii period*, 124–125. Interestingly, after his resignation he accepted a job reading lectures at the Institute of Communications Engineers, which would have been out of Delianov's jurisdiction. He abandoned these plans when an offer to work on gunpowder for the Navy materialized. Makarenia, "Maloizvestnyi fakt iz biografii D. I. Mendeleeva."

102 Mendeleev to V. Feoktistov, 29 March 1890, RGIA f. 776, op. 8, d. 625, ll. 1–1ob. Throughout his career, Mendeleev had found this type of popular periodical an especially effective way to lobby for his industrial vision. Kapustinskaia, "Khimiia v zhurnale 'Nauchnoe Obozrenie'"; and Kedrov, "Mendeleev, Dmitry Ivanovich," 286.

103 Delianov to Feoktistov, 31 March 1890, and Feoktistov to Mendeleev, 6 April 1890, in RGIA f. 776, op. 8, d. 625, ll. 3–4. There were two different censorship regimes in Imperial Russia. Most publications after the Great Reforms were subjected to punitive censorship, by which the editorial board of a journal could be held accountable for printing objectionable material. This encouraged a more permissive spectrum of publications. Troublesome periodicals were placed under regimes of preliminary censorship, in which a censor had to review each issue before it was printed. Preliminary censorship was thus substantially more restrictive. On censorship in Imperial Russia, see: Balmuth, *Censorship in Russia*; Ruud, "The Russian Empire's New Censorship Law of 1865"; Choldin, *A Fence around the Empire*; and Ferenczi, "Freedom of the Press under the Old Regime."

104 Quoted in Tishchenko, "Vospominaniia o D. I. Mendeleeve," 134. For a history of this incident, see Egorov, "'Gazeta obshchenarodnykh nauchnykh znanii.'"

CHAPTER 8

1 On Mendeleev's gunpowder research, see Gordin, "No Smoking Gun"; and idem, "A Modernization of 'Peerless Homogeneity.'"

2 Reproduced in Shchukarev and Valk, *Arkhiv D. I. Mendeleeva*, 34.

3 Ramsay to Mendeleev, 6 January 1892 [N.S.], ADIM Alb. 3/500, ll. 1–2.

4 Ramsay to Mendeleev, 20 January 1892 [N.S.], ADIM Alb. 3/501, l. 1.

5 Ramsay to Mendeleev, 7 July 1892 [N.S.], ADIM Alb. 3/502, l. 4. For the next year and a half, Ramsay continued to correspond with Mendeleev: Ramsay to Mendeleev, 19 September 1892 [N.S.], ADIM Alb. 3/503 and 4 September 1893, ADIM Alb. 3/504.

6 Ramsay to Mendeleev, 26 December 1893 [N.S.], ADIM Alb. 3/505. Mendeleev wrote on the top of this letter that he "answered [on] 16 Dec [O.S.]," but I could not locate any response.

7 Mendeleev to Ramsay, 12 February 1895, ADIM I-A–41–1–17.

8 Gilpin, "Krypton, Neon, Metargon, and Coronium," 699. See also idem, "Xenon, Etherion, and Monium"; and Piccini, "Das periodische System der Elemente von Mendelejeff und die neuen Bestandteile der atmosphärischen Luft." The best secondary source tracing attempts to accommodate the inert gases to the periodic law remains Petrov, "Prognozirovanie i razmeshchenie inertnykh elementov v periodicheskoi sisteme." See also Semishin, "Inertnye gazy i periodicheskii zakon D. I. Mendeleeva"; and Giunta, "Argon and the Periodic System."

9 *MS*, II, 401–403. Other chemists disputed kinetic theory, which was the basis for establishing argon's atomic weight as 40. On these disputes, see Hirsh, "A Conflict of Principles"; and Hiebert, "Historical Remarks on the Discovery of Argon." Bohuslav Brauner, Mendeleev's Czech acolyte, felt compelled as an "orthodox Mendeleeffian" to advocate the nitrogen hypothesis. Brauner, "Some Remarks on 'Argon,'" 79.

10 Protocol of the Russian Chemical Society, 2 March 1895, *MS*, II, 405–406. The dispute over inert gases proved very interesting to the Russian Chemical Society, which printed translations of many of the seminal Western articles on argon, such as those by Crookes, Piccini, and Brauner. An alternative approach, taken by G. Johnstone Stoney, was to argue that argon was actually a hydrocarbon polymer. He commented that "the hypothesis that argon is a compound has this great recommendation, that it does not involve any interruption of Mendeleeff's law, which, though only empirical, is probably true." Stoney, "Argon," 68.

11 Mendeleev, "Popytka khimicheskogo ponimaniia mirovogo efira (1903)," 115–116, quotation on 89.

12 Mendeleev, "Popytka khimicheskogo ponimaniia mirovogo efira," 118; *MS*, II, 451–452. See also Thomsen, "Über die mutmaßliche Gruppe inaktiver Elemente." Belgian physical chemist L. Errera proposed the zero-group formulation in 1900 while investigating the periodicity of magnetic properties. Errera, "Magnétisme et poids atomiques." For Errera's biography, see Petrov, "L. Errera i ego rol' v razvitii ucheniia o periodichnosti."

13 Friedman, *The Politics of Excellence*, 33–34; Crawford, *The Beginnings of the Nobel Institution*, 163. B. M. Kedrov has argued that Mendeleev "essentially" discovered the noble gases in 1869 by implication. See his "Periodicheskii zakon D. I. Mendeleeva i inertnye gazy." It is noteworthy, however, that Mendeleev—always eager to defend his priority—did *not* claim the inert gases. Others had predicted the noble gases from the periodic law, it turns out, but had not been taken seriously. Solov'ev, "Prognoz i otkrytie inertnykh gazov."

14 For a survey of attempts to explain radioactivity by causal mechanisms, including Mendeleev's, see Kragh, "The Origin of Radioactivity."

15 Zaitseva and Figurovskii, *Issledovaniia iavlenii radioaktivnosti v dorevoliutsionnoi Rossii*,

16–64. For more information on pre-Revolutionary radioactivity work, see Zaitseva, "Nekoto-rye neopublikovannye materialy, otnosiashchiesia k istorii ucheniia o radioaktivnosti"; and Makarenia and Pozdysheva, "Izuchenie radioaktivnosti russkimi uchenymi." Beketov used peri-odicity as an argument *for* radioactive energy transfer through the ether. See Beketov, "O khimicheskoi energii v sviazi s iavleniiami predstavliaemymi radiem"; and idem, *Rechi khimika*, 130. But he believed that the inert gases belonged with the transition metals in group VIII, and not in a 0-group. Ulanovskaia, "N. N. Beketov o periodicheskom zakone D. I. Mendeleeva."

16 N. Egorov to Mendeleev, 24 January 1896, ADIM I-V-53-1-26. After Mendeleev's death, this same Egorov became director of the Bureau of Weights and Measures, and the Min-ister of Popular Enlightenment solicited Egorov's views on radioactivity for a public pamphlet explaining the phenomena. Minister of Popular Enlightenment to Egorov, 14 December 1913, RGIA f. 28, op. 1, d. 378, ll. 22–22ob. In a newspaper interview, Mendeleev dwelt at length on the implications of radioactivity. See Gasanova, "Ob odnom interv'iu D. I. Mendeleeva."

17 See Mendeleev's commentary on M. V. Ivanov, "Nabliudeniia nad razriadnoi sposobnos-t'iu radiia," ADIM II-A-17-2-1, ll. 1–2.

18 Mendeleev to Geisel [*sic*], 7 (20) November 1902, RGIA f. 28, op. 1, d. 294, ll. 2–2ob.; Giesel to Mendeleev, 24 November 1902 (N.S), ADIM Alb. 3/536.

19 Zapisnaia knizhka, 1897–1902, l. 60, quoted in Kedrov, "D. I. Mendeleev i zarubezhnye slavianskie uchenye," 109. The secondary literature on Mendeleev's views on radioactivity is quite useful. See, primarily, Makarenia, *D. I. Mendeleev o radioaktivnosti i slozhnosti elementov*; and Vdovenko and Dobrotin, "D. I. Mendeleev i voprosy radioaktivnosti." The usually reliable Figurovskii erroneously claims that Mendeleev was persuaded by evidence of transmutation during this visit to the Curies. Figurovskii, *Dmitrii Ivanovich Mendeleev*, 18.

20 *MS*, II, 461n.

21 Mendeleev to Bogorodskii, 10 April 1905, quoted in Vozdvizhenskii, *Stranitsy iz istorii kazanskoi khimicheskoi shkoly*, 15.

22 Quoted in Morozov, *D. I. Mendeleev i znachenie ego periodicheskoi sistemy dlia khimii budushchago*, 89.

23 Dr. Tolouse to Mendeleev, 18 March 1906 (N.S.), ADIM I-V-49-1-64. Mendeleev's failing health precluded a response on this topic. Mikhail M. Filippov of the Russian counter-part *Nauchnoe Obozrenie* also solicited an article by Mendeleev on radioactivity. Filippov to Mendeleev, 14 and 17 November 1902 [*sic*: 1903], ADIM Alb. 3/663.

24 Substantial extracts of Emmens's pamphlet are reproduced, and heavily criticized, in Bolton, "Recent Progress of Alchemy in America." Emmens objected to Bolton's article as "a lit-tle modern scientific witch-finding at my expense." Emmens, "Modern Alchemy." To be fair to Emmens, Bolton did engage in spurious rhetoric such as name-calling and guilt by association. William Crookes, the editor, added a postscript saying that the sample of Argentaurum sent to him seemed to be pure gold by all chemical tests.

25 On the Emmens incident, see Kauffman, "Stephen H. Emmens and the Transmutation of Silver into Gold." For a contemporary critique of Emmens, including extracts from the Crookes correspondence, see Bary, "L'argentaurum."

26 Mendeleev, "Zoloto iz serebra (1898)," 1. Ellipses added. Mendeleev also cited Emmens as one of those who revived Prout's notion of interconversion of elements. Mendeleev, "Popytka khimicheskogo ponimaniia mirovogo efira," 85. See also his encyclopedia attack: Mendeleev, "Periodicheskaia zakonnost' khimicheskikh elementov (1895)," *MS*, II, 410n.

27 Mendeleev, "Zoloto iz serebra," 2 and 6 (quotation).

28 Mendeleev, "Comment j'ai trouvé le système périodique des éléments (1899)," 212. On the alchemy comparison, see Mendeleev, "Zoloto iz serebra," 7.

29 On Prout and his hypothesis, see the works by W. H. Brock: *From Protyle to Proton*; "Stud-ies in the History of Prout's Hypothesis, Parts I and II"; "The Life and Work of William Prout."

30 On the evidence for Prout's hypothesis, see Siegfried, "The Chemical Basis for Prout's Hypothesis." On attempts to refine Prout, see Farrar, "Nineteenth Century Speculations on the Complexity of the Chemical Elements"; Kragh, "Julius Thomsen and 19th-Century Speculations on the Complexity of Atoms"; idem, "The First Subatomic Explanations of the Periodic System"; and Farber, "The Theory of the Elements and Nucleosynthesis."

31 Mendeleev, *Izbrannye lektsii po khimii*, 16. In his lengthy German article on the periodic law (1871), Mendeleev issued one of his most famous statements on this question, often misinterpreted as an *advocacy* of Prout or, more ahistorically, as a prediction of Einstein's matter–energy equivalence. Mendeleev argued that even if all atomic weights were integral multiples of hydrogen, this would not directly prove Prout's hypothesis: "Even agreeing with the claim that the material of elements is entirely homogeneous, there is no cause to think that n massive parts of one element or n of its atoms, having produced one atom of another body, will give exactly n weighted parts, that is, that an atom of the second element will weigh exactly n times more than the first atom. I consider the law of the conservation of matter only a special case of the law of conservation of forces or motions. . . . Expressing this thought, I wish only to show that there is a certain possibility to reconcile the cherished thought of chemists on the complexity of atoms with the rejection of Prout." Mendeleev, "Periodicheskaia zakonnost' khimicheskikh elementov (1871)," in Mendeleev, *Novye materialy po istorii otkrytiia periodicheskogo zakona*, 66. Ellipses added. For an example of the "Einsteinian" misinterpretation of what was clearly a "devil's advocate" stance, see Roginskii, "D. I. Mendeleev o neizbezhnosti izmenenii massy pri protsessakh prevrashcheniia elementov."

32 Protocol of the Russian Chemical Society, 9 January 1886, *MS*, II, 311.

33 *MS*, II, 454n. See also Mendeleev, "Popytka khimicheskogo ponimaniia mirovogo efira," 164; and the second edition of *Principles* (1873), *MS*, II, 227.

34 Mendeleev and his rival Butlerov disagreed substantially on Prout. Butlerov proposed that the atomic weight of various elements was merely an average over different atomic weights of individual atoms—a view that Soviet historians would claim was a prediction of isotopy. In what was perhaps a tongue-in-cheek reference to Mendeleev's failed gas project, Butlerov suggested that just as Boyle-Mariotte was only accurate on average, so perhaps carbon had an atomic weight of twelve only on average, and actually ranged from, say, 11.5 to 12.5 for individual atoms. Butlerov, "Zametka ob atomnykh vesakh (1882)." He also suggested this at the physico-mathematical division of the Academy of Sciences on 9 February 1882, *Zapiski Akademii Nauk* 41 (1882): 59. In general, he felt this was consistent with the presumption of the complexity of chemical elements. Butlerov, "Osnovnye poniatiia khimii (1886)," in Butlerov, *Sochineniia*, III, 51–52; and as recalled in Glinka, "Aleksandr Mikhailovich Butlerov v chastnoi i domashnei zhizni," 183.

35 Mendeleev, "Popytka khimicheskogo ponimaniia mirovogo efira," 30. In his later writings, Mendeleev frequently employed astronomical analogies to chemical phenomena. His 1894 encyclopedia article on "Substance" is one notable example (*MS*, II, 376–377).

36 Ramsay's was one of the more interesting efforts at treating the electron as an element, based on Kantian speculations. Ramsay, "The Electron as Element." On the history of the electron and chemistry, see Chayut, "J. J. Thomson"; Sinclair, "J. J. Thomson and the Chemical Atom"; and Stranges, *Electrons and Valence*, Chapter 3.

37 Solov'ev, *Istoriia khimii v Rossii*, 230. Many wrote Mendeleev for his views on the composition of atoms, but mostly he remained silent. See the letter and manuscript from S. Freitag to Mendeleev, 6 May 1903, on the structure of the atom, ADIM II-V–25–1–16; and V. P. Chernyshev's undated speculations on differentiation of atomic weights, ADIM Alb. 3/624.

38 *MS*, II, 449. Ellipses added.

39 The exceptions to this historical neglect are in some cases excellent, as in the comprehensive study by Kragh, "The Aether in Late Nineteenth Century Chemistry," which contains a

substantial discussion of Mendeleev's ether proposal. This largely supercedes the prior studies: Kargon, "Mendeleev's Chemical Ether, Electrons, and the Atomic Theory"; and Bensaude-Vincent, "L'éther, élément chimique." Typically, Soviet historians have bypassed Mendeleev's ether as an embarrassing misstep in a triumphant narrative. One, however, has interpreted it as spontaneous dialectical materialism: Vasetskii, "Mirovozzrenie D. I. Mendeleeva." For a systematic discussion of the varieties of ether models in the physics of this period, see the classic account of Whittaker, *A History of the Theories of Aether and Electricity*, esp. I, Chapter 9. On the persistence of British ether theories after 1905, see Goldberg, "In Defense of Ether."

40 Kragh, "The Aether in Late Nineteenth Century Chemistry," 58.

41 On contemporary debates over the boundary of physics and chemistry, see Nye, *From Chemical Philosophy to Theoretical Chemistry*.

42 Quoted in Veinberg, *Iz vospominanii o Dmitrii Ivanoviche Mendeleeve kak lektor*, 27–28. Ellipses added.

43 Mendeleev, "Kalenie (1895)," *MS*, XVII, 464–465.

44 Mendeleev, "Kolebaniia pri istechenii (1905)," reprinted in *MS*, V, 267.

45 Mendeleev, "Elementy (khimicheskie) (1904)," *MS*, XV, 638.

46 Schinz, *Essai d'une Nouvelle Théorie Chimique*. Mendeleev's copy is stored at ADIM Bib. 4/2. For another rare precursor that attempted to treat the ether like a gas, see Wood, *The Luminiferous Aether*, 9 and 69. Another interesting example, of which Mendeleev was almost certainly unaware, was proposed by Lecoq de Boisbaudran, the discoverer of gallium (Mendeleev's eka-aluminum), in his "Sur la constitution des spectres lumineux."

47 Schinz, *Essai d'une Nouvelle Théorie Chimique*, 10–11.

48 Schinz, *Essai d'une Nouvelle Théorie Chimique*, 55 and 130.

49 For example, Mendeleev classed under his "Ether" and "Physico-Chemical Cosmogony" the following: Brester, *Essai d'une Théorie du Soleil et des Études variables* (1889), ADIM Bib. 145/1; Funk, *Aphoristischer Entwurf einer Kosmogonie* (1888), ADIM Bib. 145/3; Maxwell, *La Chaleur* (1891), ADIM Bib. 336/2; idem, *Materiia i dvizhenie* (1885), ADIM Bib. 336/3; idem, *Teoriia teploty v elementarnoi obrabotke* (1888), ADIM Bib. 331/16; and P. Dzh. Tet [P. G. Tait], *Svoistva materii* (1887), ADIM Bib. 336/1. In the last of these, Mendeleev only marked up the appended reproduction of Maxwell's "Atom" article from *Encyclopedia Britannica*. Mendeleev likewise never cited his personal acquaintance with A. A. Michelson, noted ether experimentalist, in connection to the chemical ether. See, for example, A. A. Michelson to Mendeleev, 21 April 1899, Paris, ADIM Alb. 3/417.

50 Shishkov to Mendeleev, 8 March 1899, ADIM Alb. 3/587, quoted in Figurovskii and Musabekov, "Vydaiushchiisia russkii khimik L. N. Shishkov," 62.

51 From the preface to the offprint of the ether pamphlet, published in 1905 by M. P. Frolova, reproduced in *MS*, II, 463.

52 Pamphlet preface, *MS*, II, 464.

53 See the bill from bookseller N. P. Karbasnikov, whom Mendeleev owed 52.5 rubles for 25 copies of the pamphlet, ADIM I-G–43–1–30. He also authorized a reprint in *Fizicheskoe Obozrenie*, a Kiev journal. See G. Dements to Mendeleev, 17 March 1906, ADIM I-V–22–2–63.

54 Mendeleev, "Elementy (khimicheskie) (1904)," *MS*, XV, 638n1. The English translation is Mendeléef [Mendeleev], *An Attempt towards a Chemical Conception of the Ether* (1904). Throughout, I will cite the journal form of the Russian version for greater fidelity.

55 Iv. Chetverikov to Mendeleev, 30 November 1904, ADIM II-V–25-Ch. The translation was published as *Provo de Kemia kompreno de l'monda etero de P° Mendelejev* (Paris: 1904). Mendeleev was obviously struck by the Esperanto translation, commenting in his annotated list of publications: "I wrote an article about the ether. It was then translated even into

Esperanto. . . ." "Biograficheskie zametki o D. I. Mendeleeve (1906)," in Shchukarev and Valk, *Arkhiv D. I. Mendeleeva*, 26–27.

56 On the role of the discovery of Mendeleev's predicted new elements in establishing his reputation, see Brush, "The Reception of Mendeleev's Periodic Law in America and Britain."

57 Mendeleev, "Popytka khimicheskogo ponimaniia mirovogo efira," 28.

58 Mendeleev, "Popytka khimicheskogo ponimaniia mirovogo efira," 25–26. This segment is left out of the English translation.

59 Mendeleev, "Popytka khimicheskogo ponimaniia mirovogo efira," 29n.

60 Mendeleev had held to the rarefied-gas hypothesis of the ether as late as his "Substance" encyclopedia article in 1894 (*MS*, II, 377); and in his work on the volume of air (Mendeleev, "O vese litra vozdukha [1894]," *MS*, XXII, 69). On the importance of homogeneity as a category for Mendeleev, see Gordin, "A Modernization of 'Peerless Homogeneity.'"

61 Coronium did not play a large role in Mendeleev's treatise. It had already been predicted from irregularities in the sun's spectrum. Its chief function for Mendeleev was to round out the first period of the table so that the ether could be in both the 0-group and the 0-period. He calculated that coronium should have a density of 0.2 and move 2.24 times faster than hydrogen, with a weight of 0.4. This was too heavy to be the ether. Mendeleev, "Popytka khimicheskogo ponimaniia mirovogo efira," 120–122. Coronium was eventually identified with excited states of helium and hydrogen: Gruenwald, "On Remarkable Relations between the Spectrum of Watery Vapour [. . .] (1887)." For secondary accounts of coronium's rise and fall, see Kragh, "The Aether in Late Nineteenth Century Chemistry"; and Karpenko, "The Discovery of Supposed New Elements."

62 Mendeleev, "Popytka khimicheskogo ponimaniia mirovogo efira," 89–90. Emphasis in original. Crookes's theory of the fourth state of matter was an attempt to explain cathode rays and the radiometer, and explicitly rejected the Spiritualist interpretation others attributed to him. See Whittaker, *Theories of Aether and Electricity*, I, 352; and Fournier d'Albe, *The Life of Sir William Crookes*, Chapter 14. Mendeleev wrote to Crookes about this hypothesis after a trip to England, stating in forced English that "I am very glad to declare, that it was the great pleasure to me to get acquainted with your opinion of the essential reality of this fine observation, by which is supposed possible to explain the dissociation of elements." Mendeleev to Crookes, 25 November (8 December) 1905, RGIA f. 28, op. 1, d. 1076, ll. 44–45. Mendeleev owned a German translation of Crookes's work on the fourth state: Crookes, *Strahlende Materie oder der vierte Aggregatzustand* (1879), ADIM Bib. 84/13. The two also had a standing disagreement about the interpretation of the periodic law, which Crookes saw as evidence for the evolution of the elements from a primary matter, basing his speculations primarily on spectroscopic evidence. See DeKoskey, "Spectroscopy and the Elements in the Late Nineteenth Century."

63 Mendeleev, "Popytka khimicheskogo ponimaniia mirovogo efira," 165–167.

64 Mendeleev, "Popytka khimicheskogo ponimaniia mirovogo efira," 163n; and ADIM II-A–25–2–4, l. 1ob. I would like to thank N. G. Karpilo for her assistance in transcribing this text. The connection to Newton is not entirely gratuitous. In Query 31 of his *Opticks,* Isaac Newton offered a complicated vision of matter's interaction through quantifiable forces and a variety of imponderable "ethers." There is no evidence that Mendeleev worked through Newton's intricate matter theories. There was, however, enough similarity in their visions of atomic matter and how atomic ethers could serve as mediators that president of the Soviet Academy of Sciences and Newton historian S. I. Vavilov was in his rights to see Query 31 as "the plan of all of D. I. Mendeleev's scientific work." Vavilov, "Fizika v nauchnom tvorchestve D. I. Mendeleeva," 5. On Newton's matter theory, see Thackray, *Atoms and Powers*; Carrier, "Newton's Ideas on the Structure of Matter and Their Impact on Eighteenth-Century Chemistry"; and McGuire, "Transmutation and Immutability." For the decline of Newtonian chemistry, see

Gregory, "Romantic Kantianism and the End of the Newtonian Dream in Chemistry."

65 Mendeleev, "Popytka khimicheskogo ponimaniia mirovogo efira," 92. The nature of "typical" elements as the lightest of all elements is discussed in Chapter 2.

66 Mendeleev, "Popytka khimicheskogo ponimaniia mirovogo efira," 115.

67 Mendeleev, "Popytka khimicheskogo ponimaniia mirovogo efira," 171–172. Despite the importance of the ether for Mendeleev's worldview, he championed none of the contemporary declarations of the isolation of the ether, the most famous being Charles Brush's "etherion," which Crookes definitively refuted as residual aqueous vapor. See Brush, "A New Gas"; Gilpin, "Xenon, Etherion, and Monium"; Crookes, "On the Supposed New Gas, Etherion"; and Smoluchowski de Smolan, "Etherion, a New Gas?"

68 Quoted in Makarenia, *D. I. Mendeleev o radioaktivnosti i slozhnosti elementov*, 28. Ellipses in original.

69 Mendeleev, "Popytka khimicheskogo ponimaniia mirovogo efira," 163. Readers of his pamphlet responded favorably and asked Mendeleev to be even more explicit about how the ether would eliminate Prout's hypothesis, as in Aleksandr Nemirov to Mendeleev, 25 October 1902, ADIM II-V–24-N (Nemirov).

70 Motivated by the existence of only four halogens but five alkali metals in his periodic system, he felt hydrogen should have a halogen complement, possibly with an atomic weight of 3. Mendeleev, "Popytka khimicheskogo ponimaniia mirovogo efira," 119n. Others held this now outlandish idea. Johnstone Stoney, for example, felt that argon was actually a polymer that would point the way to the six elements that connected H and Li, and even in the mid-1920s one Russian chemist still believed that there were some elements in the vicinity of hydrogen that remained to be discovered. Stoney, "Argon—a Suggestion," 67–68; and Kurbatov, *Zakon D. I. Mendeleeva*, 276. For other such predictions, see Karpenko, "The Discovery of Supposed New Elements." On possibly Mendeleev's first sketch of a periodic system, on the back of a letter from A. I. Khodnev dated 17 February 1869, Mendeleev wrote before fluorine "#3?," indicating this thought had crossed his mind even then. Kedrov, *Den' odnogo velikogo otkrytiia*, 59.

71 Andrei Litkin to Mendeleev, 30 October 1904, ADIM Alb. 3/319. There is no record of Mendeleev's response. Not all of Mendeleev's lay readers were so reasonable. For example, Ia. Reigler wrote to Mendeleev on 1 January 1901 for support in creating a perpetual motion machine harnessing the earth's energy. ADIM Alb. 2/597. Mendeleev never responded.

72 Remsen, Review of Mendeleev's *An Attempt towards a Chemical Conception of the Ether*, 519; and J. L., Review of Mendeleev's *An Attempt towards a Chemical Conception of the Ether*.

73 Allen, Review of D. Mendeléeff's *The Principles of Chemistry*, 3d. English edition.

74 Matout, "L'éther considéré comme élément chimique"; G. Eldman to Mendeleev, 22 January 1904, ADIM II-B–24–2-E, reproduced in Dobrotin, Ter-Avakova, and Volkova, "Perepiska D. I. Mendeleeva s zarubezhnymi uchenymi."

75 Mendeleev to Winkler, 10 (23) May 1904, ADIM I-A–4–2–10, l. 2. Mendeleev's refusal to attend actually had little to do with surgery. He confided in a letter to Moissan on 27 March 1904 that he could not travel "because of that attitude which the U. States displayed upon the beginning of the war of Russia with Japan." ADIM II-A–15–1-M, quoted in Dobrotin et al., *Letopis' zhizni i deiatel'nosti D. I. Mendeleeva*, 454. He reiterated this to Simon Newcomb on 7 May.

76 Quoted in Dobrotin et al., *Letopis' zhizni i deiatel'nosti D. I. Mendeleeva*, 453.

77 Ivanov article draft, October 1902, ADIM II-A–17–2–1, ll. 3–5. A series of classic contemporary studies also attempted to deduce the properties of the solar atmosphere, most importantly Lockyer, *Inorganic Evolution*. Lockyer specifically referred to Mendeleev's periodic system as the underlying mechanism for his organization (pp. 94, 176). See also Stoney, "On Atmospheres upon Planets and Satellites." In addition, a contemporary astronomer in St. Petersburg published his calculations of the properties of stellar atmospheres. It is hard to imagine that

Mendeleev was unaware of this local effort. Rogovsky, "On the Temperature and Composition of the Atmospheres of the Planets and the Sun."

78 Mendeleef [Mendeleev], *Principles of Chemistry* (1905), I, 1, n1. Ellipses added.

79 Mendeleev, "Popytka khimicheskogo ponimaniia mirovogo efira," 27n. See also Mendeleev to Witte, 15 October 1895, RGIA f. 560, op. 26, d. 29, ll. 28–30.

80 Mendeleev, *Zavetnye mysli* (1903–1905), 5–6.

81 Quoted in Mendeleeva, "Zametki o Mendeleeve," 70. Emphasis in original. Mendeleev was careful to separate himself from Slavophile or Pan-Slavist thinking: "The essence of this thought [Asians as thesis and Europeans as antithesis] has been expressed by Slavophiles, and although I cannot count myself among their good number, yet all the same I propose together with them that among us Russians more than anyone are rudiments of all types for the achievement of this synthesis. . . ." Mendeleev, *Zavetnye mysli*, 276. Kedrov has emphasized Mendeleev's "Slavic" references and his correspondence with fellow "Slavic" scientists without pointing to Mendeleev's disavowals of the political implications. Kedrov, "D. I. Mendeleev i zarubezhnye slavianskie uchenye." In doing so, he was following Czech chemist Bohuslav Brauner. See Volkova, "Ukrepiteli periodicheskogo zakona"; and Kedrov and Chentsova, *Brauner—spodvizhnik Mendeleeva*.

82 Mendeleev, *Zavetnye mysli*, 406.

83 Mendeleev, "Mirovozzrenie." For inappropriate positivist readings of Mendeleev, see Vucinich, "Mendeleev's Views on Science and Society," 345; Stackenwalt, "The Economic Thought and Work of Dmitrii Ivanovich Mendeleev," 25 and passim; and Bensaude-Vincent, "Mendeleev's Periodic System of Chemical Elements," 14.

84 Mendeleev, *Zavetnye mysli*, 394n.

85 In 1899, he referred to Popova as his "wife," although he in fact had not married her until several years after writing the article in question: "Spisok moikh sochinenii (1899)," in Shchukarev and Valk, eds., *Arkhiv D. I. Mendeleeva*, 83.

86 D. Popov [Mendeleev], "Edinitsa (1877)." It is reprinted in Mendeleev, *Zavetnye mysli*, 394–395n68.

87 Alenitsin, "Polozhitel'noe i otritsatel'noe."

88 Mendeleev, "Edinitsa," 247. Emphasis in original; ellipses added.

89 Almgren, "D. I. Mendeleev and Siberia," 57–60.

90 Reproduced in Tishchenko and Mladentsev, *Dmitrii Ivanovich Mendeleev, ego zhizn' i deiatel'nost'. Universitetskii period*, 358.

91 See the works of Gregory L. Freeze: "Handmaiden of the State?"; *The Parish Clergy in Nineteenth-Century Russia*; "Subversive Piety"; and "Bringing Order to the Russian Family."

92 Reproduced in Tishchenko and Mladentsev, *Dmitrii Ivanovich Mendeleev, ego zhizn' i deiatel'nost'. Universitetskii period*, 70. Mendeleev wrote a similar note that same year to his first daughter from his second marriage, which he gave to her on her nineteenth birthday: "The chief secret of life is this: one person is nothing, only together—they are people" (p. 72).

93 Quoted in Kapustina-Gubkina, *Semeinaia khronika*, 198. Emphasis in original.

94 Men'shikov, "Pis'ma k blizhnim," 4.

95 Mendeleev, interview reproduced in Bourdon, "Les Opinions de Professeur Mendéléeff," 3. Ellipses added.

96 Among the vast literature on the 1905 Revolution, see especially Ascher, *The Revolution of 1905*, especially volume 1: *Russia in Disarray*; Reichman, *Railwaymen and Revolution*; Engelstein, *Moscow, 1905*; and Verner, *The Crisis of Russian Autocracy*.

97 Mendeleev, *Zavetnye mysli*, 407 and 327.

98 "Biograficheskie zametki o D. I. Mendeleeve (1906)," in Shchukarev and Valk, eds., *Arkhiv D. I. Mendeleeva*, 28; Mendeleeva, *Mendeleev v zhizni*, 129; Trirogova-Mendeleeva, *Mendeleev i ego sem'ia*, 89; and Witte, *The Memoirs of Count Witte*, 400–403.

99 See the account in Visokovatov, "Iz vospominanii o D. I. Mendeleeve."

100 Mendeleev to Witte, August 1903, ADIM Alb. 1/486.

101 Visokovatov, "Iz vospominanii o D. I. Mendeleeve," 2.

102 Walden, "Dmitri Iwanowitsch Mendelejeff," 4786–4787. Ellipses in original. On Mendeleev's support of Russian forces during the war, see Men'shikov, "Pis'ma k blizhnim."

CHAPTER 9

1 Pynchon, *V.*, 449.

2 Lermontov, *Sobranie sochinenii*, I, 101.

3 Pikulev, *Ivan Ivanovich Shishkin*, 126–128 (on Mendeleev) and 168–173 (on the painting).

4 Kisch, *Alexander Blok*, 9. See also Mochulsky, *Aleksandr Blok*.

5 Pyman, *The Life of Aleksandr Blok. Volume I*, 2. See also Orlov, *Hamayun*, 48–49.

6 Pyman, *The Life of Aleksandr Blok. Volume II*, 18–20. On Blok at the University, see Iezuitova and Skvortsova, "Aleksandr Blok v Peterburgskom universitete."

7 Aleksandr Blok, "Narod i intelligentsiia (1909)," in *Sobranie sochinenii*, V, 324. As Blok wrote to his mother on 5–6 November 1908: "The closer a man is to the people (Mendeleev, Gorky, Tolstoy), the more sharply he hates the intelligentsia." Blok, *Sobranie sochinenii*, VIII, 258–259.

8 Letter of 15 May 1903, quoted in Ekimov, "D. I. Mendeleev v zhizni i tvorchestve Aleksandra Bloka," 157.

9 Blok, *Sobranie sochinenii*, VII, 111–112.

10 Blok, "Narod i intelligentsiia (1909)," *Sobranie sochinenii*, V, 324–325. Many contemporaries noted the contrasts between the two bearded figures who seemed stuck in the nineteenth century. For a comparison with respect to the Russo-Japanese War, see Men'shikov, "Pis'ma k blizhnim."

11 Entry of 26 September 1908, in Aleksandr Blok, *Zapisnye knizhki*, 114. Ellipses added. Blok often grouped the two, even temporally: ". . . in the moment of history when Tolstoy is writing *War and Peace*, Mendeleev discovers the periodic system of elements. . . ." Blok, "Stikhiia i kul'tura (1909)," *Sobranie sochinenii*, V, 354.

12 Blok, *Zapisnye knizhki*, 484. He had inherited the desk in 1907.

13 In 1918, Lenin, through V. D. Bonch-Burevich, exhorted Mendeleev's daughter to fulfill her "duty" by writing her reminiscences of her father. Trirogova-Mendeleeva, *Mendeleev i ego sem'ia*, 3. Lenin had read Mendeleev's *Principles of Chemistry* while a student at Kazan and the Ulyanov family owned several of his chemical texts. In 1905, V. Ia. Kurbatov engaged Lenin in a discussion of Mendeleev's philosophy of science. See Musabekov, Makarenia, and Pozdysheva, "Vladimir Il'ich Lenin o trudakh i ideiakh D. I. Mendeleeva." For a concise explanation of dialectical materialism, see Graham, *Science, Philosophy, and Human Behavior in the Soviet Union*, Chapter 2.

14 Stalin, *Sochineniia*, I, 301. It goes without saying that Mendeleev only referred to Marxist writings to criticize them. He did own a copy of G. Plekhanov's 1882 translation of the Communist Manifesto. Volkova, "D. I. Mendeleev i kniga," 101.

15 Ionidi, *Filosofskoe znachenie periodicheskogo zakona D. I. Mendeleeva*; idem, *Mirovozzrenie D. I. Mendeleeva*; Kedrov and Nikiforov, "K voprosu o mendeleevskikh khimicheskikh s"ezdakh."

16 Trotsky, "Dialectical Materialism and Science," in Trotsky, *Problems of Everyday Life*, 212–213. Throughout, I have modified Trotsky's transliteration of "Mendeleyev" and capitalized "Marxist."

17 Trotsky, "Dialectical Materialism and Science," 217.

18 On early Soviet attempts to preserve Mendeleev's legacy, see Volkov, "Novye dokumenty o literaturnom nasledstve D. I. Mendeleeva i o ego sem'e." For a Russian nationalist interpretation of Mendeleev's priority dispute with Lothar Meyer, see Chasovnikov, "K semidesiatipiatiletiiu so dnia opublikovaniia periodicheskoi sistemy D. I. Mendeleeva," 87.

19 Mendeleev, *Zavetnye mysli*, 9n2. For Mendeleev as endorser of five-year plans, see Veinberg, "D. I. Mendeleev," 108; Kedrov, "Ob otnoshenii k mendeleevskomu nasledstvu."

20 Quoted in Pogodin, "Otkrytie periodicheskogo zakona D. I. Mendeleevym i ego bor'ba za pervenstvo russkoi nauki," 40. Similarly, an unsubtle reference on the last page of Ionidi, *Filosofskoe znachenie periodicheskogo zakona D. I. Mendeleeva*, 48, associated the periodic law with thermonuclear power. Elsewhere, Mendeleev's balloon flight was analogized to the bravery of Soviet cosmonauts: Vol'fkovich, "D. I. Mendeleev i otechestvennoi khimii," 119.

21 Trotsky, "Dialectical Materialism and Science," 218.

22 Trotsky, "Dialectical Materialism and Science," 221. Emphasis in original.

BIBLIOGRAPHY

I.   MENDELEEV'S COLLECTED WORKS

Mendeleev, D. I. *Sochineniia*, 25 v. Leningrad: Izd. AN SSSR, 1934–1956.

v. 1: Candidate and Master's Dissertations
v. 2: Periodic Law
v. 3: *Research on Water Solutions by Specific Weight*
v. 4: Solutions
v. 5: Liquids
v. 6: Gases
v. 7: Geophysics and Hydrodynamics
v. 8: Works in the Area of Organic Chemistry
v. 9: Gunpowder
v. 10: Oil
v. 11: Fuel
v. 12: Works in the Area of Metallurgy
v. 13: *Principles of Chemistry*, volume 1 [first edition]
v. 14: *Principles of Chemistry*, volume 2 [first edition]
v. 15: "Theoretical Knowledge," Small Notes
v. 16: Agriculture and the Refining of Agricultural Products (Fertilization, Agrotechnology, Amelioration, Husbandry, Dairy Work and Cheese-Making, Wine Production and Distillation)
v. 17: Technology
v. 18: Economic Works I
v. 19: Economic Works II
v. 20: Economic Works III
v. 21: Economic Works IV
v. 22: Metrological Works
v. 23: Popular Enlightenment and Higher Education
v. 24: Articles and Materials on General Questions
v. 25: Supplementary Materials

II. OTHER WORKS BY D. I. MENDELEEV (MENDELÉEF, MENDELEJEW)

An Attempt towards a Chemical Conception of the Ether, tr. G. Kamensky. London: Longmans, Green, and Co., 1904.

"La Balance de Précision: Étude de ses Oscillations." *Revue générale de chimie pure & appliquée* 1 (1899): 100–102.

"Comment j'ai trouvé le système périodique des éléments." *Revue générale de chimie pure & appliquée* 1 (1899): 211–214, 510–512; 4 (1901): 533–546.

Dopolneniia k poznaniiu Rossii, 2d. ed. St. Petersburg: A. S. Suvorin, 1907.

"Dnevnik 1861 g." *Nauchnoe Nasledstvo* 2 (1951): 111–212.

"Des écarts dans les lois relatives aux gaz." *Comptes Rendus* 82 (1876): 412–415.

"Edinitsa [under pseudonym D. Popov]." *Svet*, no. 11 (1877): 247–249.

Granits poznaniiu predvidet' nevozmozhno, ed. Iu. I. Solov'ev. Moscow: Sovetskaia Rossiia, 1991.

Izbrannye lektsii po khimii. Moscow: Vysshaia shkola, 1968.

K poznaniiu Rossii, 6th ed. St. Petersburg: A. S. Suvorin, 1907.

"Kakaia zhe Akademiia nuzhna v Rossii." *Novyi Mir*, no. 12 (1966): 176–191.

"Kolebaniia pri istechenii." *VGPMV* 7 (1905): 167–169.

"[Letter to the editor]." *Golos*, 16 November 1880, #317: 2.

"[Letter to the editor]." *Novoe Vremia*, 23 June (5 July) 1891, #5500: 4.

"La Loi Périodique des Éléments Chimiques." *Le Moniteur Scientifique de le Dr. Quesneville* 21 (1879): 691–737.

"Materialy dlia izucheniia sovremennago sostoianiia priemov poverki mer i vesov, primeniaiushchikhsia v torgovle." *VGPMV* 3 (1896): 118–119.

Materialy dlia suzhdeniia o spiritizme. St. Petersburg: Obshchestvennaia Pol'za, 1876.

"Mirovozzrenie," ed. E. Kh. Fritsman. *Nauchnoe Nasledstvo* 1 (1948): 157–162.

Nauchnyi arkhiv: Osvoenie krainego severa, t.1: Vysokie shiroty severnogo okeana. Moscow: Izd. AN SSSR, 1960.

Nauchnyi arkhiv, t. I: Periodicheskii zakon, ed. B. M. Kedrov. Moscow: Izd. AN SSSR, 1953.

Nauchnyi arkhiv: Rastvory. Moscow: Izd. AN SSSR, 1960.

Novye materialy po istorii otkrytiia periodicheskogo zakona, ed. N. A. Figurovskii. Moscow: Izd. AN SSSR, 1950.

"O kolebanii vesov: Rech' dlia obshchago sobraniia X-go s"ezda russkikh estestvoispytatelei v Kieve (avg. 1898 g.)." *VGPMV* 4 (1899): 33–45.

O soprotivlenii zhidkostei i o vozdukhoplavanii, vol. 1. St. Petersburg: V. Demakov, 1880.

"O vese litra vozdukha." *VGPMV* 1 (1894): 57–88.

"On the Chemical Elements." *Nature* 71 (17 November 1904): 65–66.

"Opravdanie proteksionizma." *Novoe Vremia*, 11 (23) July 1897, #7675: 2–3.

Osnovy khimii, 9th ed. Moscow: Gos. Izd., 1927.

"The Periodic Law of Chemical Elements." *Journal of the Chemical Society* 55 (1889): 634–656.

"The Periodic Law of the Chemical Elements." *Chemical News* 40 (1879): 231–232, 243–244, 255–256, 279–280, 291–292, 303–304; 41 (1881): 2–3, 27–28, 39–40, 49–50, 61–62, 71–72, 83–84, 93–94, 106–108, 113–114, 125–126.

Periodicheskii zakon, ed. B. N. Menshutkin. Moscow: Gos. Izd., 1926.

Periodicheskii zakon. Dopolnitel'nye materialy. Klassiki nauki, ed. B. M. Kedrov. Moscow: Izd. AN SSSR, 1960.

Periodicheskii zakon. Klassiki nauki, ed. B. M. Kedrov. Moscow: Izd. AN SSSR, 1958.

"Die periodische Gesetzmässigkeit der chemischen Elemente." *Liebigs Annalen der Chemie und Pharmacie,* Supp. VIII (1872): 133–229.

"Podgotovka k opredeleniiu absoliutnago napriazheniia tiazhesti v Glavnoi Palate mer i vesov pri pomoshchi dlinnago maiatnika s zolotym sharom." *VGPMV* 8 (1907): 1–41.

Polozheniia, izbrannyia dlia zashchishcheniia na stepen' magistra khimii, 9 September 1856. St. Petersburg: Tip. Departamenta vneshnei torgovli, 1856.

"Popytka khimicheskogo ponimaniia mirovogo efira." *Vestnik i Biblioteka Samoobrazovaniia,* nos. 1–4 (1903): 25–32, 83–92, 113–122, 161–176.

"Pred kartinoiu A. I. Kuindzhi." *Golos,* 13 (25) November 1880, #314: 2.

"Predislovie." *VGPMV* 7 (1905): 54–57.

"Predislovie k russkomu izdaniiu." In Mohn, *Meteorologiia ili ucheniia o pogode* (1876): v-xxi.

"Predislovie. Polozhenie o Glavnoi Palate mer i vesov." *VGPMV* 1 (1894): iii-viii.

The Principles of Chemistry, 2 v., 3d. English edition, tr. G. Kamensky. London: Longmans, Green, and Co., 1905.

Problemy ekonomicheskogo razvitiia Rossii. Moscow: Izd. sotsial'no-ekonomicheskoi literatury, 1960.

"Programma dlia ispytaniia v znanii metrologicheskikh priemov dlia lits, zhelaiushchikh postupit' poveriteliami v mestnyia poverochnyia palatki." *VGPMV* 5 (1900): 179–181.

"Proiskhozhdenie nefti." *Svet,* no. 12 (1877): 277–284.

Raboty po sel'skomu khoziaistvu i lesovodstvu, eds. S. I. Vol'fkovich and F. S. Sobolev. Moscow: Izd. AN SSSR, 1954.

Rastvory. Klassiki nauki, ed. K. P. Mishchenko. Leningrad: Izd. AN SSSR, 1959.

"Remarques à propos de la découverte du gallium." *Comptes Rendus* 81 (1875): 969–972.

"Researches on Mariotte's Law." *Nature* 15 (1877): 455–457, 498–500.

S dumoiu o blage rossiiskom: Izbrannye ekonomicheskie proizvedeniia. Novosibirsk: Nauka, 1991.

"Spiriticheskie uzly." *Novoe Vremia,* 18 (31) May 1904, #10132: 3.

Tolkovyi tarif, ili izsledovanie o razvitii promyshlennosti Rossii v sviazi s eia obshchim tamozhennym tarifom 1891 goda. St. Petersburg: V. Demakov, 1892.

"Über die Stellung des Ceriums im System der Elemente." *Bulletin de l'Académie Impériale des Sciences de Saint-Pétersbourg* 16, no. 1 (1871): 45–51.

"Ueber die Anwendbarkeit des periodischen Gesetzes bei den Ceritmetallen." *Annalen der Chemie und Pharmacie* 168 (1873): 45–63.

"Vozdushnyi polet iz Klina vo vremia zatmeniia." *Severnyi Vestnik,* no. 1 (1887): 87–124; no. 12 (1887): 57–93.

Zavetnye mysli: Polnoe izdanie. Moscow: Mysl', 1995 [1903–1905].

"Zoloto iz serebra." *Zhurnal Zhurnalov i Entsiklopedicheskoe Obozrenie,* no. 1 (1898): 1–11.

III.   WORKS BY OTHER AUTHORS

A. M. Butlerov, 1828–1928. Leningrad: Izd. AN SSSR, 1929.

"Account of the Sessions of the International Congress of Chemists in Karlsruhe, on 3, 4, and 5 September 1860." In Mary Jo Nye, ed., *The Question of the Atom: From the Karlsruhe Congress to the First Solvay Conference, 1860–1911* (Los Angeles: Tomash Publishers, 1984): 5–28.

Akatov, V. S. *O pozitivnykh osnovakh noveishago spiritualizma.* Moscow: Kardek, 1909.

Aksakov, A. N. *Animisme et Spiritisme: Essai d'un examen critique des phénomènes médiumiques*, tr. Berthold Sandow. Paris: Librairie des sciences physiques, 1895.

———. *Animismus und Spiritismus: Versuch einer kritischen Prüfung der mediumistischen Phänomene mit besonderer Berücksichtigung der Hypothesen der Halluzination und des Unbewussten.* 2 v. Leipzig: Oswald Mutze, 1919.

———. "[Letter to the Commission]." *S.-Peterburgskie Vedomosti*, 11 (23) March 1876, #70: 2–3.

———. "[Letter to the editor]." *S.-Peterburgskie Vedomosti*, 7 December 1875, #329.

———. *Pamiatnik nauchnogo predubezhdeniia: Zakliuchenie mediumicheskoi kommisii Fizicheskago obshchestva pri S.-Peterburgskom Universitete s primechaniiami.* St. Petersburg: V. Bezobrazov, 1883.

———. "Pis'mo v redaktsiiu." *Novoe Vremia*, 12 March 1876, #13: 2.

———. "Po povodu odnogo iz 'pshikov' professora Mendeleeva." *Rebus*, no. 2 (1894): 15–16.

———. *Razoblacheniia: Istoriia mediumicheskoi kommisii Fizicheskago obshchestva pri S.-Peterburgskom universitete s prilozheniem vsekh protokolov i prochikh dokumentov.* St. Petersburg: V. Bezobrazov, 1883.

———, ed., tr., and pub. *Spiritualizm i nauka. Opytnyia izsledovaniia nad psikhicheskoi siloi Uil'iama Kruksa, chlena Londonskago korolevskago obshchestva. Podtverditel'nyia svidetel'stva khimika P. Gera, matematika A. De-Morgana, naturalista A. Uallesa, fizika K. Varleia i drugikh izsledovatelei. Udachnye i neudachnye seansy D. D. Iuma s angliiskimi i russkimi uchenymi.* St. Petersburg: Tip. A. M. Kotomina, 1872.

———. *Vorläufer des Spiritismus. Hervorragende Fälle willfürlicher mediumistischer Erscheinungen aus den letzten drei Jahrhunderten*, tr. F. Feilgenhauer. Leipzig: Oswald Mutze, 1898.

Aksakov, I. S. "[Editor's introduction]." *Rus'*, 13 February 1882, #7: 1–4.

Alborn, Timothy L. "Negotiating Notation: Chemical Symbols and British Society, 1831–1835." *Annals of Science* 46 (1989): 437–460.

Alder, Ken. "A Revolution to Measure: The Political Economy of the Metric System in France." In Wise, ed., *The Values of Precision* (1995): 39–71.

Alenitsin, V. "Polozhitel'noe i otritsatel'noe." *Svet* 1, nos. 6, 10 (1877): 129–134, 227–233; *Svet* 2, no. 5 (1878): 172–178.

Allen, Eugene T. Review of D. Mendeléeff's *The Principles of Chemistry*, 3d. English edition. *Journal of the American Chemical Society* 27 (1905): 789–790.

Almgren, Beverly S. "D. I. Mendeleev and Siberia." *Ambix* 45 (1998): 50–66.

———. "Mendeleev: The Third Service, 1834–1882." Ph.D. Dissertation, Brown University, 1968.

Alston, Patrick L. "The Dynamics of Educational Expansion in Russia." In Jarausch, ed., *The Transformation of Higher Learning* (1982): 89–107.

———. *Education and the State in Tsarist Russia.* Stanford: Stanford University Press, 1969.

Ambler, Effie. *Russian Journalism and Politics: The Career of Aleksei S. Suvorin, 1861–1881.* Detroit: Wayne State University Press, 1972.

Anderson, Katherine. "The Weather Prophets: Science and Reputation in Victorian Meteorology." *History of Science* 37 (1999): 179–216.

Anderson, Wilda C. *Between the Library and the Laboratory: The Language of Chemistry in Eighteenth-Century France.* Baltimore: The Johns Hopkins University Press, 1984.

Antonoshevskii, I. K. *Bibliografiia okkul'tizma: Ukazatel' sochinenii.* St. Petersburg: M. S. K., 1911.

Antonovich, M. "Predislovie." *Novoe Obozrenie*, kn. 1 (January 1881): 227–246.

Arbuzov, A. E., ed. *Materialy po istorii otechestvennoi khimii.* Moscow: Izd. AN SSSR, 1950.

———. "Pamiati D. I. Mendeleeva i A. M. Butlerova." *ZhRFKhO* 61, no. 4, ch. khim. (1929): 629–640.

Arutiunov, V. O., ed. *50 let metricheskoi reformy v SSSR.* Moscow: Izd. Standartov, 1972.

Ascher, Abraham. *The Revolution of 1905,* 2 v. Stanford: Stanford University Press, 1988–1992.

Averbukh, A. Ia., and A. A. Makarenia. *Mendeleev v Tekhnologicheskom Institute.* Leningrad: Znanie, 1976.

Azernikov, V. Z., and N. F. Vasil'kova, comps. *Mendeleev—metrolog.* Moscow: Izd. Standartov, 1969.

Balmuth, Daniel. *Censorship in Russia, 1865–1905.* Washington, D.C.: University Press of America, 1979.

Balzer, Harley D. "The Problem of Professions in Imperial Russia." In Clowes, Kassow, and West, eds., *Between Tsar and People* (1991): 183–198.

———, ed. *Russia's Missing Middle Class: The Professions in Russian History.* Armonk, N.Y.: M. E. Sharpe, 1996.

Bardin, I. P., and V. V. Rikman. "Raboty D. I. Mendeleeva v oblasti metallurgii." In Vol'fkovich, ed., *Dmitrii Ivanovich Mendeleev* (1957): 115–129.

Barrow, Logie. *Independent Spirits: Spiritualism and English Plebians, 1850–1910.* London: Routledge and Kegan Paul, 1986.

Bary, Paul. "L'argentaurum." *Revue de physique et de chimie et de leurs applications industrielles* 1, no. 11 (1896–1897): 497–503.

Barzakovskii, V. P., and R. B. Dobrotin. *Trudy D. I. Mendeleeva v oblasti khimii silikatov i stekloobraznogo sostoianiia.* Moscow: Izd. AN SSSR, 1960.

Bassin, Mark. "Inventing Siberia: Visions of the Russian East in the Early Nineteenth Century." *American Historical Review* 96 (1991): 763–794.

———. "Turner, Solov'ev, and the 'Frontier Hypothesis': The Nationalist Signification of Open Spaces." *Journal of Modern History* 65 (1993): 473–511.

Bater, James H. *St. Petersburg: Industrialization and Change.* London: Edward Arnold, 1976.

Batten, Alan H. "Johann Franz Encke, 1791–1965." *Journal of the Royal Astronomical Society of Canada* 85 (1991): 316–323.

Becker, Seymour. *Nobility and Privilege in Late Imperial Russia.* DeKalb: Northern Illinois University Press, 1985.

Bederman, Gail. *Manliness & Civilization: A Cultural History of Gender and Race in the United States, 1880–1917.* Chicago: University of Chicago Press, 1995.

Beketov, N. N. "Dmitrii Ivanovich Mendeleev, 1834–1907. Nekrolog." *Izvestiia Imperatorskoi Akademii Nauk,* Series VI (1907): 51–54.

———. "Istoriia khimicheskoi laboratorii pri Akademii nauk." In *Lomonosovskii sbornik* (1901): 1–5.

———. "O khimicheskoi energii v sviazi s iavleniiami predstavliaemymi radiem." *ZhRFKhO* 35, no. 3, otd. I (1903): 189–197.

———. *Rechi khimika, 1862–1903.* St. Petersburg: Znanie, 1908.

———. "Znachenie periodicheskoi sistemy D. I. Mendeleeva." In Tishchenko, ed., *Trudy pervago mendeleevskogo s"ezda* (1909): 33–35.

Belov, P. T. *Filosofiia vydaiushchikhsia russkikh estestvoispytatelei vtoroi poloviny XIX—nachala XX v.* Moscow: Mysl', 1970.

Bennett, Helju Aulik. "*Chiny, Ordena,* and Officialdom." In Pintner and Rowney, eds., *Russian Officialdom* (1980): 162–189.

———. "Evolution of the Meanings of Chin: An Introduction to the Russian Institution of Rank Ordering and Niche Assignment from the Time of Peter the Great's Table of Ranks to the Bolshevik Revolution." *California Slavic Studies* 10 (1977): 1–43.

Bensaude-Vincent, Bernadette. "Between History and Memory: Centennial and Bicentennial Images of Lavoisier." *Isis* 87 (1996): 481–499.

———. "L'éther, élément chimique: un essai malheureux de Mendéléev (1902)?" *British Journal for the History of Science* 15 (1982): 183–188.

———. "Mendeleev's Periodic System of Chemical Elements." *British Journal for the History of Science* 19 (1986): 3–17.

———. "A View of the Chemical Revolution through Contemporary Textbooks: Lavoisier, Fourcroy and Chaptal." *British Journal for the History of Science* 23 (1990): 435–460.

Benson, Donald R. "Facts and Fictions in Scientific Discourse: The Case of Ether." *Georgia Review* 38 (1984): 825–837.

Beretta, Marco. *The Enlightenment of Matter: The Definition of Chemistry from Agricola to Lavoisier.* Canton, Mass.: Science History Publications, 1993.

Berlin, Isaiah. *Russian Thinkers*, eds. Henry Hardy and Aileen Kelly. New York: Penguin Books, 1978 [1948].

Berry, Thomas E. "Mediums and Spiritualism in Russian Literature during the Reign of Alexander II." In Amy Mandelker and Roberta Reeder, eds., *The Supernatural in Slavic and Baltic Literature: Essays in Honor of Victor Terras* (Columbus: Slavica, 1988): 129–144.

———. *Spiritualism in Tsarist Society and Literature.* Baltimore: Edgar Allan Poe Society, 1985.

"Bezzakonniki v Akademii nauk." *Novoe Vremia*, 7 (19) December 1880, #1717: 3.

Biagioli, Mario. *Galileo, Courtier: The Practice of Science in the Culture of Absolutism.* Chicago: University of Chicago Press, 1993.

"Bibliografiia osnovnykh nauchnykh trudov B. M. Kedrova." In Markov et al., eds., *Filosofiia i estestvoznanie* (1974): 270–278.

Birkenmaier, Willy. *Das russische Heidelberg: Zur Geschichte der deutsch-russischen Beziehungen im 19. Jahrhundert.* Heidelberg: Wunderhorn, 1995.

Bischoff, C. A. [Review of Mendelejeff's *Grundlagen der Chemie*.] *Rigasche Industrie-Zeitung* 17, no. 22 (1891): 264.

Black, J. L. *G.-F. Müller and the Imperial Russian Academy.* Kingston, Ontario: McGill-Queen's University Press, 1986.

Blackwell, William. "The Russian Entrepreneur in the Tsarist Period: An Overview." In Guroff and Carstensen, eds., *Entrepreneurship in Imperial Russia and the Soviet Union* (1983): 13–26.

Blok, Aleksandr. *Sobranie sochinenii v vos'mi tomakh*, 8 v. Moscow: Khudozhestvennaia literatura, 1960–1963.

———. *Zapisnye knizhki, 1901–1920.* Moscow: Khudozhestvennaia literatura, 1965.

Blokh, M. A., ed. *Iubileinomu mendeleevskomu s"ezdu v oznamenovanie 100-letnei godovshchiny so dnia rozhdeniia D. I. Mendeleeva.* Leningrad: Goskhimtekhizdat, 1934.

Blumbakh, F. "Dannyia o vyverke mer i vesov v Sibiri; iz otcheta, predstavlennago poveritelem G. Palaty F. I. Blumbakhom." *VGPMV* 3 (1896): 124–132.

———. "Geograficheskoe polozhenie Glavnoi Palaty mer i vesov." *VGPMV* 3 (1896): 108–117.

Boborykin, P. D. "Ni vzad—ni vpered." *S.-Peterburgskie Vedomosti*, 16 (28) March 1876, #75: 1–2; 23 March (4 April) 1876, #82: 1–2; 30 March (11 April) 1876, #89: 1–2.

———. "Voskresnyi fel'eton." *S.-Peterburgskie Vedomosti*, 21 December 1875 (2 January 1876), #343: 1–2.

Boele, Otto. *The North in Russian Romantic Literature.* Amsterdam: Rodopi, 1996.

Boitsov, V. V., ed. *D. I. Mendeleev—osnovopolozhnik sovremennoi metrologii.* Moscow: Izd. Standartov, 1978.

Bolton, H. Carrington. "Recent Progress of Alchemy in America." *Chemical News* 76 (1897): 61–64.

Bondarenko, L. B. "Iz istorii russkoi spirtometrii." *VIET*, no. 2 (1999): 184–204.

Borodin, A. P. *Pis'ma A. P. Borodina: Polnoe sobranie, kriticheski sverennoe s podlinnymi tekstami*, v. 1. Moscow: Muzykal'nyi sektor, 1927–1928.

Bourdon, Georges. "Les Opinions de Professeur Mendéléeff." *Le Figaro*, 20 March 1905: 2–3.

Bowman, Linda. "Russia's First Income Taxes: The Effects of Modernized Taxes on Commerce and Industry, 1885–1914." *Slavic Review* 52 (1993): 256–282.

Bradley, Joseph. "Voluntary Associations, Civic Culture, and *Obshchestvennost'* in Moscow." In Clowes, Kassow, and West, eds., *Between Tsar and People* (1991): 131–148.

Brandon, Ruth. *The Spiritualists: The Passion for the Occult in the Nineteenth and Twentieth Centuries*. New York: Alfred A. Knopf, 1983.

Braude, Ann. *Radical Spirits: Spiritualism and Women's Rights in Nineteenth-Century America*. Boston: Beacon Press, 1989.

Brauner, Bohuslav. "Some Remarks on 'Argon.'" *Chemical News* 71 (1895): 67–68.

Brester, A. *Essai d'une Théorie du Soleil et des Études variables*. Delft: J. Waltman, Jr., 1889.

Britten, Emma Hardinge. *Nineteenth Century Miracles; or, Spirits and Their Work in Every Country of the Earth*. New York: William Britten, 1884.

Brock, W. H., ed. *The Atomic Debates: Brodie and the Rejection of the Atomic Theory. Three Studies*. Leicester: Leicester University Press, 1967.

_____. "A British Career in Chemistry: Sir William Crookes (1832–1919)." In Knight and Kragh, eds., *The Making of the Chemist* (1998): 121–129.

_____. *From Protyle to Proton: William Prout and the Nature of Matter, 1785–1985*. Bristol: Adam Hilger Ltd., 1985.

_____, ed. *H. E. Armstrong and the Teaching of Science, 1880–1930*. Cambridge: Cambridge University Press, 1973.

_____. *Justus von Liebig: The Chemical Gatekeeper*. Cambridge: Cambridge University Press, 1997.

_____. "The Life and Work of William Prout." *Medical History* 9 (1965): 101–126.

_____. "Studies in the History of Prout's Hypothesis, Parts I and II." *Annals of Science* 25 (1969): 49–80, 127–137.

Brooke, John Hedley. "Avogadro's Hypothesis and Its Fate: A Case-Study in the Failure of Case-Studies." *History of Science* 19 (1981): 235–273.

Brooks, Jeffrey. *When Russia Learned to Read: Literacy and Popular Culture, 1861–1917*. Princeton: Princeton University Press, 1985.

Brooks, Nathan M. "Alexander Butlerov and the Professionalization of Science in Russia." *Russian Review* 57 (1998): 10–24.

_____. "Developing the Periodic Law: Mendeleev's Work during 1869–1871." *Foundations of Chemistry* 4 (2002): 127–147.

_____. "The Evolution of Chemistry in Russia during the Eighteenth and Nineteenth Centuries." In Knight and Kragh, eds., *The Making of the Chemist* (1998): 163–176.

_____. "The Formation of a Community of Chemists in Russia: 1700–1870." Ph.D. Dissertation, Columbia University, 1989.

_____. "Mendeleev and Metrology." *Ambix* 45 (1998): 116–128.

_____. "Public Lectures in Chemistry in Russia." *Ambix* 44 (1997): 1–10.

_____. "Russian Chemistry in the 1850s: A Failed Attempt at Institutionalization." *Annals of Science* 52 (1995): 577–589.

Brower, Daniel R. *The Russian City between Tradition and Modernity, 1850–1900*. Berkeley: University of California Press, 1990.

_____. "Social Stratification in Russian Higher Education." In Jarausch, ed., *The Transformation of Higher Learning* (1982): 245–260.

_____. *Training the Nihilists: Education and Radicalism in Tsarist Russia*. Ithaca: Cornell University Press, 1975.

Brown, Peter. *Augustine of Hippo: A Biography.* Berkeley: University of California Press, 1967.

Brown, Slater. *The Heyday of Spiritualism.* New York: Hawthorn Books, 1970.

Brush, Charles F. "A New Gas." *Science* 8 (1898): 485–494.

Brush, Stephen G. "Prediction and Theory Evaluation in Physics and Astronomy." In A. J. Kox and Daniel M. Siegel, eds., *No Truth Except in the Details: Essays in Honor of Martin J. Klein* (Dordrecht: Kluwer Academic, 1995): 299–318.

———. "The Reception of Mendeleev's Periodic Law in America and Britain." *Isis* 87 (1996): 595–628.

Buchdahl, Gerd. "Sources of Scepticism in Atomic Theory." *British Journal for the Philosophy of Science* 10 (1960): 120–134.

Buchwald, Jed Z. "The Quantitative Ether in the First Half of the Nineteenth Century." In Cantor and Hodge, eds., *Conceptions of Ether* (1981): 215–237.

Bud, Robert, and Gerrylynn K. Roberts. *Science versus Practice: Chemistry in Victorian Britain.* Manchester: Manchester University Press, 1984.

Bulgakov, F. I. "Nemetskaia partiia v Russkoi akademii." *Istoricheskii Vestnik* 4 (1881): 421–431.

Bulgakov, M. A. *Sobranie sochinenii,* v. 3. Ann Arbor: Ardis, 1983.

Burton, Jean. *Heyday of a Wizard: Daniel Home, the Medium.* New York: Alfred A. Knopf, 1944.

Burton, Jim. "Robert FitzRoy and the Early History of the Meteorological Office." *British Journal for the History of Science* 19 (1986): 147–176.

Butlerov, A. M. "Chetvertoe izmerenie prostranstva i mediumizm." *Russkii Vestnik* (February 1878): 945–971.

———. *Empirizm i dogmatizm v oblasti mediumizma.* Moscow: Universitetskaia tipografiia (M. Katkov), 1879.

———. "Koe-chto o mediumizme." *Rebus,* no. 16 (1883): 147–148; no. 18 (1883): 165–167; no. 31 (1883): 276–278; no. 35 (1883): 309–312; no. 40 (1883): 355–356; no. 42 (1883): 376–378; no. 43 (1883): 385–387.

———. "Mediumicheskiia iavleniia." *Russkii Vestnik* 120 (November 1875): 300–348.

———. "Mediumizm i umozrenie bez opyta. (Otvet g. Strakhovu)." *Rebus,* no. 41 (1885): 367–370.

———. *Nauchnaia i pedagogicheskaia deiatel'nost'.* Moscow: Izd. AN SSSR, 1961.

———. "O 'vozmozhnom' i 'nevozmozhnom' v nauke." *Rebus,* no. 48 (1883): 435–437.

———. *Sochineniia,* 3 v. Moscow: Izd. AN SSSR, 1953–1958.

———. "[Two letters to the editor]." *Novoe Vremia,* 8 March 1876, #9: 2–3.

———. "Umstvovanie i opyt." *Novoe Vremia,* 7 February 1884, #2854.

———. "Zametka ob atomnykh vesakh." *ZhRFKhO* 14, no. 5, otd. I (1882): 208–212.

Butlerov, A. M., and A. N. Aksakov. "[Letter to the editor]." *Golos,* 29 March 1876, #89.

Bykov, G. V. *Aleksandr Mikhailovich Butlerov: Ocherk zhizni i deiatel'nosti.* Moscow: Izd. AN SSSR, 1961.

———. "Dva otzyva D. I. Mendeleeva ob A. M. Butlerove." *Uspekhi Khimii* 22, no. 1 (1953): 115–118.

———. "Materialy k istorii trekh pervykh izdanii 'Vvedeniia k polnomu izucheniiu organicheskoi khimii' A. M. Butlerova." *TIIEiT* 6 (1955): 243–291.

———. "O nauchnom metode A. M. Butlerova." *Voprosy Filosofii,* no. 6 (1955): 110–117.

———, ed. *Pis'ma russkikh khimikov k A. M. Butlerovu,* published as *Nauchnoe Nasledstvo,* v. 4. Moscow: Izd. AN SSSR, 1961.

———. "Svet i teni v nauchnoi biografii." In M. G. Iaroshenko, ed., *Chelovek nauki* (Moscow: Nauka, 1974): 68–72.

Bykov, G. V., and Z. I. Sheptunova. "Nemetskii 'Zhurnal Khimii' (1858–1871) i russkie khimiki. (K istorii khimicheskoi periodiki)." *TIIEiT* 30 (1960): 97–110.

Bykow, G. W., and L. M. Bekassowa. "Beiträge zur Geschichte der Chemie der 60-er Jahre des XIX. Jahrhunderts: II. F. Beilsteins Briefe an A. M. Butlerow." *Physis* 8 (1966): 267–285.

Cahan, David. *An Institute for an Empire: The Physikalisch-Technische Reichsanstalt, 1871–1918.* Cambridge: Cambridge University Press, 1989.

Caneva, Kenneth L. "Ampère, the Etherians, and the Oersted Connexion." *British Journal for the History of Science* 13 (1980): 121–138.

Cantor, Geoffrey. "The Theological Significance of Ethers." In Cantor and Hodge, eds., *Conceptions of Ether* (1981): 135–155.

Cantor, Geoffrey N., and M. J. S. Hodge, eds. *Conceptions of Ether: Studies in the History of Ether Theories, 1740–1900.* Cambridge: Cambridge University Press, 1981.

———. "Introduction: Major Themes in the Development of Ether Theories from the Ancients to 1900." In Cantor and Hodge, eds., *Conceptions of Ether* (1981): 1–60.

Carlson, Maria. "Fashionable Occultism: Spiritualism, Theosophy, Freemasonry, and Hermeticism in Fin-de-Siècle Russia." In Rosenthal, ed., *The Occult in Russian and Soviet Culture* (1997): 135–152.

———. *"No Religion Higher Than Truth": A History of the Theosophical Movement in Russia, 1875–1922.* Princeton: Princeton University Press, 1993.

Carnelley, Thomas. "The Influence of Atomic Weight." *Philosophical Magazine* 8, no. 49 (1879): 305–324, 369–381, 461–476.

———. "Mendelejeff's periodisches Gesetz und die magnetischen Eigenschaften der Elemente." *Berichte der Deutschen Chemischen Gesellschaft* 12 (1879): 1958–1961.

———. "Suggestions as to the Cause of the Periodic Law and the Nature of the Chemical Elements." *Chemical News* 53 (1886): 157–159, 169–172, 183–186, 197–200.

Carnes, Mark C., and Clyde Griffen, eds. *Meanings for Manhood: Constructions of Masculinity in Victorian America.* Chicago: University of Chicago Press, 1990.

Carpenter, William. "Fiziologicheskoe ob"iasnenie nekotorykh iavlenii spiritizma." *Znanie* (May 1875): 27–61.

Carrier, Martin. "Newton's Ideas on the Structure of Matter and Their Impact on Eighteenth-Century Chemistry: Some Historical and Methodological Remarks." *International Studies in the Philosophy of Science* 1 (1986): 85–105.

Carroll, Bret E. *Spiritualism in Antebellum America.* Bloomington: Indiana University Press, 1997.

Carstensen, Fred C. *American Enterprise in Foreign Markets: Studies of Singer and International Harvester in Imperial Russia.* Chapel Hill: University of North Carolina Press, 1984.

———. "Foreign Participation in Russian Economic Life: Notes on British Enterprise, 1865–1914." In Guroff and Carstensen, eds., *Entrepreneurship in Imperial Russia and the Soviet Union* (1983): 140–158.

Cassebaum, Heinz, and George B. Kauffman. "The Periodic System of the Chemical Elements: The Search for Its Discoverer." *Isis* 62 (1971): 314–327.

Cassidy, David C. "Meteorology in Mannheim: The Palatine Meteorological Society, 1780–1795." *Sudhoffs Archiv* 69 (1985): 8–25.

Cawood, John. "Terrestrial Magnetism and the Development of International Collaboration in the Early Nineteenth Century." *Annals of Science* 34 (1977): 551–587.

Cerruti, Luigi, and Eugenio Torracca. "Development of Chemistry in Italy, 1840–1910." In Knight and Kragh, eds., *The Making of the Chemist* (1998): 233–262.

Chasovnikov, N. L. "K semidesiatipiatiletiiu so dnia opublikovaniia periodicheskoi sistemy D. I. Mendeleeva." *Uspekhi Khimii* 13, no. 2 (1944): 85–89.

Chayut, Michael. "J. J. Thomson: The Discovery of the Electron and the Chemists." *Annals of Science* 48 (1991): 527–544.

Choldin, Marianna Tax. *A Fence around the Empire: Russian Censorship of Western Ideas under the Tsars*. Durham, N.C.: Duke University Press, 1985.

Christian, David. *"Living Water": Vodka and Russian Society on the Eve of Emancipation*. Oxford: Clarendon Press, 1990.

————. "A Neglected Great Reform: The Abolition of Tax Farming in Russia." In Eklof, Bushnell, and Zakharova, eds., *Russia's Great Reforms* (1994): 102–114.

Christie, J. R. R. "Ether and the Science of Chemistry: 1740–1790." In Cantor and Hodge, eds., *Conceptions of Ether* (1981): 85–110.

Chugaev, L. A. *Dmitrii Ivanovich Mendeleev: Zhizn' i deiatel'nost'*. Leningrad: Nauchnoe khimiko-tekhnicheskoe izd., 1924.

Cleve, P. T. "Om Skandium." *Ofversigt af Kongl. Vetenskaps-Akademiens Förhandlingar*, no. 7 (1879): 3–10.

————. "Sur le scandium." *Comptes Rendus* 89 (1879): 419–422.

Clowes, Edith W., Samuel D. Kassow, and James L. West, eds. *Between Tsar and People: Educated Society and the Quest for Public Identity in Late Imperial Russia*. Princeton: Princeton University Press, 1991.

Cole, Theron M., Jr. "Early Atomic Speculations of Marc Antoine Gaudin: Avogadro's Hypothesis and the Periodic System." *Isis* 66 (1975): 334–360.

Confino, Michael. "On Intellectuals and Intellectual Traditions in Eighteenth- and Nineteenth-Century Russia." *Daedalus* 101, no. 2 (1972): 117–149.

"The Congress of Chemists at Carlsruhe." *Chemical News* 2 (1860): 226–227.

Cooter, Roger. *The Cultural Meaning of Popular Science: Phrenology and the Organization of Consent in Nineteenth-Century Britain*. Cambridge: Cambridge University Press, 1984.

Crawford, Elisabeth. *The Beginnings of the Nobel Institution: The Science Prizes, 1901–1915*. Cambridge: Cambridge University Press, 1984.

Crisp, Olga. *Studies in the Russian Economy before 1914*. London: Macmillan, 1976.

Crisp, Olga, and Linda Edmondson, eds. *Civil Rights in Imperial Russia*. Oxford: Clarendon Press, 1989.

Cromartie, Alan. *Sir Matthew Hale, 1609–1676: Law, Religion, and Natural Philosophy*. Cambridge: Cambridge University Press, 1992.

Crookes, William. "The Chemistry of the Future." *Quarterly Journal of Science* 7 (1877): 289–306.

————. "On the Supposed New Gas, Etherion." *Chemical News* 78 (1898): 221–222.

————. *Strahlende Materie oder der vierte Aggregatzustand*. Leipzig: Quandt und Händel, 1879.

Crosland, Maurice. *Gay-Lussac: Scientist and Bourgeois*. Cambridge: Cambridge University Press, 1978.

————. "'Nature' and Measurement in Eighteenth-Century France." *Studies on Voltaire and the Eighteenth Century* 87 (1972): 277–309.

————. "The Organisation of Chemistry in Nineteenth-Century France." In Knight and Kragh, eds., *The Making of the Chemist* (1998): 3–14.

————. *The Society of Arcueil: A View of French Science at the Time of Napoleon I*. Cambridge, Mass.: Harvard University Press, 1967.

Crouch, Tom D. *The Eagle Aloft: Two Centuries of the Balloon in America*. Washington, D.C.: Smithsonian Institution Press, 1983.

D. I. Mendeleev. Ego nauchnoe tvorchestvo i raboty v Glavnoi Palate mer i vesov. Moscow: Tsentral'noe upravlenie prompropagandy i pechati VSNKh SSSR, 1926.

Daly, Jonathan W. *Autocracy under Siege: Security Police and Opposition in Russia, 1866–1905*. DeKalb: Northern Illinois University Press, 1998.

Darnton, Robert. *Mesmerism and the End of the Enlightenment in France*. Cambridge, Mass.: Harvard University Press, 1968.

Davis, Andrew Jackson. *The Magic Staff: An Autobiography.* New York: J. S. Brown, 1857.

Davis, John L. "Weather Forecasting and the Development of Meteorological Theory at the Paris Observatory, 1853–1878." *Annals of Science* 41 (1984): 359–382.

Dawson, Graham. "The Blond Bedouin: Lawrence of Arabia, Imperial Adventure and the Imagining of English-British Masculinity." In Roper and Tosh, eds., *Manful Assertions* (1991): 113–144.

DeKoskey, Robert K. "Spectroscopy and the Elements in the Late Nineteenth Century: The Work of Sir William Crookes." *British Journal for the History of Science* 6 (1973): 400–423.

"Delo Mendeleeva. Predstavlenie akademikov A. Butlerova, P. Chebysheva, F. Ovsiannikova i N. Koksharova v 1-e otdelenie Imperatorskoi Akademii nauk." *Novoe Vremia*, 21 November (3 December) 1880, #1701: 3; 23 November (5 December) 1880, #1703: 4; 24 November (6 December) 1880, #1704: 2–3; 25 November (7 December) 1880, #1705: 3.

DeMilt, Clara. "The Congress at Karlsruhe." *Journal of Chemical Education* (August 1951): 421–425.

Departament Torgovli i Manufaktur Ministerstva Finansov. *Fabrichno-zavodskaia promyshelennost' i torgovlia Rossii. Vsemirnaia Kolumbova vystavka 1893 g. v Chikago.* St. Petersburg: V. S. Balashev and V. F. Demakov, 1893.

"Derptskie otgoloski akademicheskogo skandala." *Novoe Vremia*, 15 (27) December 1880, #1725: 1.

Desmond, Adrian, and James Moore. *Darwin.* London: Michael Joseph, 1991.

Dmitriev, I. S. "Natsional'naia legenda: Byl li D. I. Mendeleev sozdatelem russkoi 'monopol'noi' vodki?" *VIET*, no. 2 (1999): 177–183.

_____. "Nauchnoe otkrytie in statu nascendi: Periodicheskii zakon D. I. Mendeleeva." *VIET*, no. 1 (2001): 31–82.

_____. "Problema razmeshcheniia indiia v periodicheskoi sisteme elementov." *VIET*, no. 2 (1984): 3–14.

_____. "Skuchnaia istoriia (o neizbranii D. I. Mendeleeva v Imperatorskuiu akademiiu nauk v 1880 g.)." *VIET*, no. 2 (2002): 231–280.

_____. "Teoreticheskie issledovaniia P. E. Lekoka de Buabodrana po klassifikatsii khimicheskikh elementov i sistematike spektrov." In Trifonov, ed., *Uchenie o periodichnosti* (1981): 19–36.

Dobrokhotov, A. "Otchet o komandirovke v Tver', Moskvu, Maloiaroslavets, Kalugu, Tulu, Vladimir, Nizhnii-Novgorod, Kostromu, Iaroslavl' i Rybinsk." *VGPMV* 6 (1903): 1–24.

Dobrotin, R. B. "K istorii otkrytiia germaniia (ekasilitsiia)." *Vestnik Leningradskogo universiteta*, no. 10 (1956): 55–59.

_____. "Rannii period nauchnoi deiatel'nosti D. I. Mendeleeva kak etap na puti k otkrytiiu periodicheskogo zakona." Candidate Dissertation, Leningrad State University, 1953.

Dobrotin, R. B., et al. *Letopis' zhizni i deiatel'nosti D. I. Mendeleeva.* Leningrad: Nauka, 1984.

Dobrotin, R. B., and N. G. Karpilo. *Biblioteka D. I. Mendeleeva.* Leningrad: Nauka, 1980.

_____. "D. I. Mendeleev o L. N. Tolstom." *Priroda*, no. 9 (1978): 11–13.

Dobrotin, R. B., and L. S. Kerova. "Logicheskii analiz tvorcheskogo puti D. I. Mendeleeva." *Voprosy Istorii i Metodologii Khimii* 1 (1976): 5–23.

Dobrotin, R. B., and A. A. Makarenia. "Prognozirovanie svoistv skandiia i germaniia v rabotakh D. I. Mendeleeva." In Kedrov and Trifonov, eds., *Prognozirovanie v uchenii o periodichnosti* (1976): 53–70.

Dobrotin, R. B., M. G. Ter-Avakova, and T. V. Volkova. "Perepiska D. I. Mendeleeva s zarubezhnymi uchenymi." *VIET*, no. 3 (1957): 176–189.

"Doklady gen.-ad. A. R. Drentel'na Aleksandru II. (Aprel'-Noiabr' 1879 g.)." *Krasnyi Arkhiv* 3 (40) (1930): 124–175.

Dolby, R. G. A. "Debates over the Theory of Solution: A Study of Dissent in Physical Chemistry in the English-Speaking World in the Late Nineteenth and Early Twentieth Centuries." *Historical Studies in the Physical Sciences* 7 (1976): 297–404.

Donovan, A. L. *Philosophical Chemistry in the Scottish Enlightenment: The Doctrines and Discoveries of William Cullen and Joseph Black.* Edinburgh: Edinburgh University Press, 1975.

Dostoevskii, F. M. *Polnoe sobranie sochinenii v tridtsati tomakh,* 30 v. Leningrad: Nauka, 1972–1990.

Douglas, Ann. *The Feminization of American Culture.* New York: Alfred A. Knopf, 1977.

Dowler, Wayne. *Dostoevsky, Grigor'ev, and Native Soil Conservatism.* Toronto: University of Toronto Press, 1982.

Doyle, Arthur Conan. *The History of Spiritualism,* 2 v. New York: George H. Doran, 1926.

Drzewiecki, S. *Les Ouiseaux Considérés comme des Aéroplanes Animés: Essai d'une Nouvelle Théorie du Vol.* Clermont (Oise): Imprimerie Daix Frères, 1889.

Dubravin, A. I. "O laboratornykh priborakh D. I. Mendeleeva dlia issledovaniia uprugosti gazov." *Voprosy Istorii i Metodologii Khimii* 2 (1978): 97–103.

Duhem, Pierre. *The Aim and Structure of Physical Theory,* tr. Philip P. Wiener. Princeton: Princeton University Press, 1982 [1906].

Dunraven, Windham Thomas Wyndham-Quin, Earl of. *Experiences in Spiritualism with D. D. Home.* Glasgow: Robert Maclehose and Co., 1924.

E. K. "Spiritizm i nasha intelligentsia." *Nedelia,* 25 March 1875, #21: 686–701.

Eastburn, George. *The Metric System.* New York: The American Metrological Society, 1892.

Edmondson, Linda, and Peter Waldron, eds. *Economy and Society in Russia and the Soviet Union, 1860–1930: Essays for Olga Crisp.* New York: St. Martin's Press, 1992.

Egorov, K. "Otchet o komandirovke v goroda: Riazan', Syzran', Penzu, Samaru, Ufu, Zlatoust', Cheliabinsk, Ekaterinburg, Perm' i Irbit." *VGPMV* 6 (1903): 46–83.

———. "Otchet o komandirovke v goroda: Smolensk, Chernigov i Kiev." *VGPMV* 6 (1903): 99–108.

———. "Otchet po komandirovke v goroda Varshavu, Lodz' i Liublin." *VGPMV* 5 (1900): 74–144.

Egorov, K., A. Dobrokhotov, and V. Müller. "Reviziia vesov i gir' v Pochtamte i pochtovykh otdeleniiakh g. S. Peterburga." *VGPMV* 5 (1900): 63–73.

Egorov, N. G. "Dmitrii Ivanovich Mendeleev (nekrolog)." *VGPMV* 8 (1907): 3–18.

———. "O pravitel'stvennoi vyverke elektricheskikh izmeritel'nykh priborov v zapadnoevropeiskikh gosudarstvakh." *VGPMV* 4 (1899): 81–121.

Egorov, N., and A. Dobrokhotov. "Reviziia vesov i gir' v Gosudarstvennom Banke." *VGPMV* 5 (1900): 60–63.

Egorov, V. M. "'Gazeta obshchenarodnykh nauchnykh znanii.' (Neosushchestvennyi zamysel D. I. Mendeleeva)." *Ocherki po istorii Sankt-Peterburgskogo Universiteta* 7 (1998): 124–131.

Eimontova, R. G. *Russkie universiteti na grani dvukh epokh: Ot Rossii krepostnoi k Rossii kapitalisticheskoi.* Moscow: Nauka, 1985.

Ekimov, A. "D. I. Mendeleev v zhizni i tvorchestve Aleksandra Bloka." *Russkaia literatura,* no. 1 (1960): 156–160.

Eklof, Ben. "Introduction." In Eklof, Bushnell, and Zakharova, eds., *Russia's Great Reforms* (1994): vii–xvi.

Eklof, Ben, John Bushnell, and Larissa Zakharova, eds. *Russia's Great Reforms, 1855–1881.* Bloomington: Indiana University Press, 1994.

Emmens, Stephen H. "Modern Alchemy." *Chemical News* 76 (1897): 117–118.

Engels, Frederick. *Dialectics of Nature,* ed. and tr. Clemens Dutt. New York: International Publishers, 1940.

Engelstein, Laura. "Combined Underdevelopment: Discipline and the Law in Imperial and Soviet Russia." In Jan Goldstein, ed., *Foucault and the Writing of History* (Oxford: Blackwell,
1994): 220–236.

———. *The Keys to Happiness: Sex and the Search for Modernity in Fin-de-Siècle Russia*. Ithaca:
Cornell University Press, 1992.

———. *Moscow, 1905: Working-Class Organization and Political Conflict*. Stanford: Stanford
University Press, 1982.

Errera, L. "Magnétisme et poids atomiques." *Académie Royale de Belgique. Bulletin de la Classe
des Sciences* (1900): 152–161.

Farber, Eduard. "The Theory of the Elements and Nucleosynthesis in the Nineteenth Century."
Chymia 9 (1964): 181–200.

Farrar, W. V. "Nineteenth Century Speculations on the Complexity of the Chemical Elements."
British Journal for the History of Science 2 (1965): 297–323.

———. "Sir B. C. Brodie and His Calculus of Chemical Operation." *Chymia* 9 (1964):
169–179.

Favre, Adrien. *Les Origines du Système Métrique*. Paris: Les Presses Universitaires de France,
1931.

Fell, Ulrike. "The Chemistry Profession in France: The Société Chimique de Paris/de France,
1870–1914." In Knight and Kragh, eds., *The Making of the Chemist* (1998): 15–38.

Fenin, Aleksandr I. *Coal and Politics in Late Imperial Russia: Memoirs of a Russian Mining Engineer*, tr. Alexandre Fediaevsky, ed. Susan P. McCaffray. DeKalb: Northern Illinois University
Press, 1990.

Ferenczi, Caspar. "Freedom of the Press under the Old Regime, 1905–1914." In Crisp and Edmondson, eds., *Civil Rights in Imperial Russia* (1989): 191–214.

Field, Daniel. *Rebels in the Name of the Tsar*. Boston: Unwin Hyman, 1989.

———. "The Reforms of the 1860s." In Samuel H. Baron and Nancy W. Heer, eds., *Windows
on the Russian Past: Essays on Soviet Historiography Since Stalin* (Columbus, Ohio: American
Association for the Advancement of Slavic Studies, 1977): 89–104.

Fierro, Alfred. *Histoire de la Météorologie*. Paris: Denoël, 1991.

Figurovskii, N. A. *Dmitrii Ivanovich Mendeleev, 1834–1907*. Moscow: Izd. AN SSSR, 1961.

———. "Dnevniki D. I. Mendeleeva 1861 i 1862 gg." *Nauchnoe Nasledstvo* 2 (1951): 95–110.

———. "Khimiia v dopetrovskoi Rusi." In Solov'ev, *Ocherki po istorii khimii* (1963): 370–398.

———. "Triumf periodicheskogo zakona D. I. Mendeleeva." *VIET*, no. 3 (1957): 3–13.

Figurovskii, N. A., et al., eds. *Materialy po istorii otechestvennoi khimii*. Moscow: Izd. AN SSSR,
1953.

Figurovskii, N. A., and K. Ts. Elagina. "Aleksandr Abramovich Voskresenskii (1809–1880)."
TIIEiT 18 (1958): 213–235.

Figurovskii, N. A., and Iu. S. Musabekov. "Vydaiushchiisia russkii khimik L. N. Shishkov." *TI-
IEiT* 2 (1954): 46–66.

Figurovskii, N. A., and Iu. I. Solov'ev. *Aleksandr Porfir'evich Borodin*. Moscow: Izd. AN SSSR,
1950.

———. *Aleksandr Porfir'evich Borodin: A Chemist's Biography*, tr. Charlene Steinberg and
George B. Kauffman. Berlin: Springer-Verlag, 1988.

———. *Nikolai Nikolaevich Zinin: Biograficheskii ocherk*. Moscow: Izd. AN SSSR, 1957.

Fisher, N. W. "Avogadro, the Chemists, and Historians of Chemistry: Parts I and II." *History of
Science* 20 (1982): 77–102, 212–231.

Flammarion, Camille. *Mysterious Psychic Forces: An Account of the Author's Investigations in Psychical Research, Together with Those of Other European Savants*. Boston: Small, Maynard, and
Company, 1907.

Flaubert, Gustave. *Madame Bovary*, tr. W. Blaydes. New York: P. F. Collier & Son, 1902.

Fleck, Ludwik. *The Genesis and Development of a Scientific Fact*, eds. Thaddeus J. Treun and Robert K. Merton, trs. Fred Bradley and Thaddeus J. Treun. Chicago: University of Chicago Press, 1979 [1935].

Fleming, James Rodger. "Meteorological Observing Systems before 1870 in England, France, Germany, Russia and the USA: A Review and Comparison." *Bulletin of the World Meteorological Organization* 46 (1997): 249–258.

_____. *Meteorology in America, 1800–1870*. Baltimore: The Johns Hopkins University Press, 1990.

Foucault, Michel. *The Order of Things: An Archaeology of the Human Sciences*. New York: Vintage Books, 1970.

Fournier d'Albe, E. E. *The Life of Sir William Crookes, O. M., F. R. S.* London: T. Fisher Unwin, 1923.

Fox, Robert. *The Caloric Theory of Gases: From Lavoisier to Regnault*. Oxford: Clarendon Press, 1971.

Freeze, Gregory L. "Bringing Order to the Russian Family: Marriage and Divorce in Imperial Russia, 1760–1860." *Journal of Modern History* 62 (1990): 709–746.

_____. "Handmaiden of the State? The Church in Imperial Russia Reconsidered." *Journal of Ecclesiastical History* 36 (1985): 82–102.

_____. *The Parish Clergy in Nineteenth-Century Russia: Crisis, Reform, Counter-Reform*. Princeton: Princeton University Press, 1983.

_____. "The *Soslovie* (Estate) Paradigm and Russian Social History." *American Historical Review* 91 (1986): 11–36.

_____. "Subversive Piety: Religion and the Political Crisis in Late Imperial Russia." *Journal of Modern History* 68 (1996): 308–350.

Freund, Ida. *The Study of Chemical Composition: An Account of Its Method and Historical Development*. Cambridge: Cambridge University Press, 1904.

Fribes, Evgeniia. *Po voprosam o vospitanii detei, nauchnom obuchenii, zhenskom trude, brake i ob otnoshenii k padshim zhenshchinam. V prilozhenii: O spiritizme (razbor soch. g-zhi Oduar "Le Monde des Esprits ou la Vie Après la Mort"). Sotvorenie mira, chto takoe priroda, rai i ad po Svedenborgu. Dusha i Telo. Svoboda voli. Otkuda ver v bezsmertie dushi?* St. Petersburg: V. Bezobrazov, 1877.

Frieden, Nancy Mandelker. *Russian Physicians in an Era of Reform and Revolution, 1856–1905*. Princeton: Princeton University Press, 1981.

Friedman, Robert Marc. *Appropriating the Weather: Vilhelm Bjerknes and the Construction of a Modern Meteorology*. Ithaca: Cornell University Press, 1989.

_____. *The Politics of Excellence: Behind the Nobel Prize in Science*. New York: W. H. Freeman, 2001.

Frisinger, H. Howard. *The History of Meteorology: To 1800*. New York: Science History Publications, 1977.

Fritsman, E. Kh. "D. I. Mendeleev. Ego nauchnaia i obshchestvennaia deiatel'nost'." *Priroda* (March 1937): 103–116.

_____. "D. I. Mendeleev i problema Arktiki." *Nauchnoe Nasledstvo* 1 (1948): 163–168.

Funk, Carl. *Aphoristischer Entwurf einer Kosmogonie: Entstehung der directen und retrograden Kometen und Beweisführung, dass die Planeten Metallugeln sind, welche sich im Aether oxydiren und hierdurch der Sonne sich nähern, sowie Nachweis einer Statik der Himmelskörper im Aether.* Helmstedt: J. C. Schmidt, 1888.

Galison, Peter. *Image and Logic: A Material Culture of Microphysics*. Chicago: University of Chicago Press, 1997.

_____. "The Many Faces of Big Science." In Galison and Hevly, eds., *Big Science* (1992): 1–17.

Galison, Peter, and Bruce Hevly, eds. *Big Science: The Growth of Large-Scale Research*. Stanford: Stanford University Press, 1992.

Galison, Peter, Bruce Hevly, and Rebecca Lowen. "Controlling the Monster: Stanford and the Growth of Physics Research, 1935–1962." In Galison and Hevly, eds., *Big Science* (1992): 46–77.

Garber, Elizabeth. "Thermodynamics and Meteorology (1850–1900)." *Annals of Science* 33 (1976): 51–65.

Garshin, V. M. *Polnoe sobranie sochinenii v trekh tomakh*, 3 v. Moscow: Academia, 1934.

Gasanova, N. V. "Ob odnom interv'iu D. I. Mendeleeva." *VIET*, no. 4 (1983): 17–19.

Gatrell, Peter. "The Meaning of the Great Reforms in Russian Economic History." In Eklof, Bushnell, and Zakharova, eds., *Russia's Great Reforms* (1994): 84–101.

Gauld, Alan. *The Founders of Psychical Research*. London: Routledge and Kegan Paul, 1968.

Geison, Gerald L. *The Private Science of Louis Pasteur*. Princeton: Princeton University Press, 1995.

Georgievskii, V. G., ed. "Literaturnoe nasledstvo D. I. Mendeleeva (neopublikovannaia perepiska). Pis'ma D. I. Mendeleeva P. P. Alekseevu." *Soobshcheniia o nauchnykh rabotakh chlenov Vsesoiuznogo khimicheskogo obshchestva im. D. I. Mendeleeva*, no. 3 (1948): 34–40.

Gerschenkron, Alexander. *Economic Backwardness in Historical Perspective: A Book of Essays*. Cambridge, Mass.: Belknap Press, 1962.

_____. *Europe in the Russian Mirror: Four Lectures in Economic History*. Cambridge: Cambridge University Press, 1970.

Gerstein, Linda. *Nikolai Strakhov*. Cambridge, Mass.: Harvard University Press, 1971.

Gessen, Sergei. "Peterburgskii universitet osen'iu 1861 g." In Goreva and Koz'min, *Revoliutsionnoe dvizhenie 1860-kh godov* (1932): 9–21.

Geyer, Dietrich. *Russian Imperialism: The Interaction of Domestic and Foreign Policy 1860–1914*, tr. Bruce Little. New Haven: Yale University Press, 1987.

Gilpin, J. E. "Krypton, Neon, Metargon, and Coronium—Recently Discovered Constituents of the Atmosphere." *American Chemical Journal* 20 (1898): 696–699.

_____. "Xenon, Etherion, and Monium." *American Chemical Journal* 20 (1898): 872–875.

Gindin, I. F. "D. I. Mendeleev o razvitii promyshlennosti v Rossii." *Voprosy Istorii*, no. 9 (1976): 210–215.

Ginzburg, Carlo. *The Cheese and the Worms: The Cosmos of a Sixteenth-Century Miller*, tr. John Tedeschi and Anne Tedeschi. Baltimore: The Johns Hopkins University Press, 1980.

Giunta, Carmen J. "Argon and the Periodic System: The Piece that Would Not Fit." *Foundations of Chemistry* 3 (2001): 105–128.

Gleason, Abbott. "The Great Reforms and the Historians Since Stalin." In Eklof, Bushnell, and Zakharova, eds., *Russia's Great Reforms* (1994): 1–16.

Glinka, S. F. "Aleksandr Mikhailovich Butlerov v chastnoi i domashnei zhizni. (Po lichnym vospominaniiam)." *TIIEiT* 12 (1956): 182–199.

Goldberg, Stanley. "In Defense of Ether: The British Response to Einstein's Special Theory of Relativity, 1905–1911." *Historical Studies in the Physical Sciences* 2 (1970): 89–125.

Goldfarb, Russell M., and Clare R. Goldfarb. *Spiritualism and Nineteenth-Century Letters*. Rutherford, N.J.: Fairleigh Dickinson University Press, 1978.

Golinski, Jan. "Barometers of Change: Meteorological Instruments as Machines of Enlightenment." In William Clark, Jan Golinski, and Simon Schaffer, eds., *The Sciences in Enlightened Europe* (Chicago: University of Chicago Press, 1999): 69–93.

_____. *Science as Public Culture: Chemistry and Enlightenment in Britain, 1760–1820*. Cambridge: Cambridge University Press, 1992.

Goloushkin, V. N. "Uravnenie sostoianiia ideal'nogo gaza D. I. Mendeleeva." *Uspekhi Fizicheskikh Nauk* 45, no. 4 (1951): 616–621.

González, Wenceslao J. "Reichenbach's Concept of Prediction." *International Studies in the Philosophy of Science* 9 (1995): 37–58.

Gorbatsevich, S. V. "Raboty D. I. Mendeleeva v oblasti fizicheskikh konstant." In *Mendeleev i metrologiia* (1969): 38–50.

Gorbov, A. "Dmitrii Ivanovich Mendeleev i 'Osnovy khimii.'" *Zhurnal Prikladnoi Khimii* 7, no. 8 (1934): 1544–1554.

Gordin, Michael D. "The Importation of Being Earnest: The Early St. Petersburg Academy of Sciences." *Isis* 91 (2000): 1–31.

———. "Loose and Baggy Spirits: Reading Dostoevskii and Mendeleev." *Slavic Review* 60 (2001): 756–780.

———. "Making Newtons: Mendeleev, Metrology, and the Chemical Ether." *Ambix* 45 (1998): 96–115.

———. "Measure of All the Russias: Metrology and Governance in the Russian Empire." *Kritika* 4 (2003): 783–815.

———. "A Modernization of 'Peerless Homogeneity': The Creation of Russian Smokeless Gunpowder." *Technology and Culture* 44 (2003): 677–702.

———. "No Smoking Gun: D. I. Mendeleev and Pyrocollodion Gunpowder." In *Troisièmes Journées Scientifiques Paul Vieille: Instrumentation, expérimentation et expertise des matériaux énergétiques (poudres, explosifs et pyrotechnie), du XVIe siècle à nos jours* (Paris: A3P, 2000): 73–96.

———. "The Organic Roots of Mendeleev's Periodic Law." *Historical Studies in the Physical and Biological Sciences* 32 (2002): 263–290.

———. "The Science of Vodka [letter to the editor]." *The New Yorker* (13 January 2003): 7.

Goreva, B. I., and B. P. Koz'min, eds. *Revoliutsionnoe dvizhenie 1860-kh godov.* Moscow: Izd. Vsesoiuznogo obshchestva politkatorzhan i ssyl'no-posentsev, 1932.

Gorlin, Robert H. "Problems of Tax Reform in Imperial Russia." *Journal of Modern History* 49 (1977): 246–265.

Graham, Loren R. *Science, Philosophy, and Human Behavior in the Soviet Union.* New York: Columbia University Press, 1987.

———. "Textbook Writing and Scientific Creativity: The Case of Mendeleev." *National Forum* (Winter 1983): 22–23.

de Gramont, A. "Lecoq de Boisbaudran: Son œuvre et ses idées." *Revue Scientifique* 51, no. 4 (1913): 97–109.

Gramsci, Antonio. *Selections from the Prison Notebooks of Antonio Gramsci,* eds. and trs. Quintin Hoare and Geoffrey Nowell Smith. New York: International Publishers, 1971.

Gray, Camilla. *The Russian Experiment in Art, 1863–1922,* rev. ed. by Marian Burleigh-Motley. New York: Thames and Hudson, 1986 [1962].

Gregory, Frederick. "Romantic Kantianism and the End of the Newtonian Dream in Chemistry." *Archives Internationales d'Histoire des Sciences* 34 (1984): 108–123.

Gregory, Paul R. *Russian National Income, 1885–1913.* Cambridge: Cambridge University Press, 1982.

Gregory, Paul R., and Joel W. Sailors. "Russian Monetary Policy and Industrialization, 1861–1913." *Journal of Economic History* 36 (1976): 836–851.

Grossberg, Michael. "Institutionalizing Masculinity: The Law and a Masculine Profession." In Carnes and Griffen, eds., *Meanings for Manhood* (1990): 133–151.

Grossman, Joan Delaney. "Rise and Decline of the 'Literary' Journal, 1880–1917." In Martinsen, ed., *Literary Journals in Imperial Russia* (1997): 171–196.

Gruenwald, A. "On Remarkable Relations between the Spectrum of Watery Vapour and the Line Spectra of Hydrogen and Oxygen, and on the Chemical Structure of the Two Latter

Bodies and their Dissociation in the Atmosphere of the Sun." *Chemical News* 56 (1887): 186–188, 201–202, 223–224, 232.

Guillaume, Charles-Edouard. *La Convention du Mètre et le Bureau International des Poids et Mesures*. Paris: Gauthier-Villars, 1902.

Gumilevskii, Lev. *Aleksandr Mikhailovich Butlerov, 1828–1886*. Moscow: Molodaia Gvardia, 1952.

Guroff, Gregory, and Fred V. Carstensen, eds. *Entrepreneurship in Imperial Russia and the Soviet Union*. Princeton: Princeton University Press, 1983.

Haimson, Leopold H. "The Problem of Social Identities in Early Twentieth Century Russia." *Slavic Review* 47 (1988): 1–20.

Hallock, William, and Herbert T. Wade. *Outlines of the Evolution of Weights and Measures and the Metric System*. New York: MacMillan, 1906.

Haltzel, Michael Harris. "The Reaction of the Baltic Germans to Russification during the Nineteenth Century." Ph.D. Dissertation, Harvard University, 1971.

Hamburg, G. M. *Politics of the Russian Nobility, 1881–1905*. New Brunswick, N.J.: Rutgers University Press, 1984.

Hamm, Michael F. "The Breakdown of Urban Modernization: A Prelude to the Revolutions of 1917." In Michael F. Hamm, ed., *The City in Russian History* (Lexington: University Press of Kentucky, 1976): 182–200.

Hannaway, Owen. *The Chemists and the Word: The Didactic Origins of Chemistry*. Baltimore: The Johns Hopkins University Press, 1975.

Harkness, Wm. *On the Progress of Science as Exemplified in the Art of Weighing and Measuring*. Washington, D.C.: Judd & Detweiler, 1888.

Heimann, P. M. "Ether and Imponderables." In Cantor and Hodge, eds., *Conceptions of Ether* (1981): 61–83.

Henriksson, Anders. "Nationalism, Assimilation and Identity in Late Imperial Russia: The St. Petersburg Germans, 1906–1914." *Russian Review* 52 (1993): 341–353.

———. "The *St. Petersburger Zeitung:* Tribune of Baltic German Conservatism in Late-Nineteenth-Century Russia." *Journal of Baltic Studies* 20 (1989): 365–378.

Hess, David J. *Spirits and Scientists: Ideology, Spiritism, and Brazilian Culture*. University Park: The Pennsylvania State University Press, 1991.

Hiebert, Erwin N. "Developments in Physical Chemistry at the Turn of the Century." In Carl Gustaf Bernhard et al., eds., *Science, Technology, and Society in the Time of Alfred Nobel* (Oxford: Pergamon, 1982): 97–115.

———. "The Energetics Controversy and the New Thermodynamics." In Duane H. D. Roller, ed., *Perspectives in the History of Science and Technology* (Norman: University of Oklahoma Press, 1971): 67–86.

———. "Historical Remarks on the Discovery of Argon, the First Noble Gas." In Herbert H. Hyman, ed., *Noble-Gas Compounds* (Chicago: University of Chicago Press, 1963): 3–20.

Hirsh, Richard F. "A Conflict of Principles: The Discovery of Argon and the Debate over Its Existence." *Ambix* 28 (1981): 121–130.

Hjelt, Edv. "Friedrich Konrad Beilstein." *Berichte der Deutschen Chemischen Gesellschaft* 40 (1907): 5041–5078.

Hogan, Heather. *Forging Revolution: Metalworkers, Managers, and the State in St. Petersburg, 1890–1914*. Bloomington: Indiana University Press, 1993.

Homburg, Ernst. "Two Factions, One Profession: The Chemical Profession in German Society, 1780–1870." In Knight and Kragh, eds., *The Making of the Chemist* (1998): 39–76.

Home, D. D. *Incidents in My Life*. London: Longman, Green, Longman, Roberts & Green, 1863.

———. *Lights and Shadows of Spiritualism*, 2d. ed. London: Virtue & Co., 1878.

Home, Roderick W. "The Third Law in Newton's Mechanics." *British Journal for the History of Science* 4 (1968): 39–51.

Home, R. W., and K. T. Livingston. "Science and Technology in the Story of Australian Federation: The Case of Meteorology, 1876–1908." *Historical Records of Australian Science* 10 (1994): 109–127.

Hufbauer, Karl. *The Formation of the German Chemical Community (1720–1795)*. Berkeley: University of California Press, 1982.

Hughes, Agatha C., and Thomas P. Hughes, eds. *Systems, Experts, and Computers: The Systems Approach to Management and Engineering, World War II and After*. Cambridge, Mass.: MIT Press, 2000.

Hunczak, Taras. "Pan-Slavism or Pan-Russianism." In Taras Hunczak, ed., *Russian Imperialism from Ivan the Great to the Revolution* (New Brunswick, N.J.: Rutgers University Press, 1974): 82–105.

Hutchinson, John F. *Politics and Public Health in Revolutionary Russia, 1890–1918*. Baltimore: The Johns Hopkins University Press, 1990.

Iezuitova, L. A., and N. V. Skvortsova. "Aleksandr Blok v Peterburgskom universitete." *Ocherki po istorii Leningradskogo universiteta* 4 (1982): 52–86.

Ihde, Aaron J. "The Karlsruhe Congress: A Centennial Retrospect." *Journal of Chemical Education* 38 (February 1961): 83–86.

Inglis, Brian. *Natural and Supernatural: A History of the Paranormal from Earliest Times to 1914*. London: Hodder and Stoughton, 1977.

Inostrantsev, A. A. *Vospominaniia (Avtobiografiia)*, eds. V. A. Prozorovskii and I. L. Tikhonov. St. Petersburg: Peterburgskoe vostokovedenie, 1998.

"Instruktsiia dlia proizvodstva vnezapnykh revizii." *VGPMV* 7 (1905): 177–178.

Ionidi, P. P. *Filosofskoe znachenie periodicheskogo zakona D. I. Mendeleeva*. Moscow: Znanie, 1958.

_____. *Mirovozzrenie D. I. Mendeleeva*. Moscow: Izd. AN SSSR, 1959.

_____. "O mendeleevskom nasledstve." *Sovetskaia Nauka*, no. 3 (1938): 99–106.

Iunge, E. F. *Vospominaniia (1843–1860 g. g.)*. [Moscow]: Sfinks, [1914].

Ivan Ivanovich Shishkin: Mastera russkoi zhivopisi. Leningrad: Khudozhnik RSFSR, 1990.

Ivanov, S. Z. "D. I. Mendeleev i pishchevaia promyshlennost'." *Trudy Leningradskogo tekhnologicheskogo instituta pishchevoi promyshlennosti* 2 (10) (1951): 40–46.

_____. "D. I. Mendeleev—sozdatel' nauchnykh osnov sovremennoi spirtometrii." *Trudy Leningradskogo tekhnologicheskogo instituta pishchevoi promyshlennosti* 12 (1955): 195–207.

J. L. Review of Mendeleev's *An Attempt towards a Chemical Conception of the Ether. Nature* 69 (1904): 558.

Jarausch, Konrad H., ed. *The Transformation of Higher Learning, 1860–1930: Expansion, Diversification, Social Opening and Professionalization in England, Germany, Russia and the United States*. Stuttgart: Klett-Cotta, 1982.

Jaubert, George F. "Mendeléeff." *Revue générale de chimie pure & appliquée* 1 (1899): 97–99.

Jenkins, Elizabeth. *The Shadow and the Light: A Defence of Daniel Dunglas Home, the Medium*. London: Hamish Hamilton, 1982.

Jennings, Edward M. "The Consequences of Prediction." *Studies on Voltaire and the Eighteenth Century* 153 (1976): 1131–1150.

Johanson, Christine. *Women's Struggle for Higher Education in Russia, 1855–1900*. Kingston, Ontario: McGill-Queen's University Press, 1987.

Johnson, Jeffrey A. "Academic Chemistry in Imperial Germany." *Isis* 76 (1985): 500–524.

Johnson, Samuel. "Spiritualism Tested by Science." *New Englander* (May 1858): 1–46.

Jolly, W. P. *Sir Oliver Lodge*. London: Constable, 1974.

"K voprosu o peresmotre ustava Akademii nauk." *Novoe Vremia*, 12 (24) December 1880, #1722: 1–2.

Kablukov, I. A. "Dmitrii Ivanovich Mendeleev." *Nauchnyi Rabotnik*, no. 7–8 (1927): 97–104.

———. "Dmitrii Ivanovich Mendeleev." *Vestnik Vospitaniia* 18, no. 2 (1907): 95–100.

———. "Iz vospominanii o khimii v Moskovskom universitete s semidesiatykh godov XIX veka." *Uspekhi Khimii* 9, no. 6 (1940): 727–733.

———. "Obzor izdanii 'Osnov khimii' D. I. Mendeleeva." In Mendeleev, *Osnovy khimii*, 9th ed., v. 1 (1927): xli–xlvi.

Kahan, Arcadius. *Russian Economic History: The Nineteenth Century*, ed. Roger Weiss. Chicago: University of Chicago Press, 1989.

Kaiser, Friedhelm Berthold. *Die russische Justizreform von 1864: Zur Geschichte der russischen Justiz von Katherina II. bis 1917*. Leiden: E. J. Brill, 1972.

Kaji, Masanori. "Mendeleev's Discovery of the Periodic Law of the Chemical Elements: The Scientific and Social Context of His Discovery (English summary)." In *Mendeleev's Discovery of the Periodic Law of the Chemical Elements—The Scientific and Social Context of His Discovery* (in Japanese) (Sapporo: Hokkaido University Press, 1997): 365–380.

Kalifati, D. D. "Otkrytie D. I. Mendeleevym universal'noi gazovoi postoiannoi." *VIET*, no. 2 (1984): 15–18.

Kalifati, D. D., and V. V. Sychev. "K 100-letiiu universal'noi gazovoi postoiannoi D. I. Mendeleeva." *Teploenergetika*, no. 11 (1975): 87.

Kamenogradskaia, O., ed. *Deiatel'nost' D. I. Mendeleeva v S.-Peterburgskom universitete i nauchnykh obshchestvakh. Uchastie v rabote mezhdunarodnykh general'nykh konferentsii po meram i vesam i mezhdunarodnogo komiteta mer i vesov v Parizhe*. Leningrad: BAN, 1985.

Kamentseva, E. I., and N. V. Ustiugov. *Russkaia metrologiia*, 2d. ed. Moscow: Vysshaia shkola, 1975.

Kapustin, F. Ia. "O trudakh D. I. Mendeleeva po voprosam ob izmenenii ob"ema gazov i zhidkostei." In Tishchenko, ed., *Trudy pervago mendeleevskogo s"ezda* (1909): 107–115.

Kapustina-Gubkina, N. Ia. *Semeinaia khronika v pis'makh materi, ottsa, brata, sester, diadi D. I. Mendeleeva, vospominaniia o D. I. Mendeleeve*. St. Petersburg: M. Frolova, 1908.

Kapustinskaia, K. A. "Khimiia v zhurnale 'Nauchnoe Obozrenie.'" *TIIEiT* 35 (1961): 380–385.

Kapustinskii, A. F., ed. *D. I. Mendeleev—Velikii russkii khimik*. Moscow: Sovetskaia nauka, 1949.

Kargon, Robert. "Mendeleev's Chemical Ether, Electrons, and the Atomic Theory." *Journal of Chemical Education* 42 (July 1965): 388–389.

Karol', V. P. *D. I. Mendeleev i meteorologiia*. Leningrad: Gidrometeoizdat, 1950.

Karpenko, V. "The Discovery of Supposed New Elements: Two Centuries of Errors." *Ambix* 27 (1980): 77–102.

Kassow, Samuel D. *Students, Professors, and the State in Tsarist Russia*. Berkeley: University of California Press, 1989.

———. "The University Statute of 1863: A Reconsideration." In Eklof, Bushnell, and Zakharova, eds., *Russia's Great Reforms* (1994): 247–263.

Kauffman, George B. "Stephen H. Emmens and the Transmutation of Silver into Gold." *Endeavour* 7 (1983): 150–154.

Kedrov, B. M. "D. I. Mendeleev i zarubezhnye slavianskie uchenye." In Arbuzov, ed., *Materialy po istorii otechestvennoi khimii* (1950): 83–115.

———. *Den' odnogo velikogo otkrytiia*. Moscow: Izd. sotsial'no-ekonomicheskoi literatury, 1958.

———. *Evoliutsiia poniatiia elementa v khimii*. Moscow: Izd. Akademii pedagogicheskikh nauk RSFSR, 1956.

_____. "Etapy razrabotki D. I. Mendeleevym periodicheskogo zakona." *TIIEiT* 2 (1948): 288–322.

_____. *Filosofskii analiz pervykh trudov D. I. Mendeleeva o periodicheskom zakone.* Moscow: Izd. AN SSSR, 1959.

_____. "K istorii otkrytiia periodicheskogo zakona D. I. Mendeleevym." In Mendeleev, *Novye materialy po istorii otkrytiia periodicheskogo zakona* (1950): 87–145.

_____, comp. "Kratkie svedeniia o zhizni i nauchnoi deiatel'nosti D. I. Mendleeva i o ego rabote nad periodicheskim zakonom." In Mendeleev, *Periodicheskii zakon* (1958): 746–770.

_____. "Mendeleev, Dmitry Ivanovich." In Charles Coulston Gillispie, ed., *Dictionary of Scientific Biography* (New York: Charles Scribner's Sons, 1980): ix, 286–295.

_____. *Mikroanatomiia velikogo otkrytiia: K 100-letiiu zakona Mendeleeva.* Moscow: Nauka, 1970.

_____. "Nad mendeleevskimi rukopisiami (o tom, kak izuchalas' istoriia otkrytiia razrabotki periodicheskogo zakona)." *VIET*, no. 4 (1983): 41–64.

_____. "Nauchnyi metod D. I. Mendeleeva." *Voprosy Filosofii*, no. 3 (1957): 19–34.

_____. "O khode otkrytiia D. I. Mendeleevym periodicheskoi sistemy elementov." In Figurovskii et al., *Materialy po istorii otechestvennoi khimii* (1953): 119–141.

_____. "Ob otnoshenii k mendeleevskomu nasledstvu." *Pod znamenem marksizma*, no. 4 (1939): 194–207.

_____. "Otkrytie galliia—pervoe khimicheskoe otkrytie novogo tipa (D. I. Mendeleev i P. E. Lekok de Buabodran)." In Kedrov and Trifonov, eds., *Prognozirovanie v uchenii o periodichnosti* (1976): 5–19.

_____. "Otkrytie periodicheskogo zakona D. I. Mendeleevym." In Solov'ev, ed., *Ocherki po istorii khimii* (1963): 36–80.

_____. "Periodicheskii zakon D. I. Mendeleeva i inertnye gazy." *Uspekhi Fizicheskikh Nauk* 47, no. 1 (1952): 95–114.

_____, comp. "Podrobnaia kommentirovannaia khronologiia otkrytiia galliia." In Kedrov and Trifonov, eds., *Prognozirovanie v uchenii o periodichnosti* (1976): 333–356.

_____. "Predvidenie i poiski D. I. Mendeleevym ekasilitsiia (budushchego germaniia)." *Khimiia redkikh metallov*, no. 1 (1954): 7–17.

_____. "Razvitie D. I. Mendeleevym estestvennoi ('korotkoi') sistemy elementov." In Mendeleev, *Nauchnyi arkhiv* (1953): 771–858.

_____. "Ser'eznye oshibki i upushcheniia (o polnom akademicheskom sobranii sochinenii D. I. Mendeleeva)." *Vestnik Akademii Nauk SSSR*, no. 1 (1957): 122–133.

_____. "Die Vorhersage des Ekasiliziums und die Entdeckung des Germaniums." *NTM* 3, no. 8 (1966): 11–37.

Kedrov, B. M., and T. Chentsova. *Brauner—spodvizhnik Mendeleeva: K stoletiiu so dnia rozhdeniia Boguslava Braunera.* Moscow: Izd. AN SSSR, 1955.

Kedrov, B. M., and V. Nikiforov. "K voprosu o mendeleevskikh khimicheskikh s"ezdakh." *Za Marksistsko-Leninskoe Estestvoznanie*, no. 2 (1931): 11–33.

Kedrov, B. M., and D. N. Trifonov, eds. *Prognozirovanie v uchenii o periodichnosti.* Moscow: Nauka, 1976.

_____. *Zakon periodichnosti i khimicheskie elementy. Otkrytiia i khronologiia.* Moscow: Nauka, 1969.

Kennelly, Arthur E. *Vestiges of Pre-Metric Weights and Measures Persisting in Metric-System Europe 1926–1927.* New York: MacMillan, 1928.

Kerova, L. S., and N. V. Gasanova. *Boblovo—odna iz tvorcheskikh laboratorii D. I. Mendeleeva.* Leningrad: Izd. AN SSSR, 1990.

Kerova, L. S., V. A. Krotikov, and R. B. Dobrotin. "Issledovaniia D. I. Mendeleeva v oblasti fiziki gazov." *Voprosy Istorii i Metodologii Khimii* 2 (1978): 73–96.

Kerr, Howard. *Mediums, and Spirit-Rappers, and Roaring Radicals: Spiritualism in American Literature, 1850–1900.* Urbana: University of Illinois Press, 1972.

Kharichkova, K. V. "Zaslugi Dmitriia Ivanovicha Mendeleeva v oblasti izucheniia nefti i neftianoi tekhniki." In Tishchenko, ed., *Trudy pervago mendeleevskogo s"ezda* (1909): 147–151.

Khlopin, V. G., A. A. Balandin, and S. A. Pogodin. *Ocherki po istorii Akademii nauk: Khimicheskie nauki.* Moscow: Izd. AN SSSR, 1945.

Khomiakov, K. G. "K istorii otkrytiia periodicheskogo zakona D. I. Mendeleeva." *Vestnik Moskovskogo universiteta*, no. 12 (1953): 17–23.

Khrgian, A. Kh. "Istoriia meteorologii v Rossii." *TIIEiT* 2 (1948): 71–104.

_____. *Ocherki razvitiia meteorologii*, vol. 1, 2d. ed. Leningrad: Gidrometeorologicheskoe Izd., 1959.

Kington, John. "The Development of Meteorology during the Enlightenment and the History of Weather in the 1780s." *Studies on Voltaire and the Eighteenth Century* 264 (1989): 811–814.

Kipins, A. Ia. "K istorii ustanovleniia uravneniia sostoianiia ideal'nogo gaza." *VIET*, no. 4 (29) (1969): 91–94.

Kireev, V. A. "Rabota D. I. Mendeleeva po uravneniiu sostoianiia ideal'nogo gaza." *Uspekhi Khimii* 20, no. 1 (1951): 132–134.

Kisch, Cecil. *Alexander Blok: Prophet of Revolution.* London: Weidenfeld and Nicolson, 1960.

Kiseleva, M. N. "K istorii otkrytiia uravneniia sostoianiia ideal'nogo gaza." In I. S. Dmitriev, ed., *Mendeleevskii sbornik* (St. Petersburg: Izd. S.-Peterburgskogo universiteta, 1999): 85–97.

_____. "K voprosu o neizbranii D. I. Mendeleeva v Rossiiskuiu Akademiiu nauk." *Ocherki po istorii Sankt-Peterburgskogo universiteta* 8 (2000): 179–195.

Kniazev, G. A. "D. I. Mendeleev i imperatorskaia akademiia." *Vestnik Akademii Nauk SSSR*, no. 3 (1931): 27–34.

_____. "D. I. Mendeleev i tsarskaia Akademiia nauk (1858–1907 gg.)." *Arkhiv istorii nauk i tekhniki* 6 (1935): 299–331.

Knight, David M. *Atoms and Elements: A Study of Theories of Matter in England in the Nineteenth Century.* London: Hutchinson, 1967.

Knight, David, and Helge Kragh. *The Making of the Chemist: The Social History of Chemistry in Europe, 1789–1914.* Cambridge: Cambridge University Press, 1998.

Koblitz, Ann Hibner. "Science, Women, and the Russian Intelligentsia: The Generation of the 1860s." *Isis* 79 (1988): 208–226.

Kollerov, D. K. "Raboty D. I. Mendeleeva po issledovaniiu svoistv i sostava gazov." In *Mendeleev i metrologiia* (1969): 64–74.

Kottler, Malcolm Jay. "Alfred Russel Wallace, the Origin of Man, and Spiritualism." *Isis* 65 (1974): 145–192.

Kozlov, N. S. "Nauchnaia i obshchestvennaia deiatel'nost' A. N. Engel'gardta." *TIIEiT* 30 (1960): 111–134.

Kozlov, V. V. *Ocherki istorii khimicheskikh obshchestv SSSR.* Moscow: Izd. AN SSSR, 1958.

_____. *Vsesoiuznoe khimicheskoe obshchestvo imeni D. I. Mendeleeva, 1868–1968.* Moscow: Nauka, 1971.

Kozlov, V. V., and A. I. Lazarev. "Tri chetverti veka Russkogo khimicheskogo obshchestva." In Vol'fkovich and Kiselev, eds., *75 let periodicheskogo zakona D. I. Mendeleeva i Russkogo khimicheskogo obshchestva* (1947): 97–114.

Kragh, Helge. "The Aether in Late Nineteenth Century Chemistry." *Ambix* 36 (1989): 49–65.

_____. "The First Subatomic Explanations of the Periodic System." *Foundations of Chemistry* 3 (2001): 129–143.

_____. "Julius Thomsen and 19th-Century Speculations on the Complexity of Atoms." *Annals of Science* 39 (1982): 37–60.

_____. "The Origin of Radioactivity: From Solvable Problem to Unsolved Non-Problem." *Archive for History of the Exact Sciences* 50 (1997): 331–358.

Kratkoe istoricheskoe obozrenie deistviia Glavnago pedagogicheskago instituta, 1828–1859 gg. St. Petersburg: Tip. Akademii nauk, 1859.

Krätz, Otto, ed. *Beilstein–Erlenmeyer: Briefe zur Geschichte der chemischen Dokumentation und des chemischen Zeitschriftenwesens.* Munich: Werner Fritsch, 1972.

Krestovnikov, N. K. *"Tolkovyi Tarif" D. I. Mendeleeva.* Moscow: Sovet Moskovskago otdeleniia Obshchestva dlia sodeistviia russkoi promyshlennosti i torgovli, 1891.

Krotikov, V. A. "K istorii organizatsii russkogo khimicheskogo obshchestva." *VIET*, no. 13 (1962): 83–88.

_____. "The Mendeleev Archives and Museum of Leningrad University." *Journal of Chemical Education* 37 (December 1960): 625–628.

Krotikov, V. A., and I. N. Filimonova. "Ocherk pedagogicheskoi deiatel'nosti D. I. Mendeleeva v Peterburgskom universitete (1856–1867 gg.)." *Vestnik Leningradskogo universiteta*, no. 10 (1958): 126–132.

_____. "Ocherk pedagogicheskoi deiatel'nosti D. I. Mendeleeva v Peterburgskom universitete (1867–1881 gg.)." *Vestnik Leningradskogo universiteta*, no. 16 (1958): 140–148.

_____. "Ocherk pedagogicheskoi deiatel'nosti D. I. Mendeleeva v Peterburgskom universitete (1881–1890 gg.)." *Vestnik Leningradskogo universiteta*, no. 4 (1959): 112–119.

Kudriavtseva, T. S., and M. E. Shekhter. *D. I. Mendeleev i ugol'naia promyshlennost' Rossii.* Moscow: Ugletekhizdat, 1952.

Kuhn, Thomas S. *The Structure of Scientific Revolutions*, 3d. ed. Chicago: University of Chicago Press, 1996 [1962].

Kula, Witold. *Measures and Men*, tr. R. Szreter. Princeton: Princeton University Press, 1986.

Kurbatov, V. Ia. *Zakon D. I. Mendeleeva.* Leningrad: Nauchnoe Khimiko-Tekhnicheskoe Izd., 1925.

Kurinnoi, V. I. "Vozniknovenie sovremennoi sistemy atomnykh vesov." In Solov'ev, *Ocherki po istorii khimii* (1963): 19–35.

Kutzbach, Gisela. *The Thermal Theory of Cyclones: A History of Meteorological Thought in the Nineteenth Century.* Boston: American Meteorological Society, 1979.

Lamanskii, S. "Iz otcheta, predstavlennago i. o. inspektora Glavnoi Palaty mer i vesov S. I. Lamanskii." *VGPMV* 3 (1896): 119–124.

Lamanskii, Vladimir. "Eshche plemiannik i sanskritolog." *Novoe Vremia*, 3 (15) May 1879, #1140: 2–3.

_____. "Otkrytoe pis'mo ordinarnago professora Imperatorskogo S.-Peterburgskogo universiteta Lamanskago k adiunktu astronomii v Pulkovskoi observatorii g. Baklundu." *Novoe Vremia*, 23 November (5 December) 1880, #1703: 2.

_____. "Otkrytoe pis'mo ordinarnago professora Imperatorskogo SPb. universiteta k g. dirertoru [sic] Pulkovskoi observatorii ordinarnomu O. V. Struve." *Novoe Vremia*, 23 November (5 December) 1880, #1703: 2.

Lapshin, I. I. *Filosofiia izobreteniia i izobreteniia v filosofii*, v. 2. Moscow: Nauka i shkola, 1922.

Larman, E. K. *Aksel' Vil'gel'movich Gadolin.* Moscow: Nauka, 1969.

Lears, T. J. Jackson. "The Concept of Cultural Hegemony: Problems and Possibilities." *American Historical Review* 90 (1985): 567–593.

Lebedev, I. "Elektricheskoe otdelenie Glavnoi Palaty mer i vesov." *VGPMV* 7 (1905): 1–22.

Lecoq de Boisbaudran, P. "Caractères chimiques et spectroscopiques d'un nouveau métal, le *Gallium*, découvert dans une blende de la mine de Pierrefitte, vallée d'Argelès (Pyrénées)." *Comptes Rendus* 81 (1875): 493–495.

_____. "Nouveau procédé d'extraction du gallium." *Comptes Rendus* 83 (1876): 636–638.

_____. "Nouvelles recherches sur le gallium." *Comptes Rendus* 82 (1876): 1036–1039.

_____. "Réactions chimiques du gallium." *Comptes Rendus* 83 (1876): 824–825.

_____. "Sur la constitution des spectres lumineux." *Comptes Rendus* 73 (1871): 658–660.

_____. "Sur l'équivalent du gallium." *Comptes Rendus* 86 (1878): 941–943.

_____. "Sur le spectre du gallium." *Comptes Rendus* 82 (1876): 168–170.

_____. "Sur les propriétés physiques du gallium." *Comptes Rendus* 82 (1876): 611–613.

_____. "Sur quelques propriétés du gallium." *Comptes Rendus* 81 (1875): 1100–1105.

LeDonne, John P. *Absolutism and Ruling Class: The Formation of the Russian Political Order, 1700–1825*. New York: Oxford University Press, 1991.

_____. *The Russian Empire and the World, 1700–1917: The Geopolitics of Expansion and Containment*. New York: Oxford University Press, 1997.

Leicester, Henry M. "Mendeleev and the Russian Academy of Sciences." *Journal of Chemical Education* 25 (1948): 439–441, 444.

_____. "Mendeleev's Visit to America." *Journal of Chemical Education* 34 (1957): 331–333.

Lermontov, M. Iu. *Sobranie sochinenii v chetyrekh tomakh*, 4 v. Moscow: Khudozhestvennaia literatura, 1975.

Letanin, N. "Pervaia lektsiia g. Mendeleeva o spiritizme." *Novoe Vremia*, 26 April 1876, #56: 3.

Levere, Trevor H. "The Rich Economy of Nature: Chemistry in the Nineteenth Century." In U. C. Knoepflmacher and G. B. Tennyson, eds., *Nature and the Victorian Imagination* (Berkeley: University of California Press, 1977): 189–200.

Levi, Primo. *The Periodic Table*, tr. Raymond Rosenthal. New York: Schocken Books, 1984.

Levin-Stankevich, Brian L. "Cassation, Judicial Interpretation and the Development of Civil and Criminal Law in Russia, 1864–1917." Ph.D. Dissertation, SUNY-Buffalo, 1984.

_____. "The Transfer of Legal Technology and Culture: Law Professionals in Tsarist Russia." In Balzer, ed., *Russia's Missing Middle Class* (1996): 223–249.

Levitskii, G. "O sushchestvovanii soprotivliaiushcheisia sredy v nebesnom prostranstve." *ZhRFKhO* 9, no. 1, otd. I, ch. fiz. (1877): 6–25.

Lieven, D. C. B. "The Security Police, Civil Rights, and the Fate of the Russian Empire, 1855–1917." In Crisp and Edmondson, eds., *Civil Rights in Imperial Russia* (1989): 235–262.

Lincoln, W. Bruce. "The Daily Life of St. Petersburg Officials in the Mid Nineteenth Century." *Oxford Slavonic Papers*, N.S. 3 (1975): 82–100.

_____. *The Great Reforms: Autocracy, Bureaucracy, and the Politics of Change in Imperial Russia*. DeKalb: Northern Illinois University Press, 1990.

_____. *In the Vanguard of Reform: Russia's Enlightened Bureaucrats, 1825–1861*. DeKalb: Northern Illinois University Press, 1982.

_____. *Nikolai Miliutin: An Enlightened Russian Bureaucrat*. Newtonville, Mass.: Oriental Research Partners, 1977.

_____. "Reform and Reaction in Russia: A. V. Golovnin's Critique of the 1860's." *Cahiers du Monde Russe et Soviétique* 16 (1975): 167–179.

_____. "Russia on the Eve of Reform: A *Chinovnik*'s View." *Slavonic and East European Review* 59 (1981): 264–271.

Lissner, A. "Sviazi D. I. Mendeleeva s gornoi akademiei vo Freiberge." *VIET*, no. 5 (1957): 50–55.

Lockyer, Norman. *Inorganic Evolution, as Studied by Spectrum Analysis*. London: MacMillan and Co., 1900.

Lomonosov, M. V. *Polnoe sobranie sochinenii*, 11 v. Moscow: Izd. AN SSSR, 1959.

Lomonosovskii sbornik: Materialy dlia istorii razvitiia khimii v Rossii. Moscow: A. I. Mamontov, 1901.

Lukashevich, Stephen. *Ivan Aksakov, 1823–1886: A Study in Russian Thought and Politics*. Cambridge, Mass.: Harvard University Press, 1965.

Lundren, Anders, and Bernadette Bensaude-Vincent, eds. *Communicating Chemistry: Textbooks and Their Audiences, 1789–1939*. Canton, Mass.: Science History Publications, 2000.

MacKenzie, John M. "The Imperial Pioneer and Hunter and the British Masculine Stereotype in Late Victorian and Edwardian Times." In Mangan and Walvin, eds., *Manliness and Morality* (1987): 176–198.

Makarenia, A. A. "A. A. Voskresenskii i ego nauchnaia shkola." *VIET*, no. 11 (1961): 141–144.

———. *D. I. Mendeleev i fiziko-khimicheskie nauki: Opyt nauchnoi biografii D. I. Mendeleeva*, 2d. ed. Moscow: Energoizdat, 1982.

———. *D. I. Mendeleev o radioaktivnosti i slozhnosti elementov*. Moscow: Atomizdat, 1965.

———. "Maloizvestnyi fakt iz biografii D. I. Mendeleeva." *VIET*, no. 54 (1976): 58–59.

———. "Voploshchenie mechty: D. I. Mendeleev o ratsional'nom vedenii sel'skogo khoziaistva." *Priroda*, no. 6 (1964): 107–111.

Makarenia, A. A., and I. N. Filimonova. *D. I. Mendeleev i Peterburgskii universitet*. Leningrad: Izd. Leningradskogo universiteta, 1969.

———. "D. I. Mendeleev na s"ezde Britanskoi assotsiatsii v 1887 godu." *VIET*, no. 22 (1967): 64–67.

Makarenia, A. A., I. N. Filimonova, and N. G. Karpilo, eds. *D. I. Mendeleev v vospominaniiakh sovremennikov*, 2d. ed. Moscow: Atomizdat, 1973.

Makarenia, A. A., and A. I. Nutrikhin. *Mendeleev v Peterburge*. Leningrad: Lenizdat, 1982.

Makarenia, A. A., and V. A. Pozdysheva. "Izuchenie radioaktivnosti russkimi uchenymi (Iz istorii organizatsii novykh nauchnykh issledovanii)." In B. M. Kedrov, ed., *Uchenie o radioaktivnosti: Istoriia i sovremennost'* (Moscow: Nauka, 1973): 144–154.

Mangan, J. A., and James Walvin, eds. *Manliness and Morality: Middle-Class Masculinity in Britain and America, 1800–1940*. New York: St. Martin's Press, 1987.

Manin, V. *Kuindzhi*. Moscow: Izobrazit. isskusstvo, 1976.

Manning, Roberta Thompson. *The Crisis of the Old Order in Russia: Gentry and Government*. Princeton: Princeton University Press, 1982.

Markov, Evgenii. "Magiia pod krylom nauki." *Golos*, 3 (15) June 1875, #152: 1–2.

Markov, M. A., et al., eds. *Filosofiia i estestvoznanie: K semidesiatiletiiu akademika Bonifatiia Mikhailovicha Kedrova*. Moscow: Nauka, 1974.

Markovnikov, V. V. "[Jubilee speech]." In *Russkoe khimicheskoe obshchestvo. XXV (1868–1893)* (1894): 56–69.

———. "Moskovskaia rech' o Butlerove," ed. Iu. S. Musabekov. *TIIEiT* 12 (1956): 135–181.

———. "Vospominaniia i cherty iz zhizni i deiatel'nosti A. M. Butlerova." *ZhRFKhO* 19, no. 1 (1877): 69–96.

Marks, Steven G. *Road to Power: The Trans-Siberian Railroad and the Colonization of Asian Russia, 1850–1917*. Ithaca: Cornell University Press, 1991.

Martinsen, Deborah A. *Literary Journals in Imperial Russia*. Cambridge: Cambridge University Press, 1997.

Materialy dlia istorii akademicheskikh uchrezhdenii za 1889–1914 gg., ch. 1. Petrograd: Izd. Imp. Akademii nauk, 1917.

Mathes, William L. "The Origins of Confrontation Politics in Russian Universities: Student Activism, 1855–1861." *Canadian Slavic Studies* 2 (1968): 28–45.

———. "University Courts in Imperial Russia." *Slavonic and East European Review* 52 (1974): 366–381.

Matout, L. "L'éther considéré comme élément chimique." *Le Radium* 1 (1904): 116–117.

Matthey, George. "The Preparation in a State of Purity of the Group of Metals Known as the Platinum Series, and Notes upon the Manufacture of Iridio-Platinum." *Chemical News* 39 (1879): 175–177.

Mauskopf, Seymour H. "The Atomic Structural Theories of Ampère and Gaudin: Molecular Speculation and Avogadro's Hypothesis." *Isis* 60 (1969): 61–74.

Maxwell, J. Clerk. *La Chaleur: Leçons élémentaires sur la thermométrie, la calorimétrie, la thermo-dynamique, et la dissipation de l'énergie.* Paris: B. Tignol, 1891.

———. *Materiia i dvizhenie,* tr. M. A. Antonovich. St. Petersburg: L. F. Panteleev, 1885.

———. *Teoriia teploty v elementarnoi obrabotke,* tr. from 7th English ed. by A. L. Korol'kov. Kiev: I. N. Kushnerev, 1888.

Mazurs, Edward G. *Graphic Representations of the Periodic System during One Hundred Years.* University: University of Alabama Press, 1974 [1957].

McCabe, James Dabney. *The Illustrated History of the Centennial Exhibition.* Philadelphia: National Publishing Co., [1876].

McClelland, James C. "Diversification in Russian-Soviet Education." In Jarausch, ed., *The Transformation of Higher Learning* (1982): 180–195.

McGuire, J. E. "Transmutation and Immutability: Newton's Doctrine of Physical Qualities." *Ambix* 14 (1967): 69–95.

McKay, John P. "Baku Oil and Transcaucasian Pipelines, 1883–1891: A Study in Tsarist Economic Policy." *Slavic Review* 43 (1984): 604–623.

———. "Entrepreneurship and the Emergence of the Russian Petroleum Industry, 1813–1883." *Research in Economic History* 8 (1983): 47–91.

———. *Pioneers for Profit: Foreign Entrepreneurship and Russian Industrialization, 1885–1913.* Chicago: University of Chicago Press, 1970.

———. "Restructuring the Russian Petroleum Industry in the 1890s: Government Policy and Market Forces." In Edmondson and Waldron, eds., *Economy and Society in Russia and the Soviet Union* (1992): 85–106.

McReynolds, Louise. "Imperial Russia's Newspaper Reporters: Profile of a Society in Transition, 1865–1914." *Slavonic and East European Review* 68 (1990): 277–293.

———. *The News under Russia's Old Regime: The Development of a Mass Circulation Press.* Princeton: Princeton University Press, 1991.

———. "V. M. Doroshevich: The Newspaper Journalist and the Development of Public Opinion in Civil Society." In Clowes, Kassow, and West, eds., *Between Tsar and People* (1991): 233–247.

Medhurst, R. G., ed. *Crookes and the Spirit World: A Collection of Writings by or Concerning the Work of Sir William Crookes, O.M., F.R.S., in the Field of Psychical Research.* New York: Taplinger, 1972.

Meilakh, B. "Posleslovie." *Novyi Mir,* no. 12 (1966): 191–198.

Melhado, Evan M. "Mitscherlich's Discovery of Isomorphism." *Historical Studies in the Physical Sciences* 11 (1980): 87–123.

Mendeleev i metrologiia. Moscow: Izd. komiteta standartov, mer i izmeritel'nykh priborov pri Sovete ministrov SSSR, 1969.

Mendeleeva, A. I. *Mendeleev v zhizni.* Moscow: M. and S. Sabashnikov, 1928.

Mendeleeva, M. D. "Novye materialy o zhizni i tvorchestve D. I. Mendeleeva v nachale 60-kh godov." *Nauchnoe Nasledstvo* 2 (1951): 85–94.

———. "Zametki o Mendeleeve." *Vestnik Akademii Nauk SSSR,* no. 2 (1957): 69–73.

"Mendeleevskii obed." *Novoe Vremia,* 8 (20) December 1880, #1718: 2–3.

Menning, Bruce W. *Bayonets before Bullets: The Imperial Russian Army, 1861–1914.* Bloomington: Indiana University Press, 1992.

Men'shikov, M. "Pis'ma k blizhnim: Lev Tolstoi, Mendeleev, Vereshchagin." *Novoe Vremia,* 4 (17) July 1904, #10179: 3–4.

Menshutkin, B. N. "Glavnye momenty v razvitii periodicheskoi sistemy elementov (1868–1937)." *Priroda* (March 1937): 116–126.

_____. *Khimiia i puti ee razvitiia*. Moscow: Izd. AN SSSR, 1937.

_____. *Zhizn' i deiatel'nost' Nikolaia Aleksandrovicha Menshutkina*. St. Petersburg: M. Frolova, 1908.

Metricheskaia reforma v SSSR. Moscow: Tsentral'naia Metricheskaia Komissiia, 1928.

Ministerstvo Finansov, Departament Torgovli i Manufaktur. *Zhurnaly zasedanii kommisii po peresmotru deistvuiushchikh o merakh i vesakh uzakonenii*. St. Petersburg: V. Kirshbaum, 1897.

Ministerstvo finansov, 1802–1902. St. Petersburg: Ekspeditsiia zagotovleniia gosudarstvennykh bumag, 1902.

Mladentsev, M. N. "Uchrezhdenie Glavnoi Palaty mer i vesov i eia deiatel'nost'." *VGPMV* 8 (1907): 42–90.

Mladentsev, M. N., and V. E. Tishchenko. *Dmitrii Ivanovich Mendeleev, ego zhizn' i deiatel'nost'*, v. 1. Moscow: Izd. AN SSSR, 1938.

Mochulsky, Konstantin. *Aleksandr Blok*, tr. Doris V. Johnson. Detroit: Wayne State University Press, 1983.

Modestov, V. "Russkaia nauka i obshchestvo." *Golos*, 4 (16) December 1880, #335: 1–2.

Mohn, Heinrich. *Meteorologiia ili uchenie o pogode*, trs. N. Iordanskii and F. Kapustin, ed. D. Mendeleev. St. Petersburg: Obshchestvennaia Pol'za, 1876.

Monmonier, Mark. "Telegraphy, Iconography, and the Weather Map: Cartographic Weather Reports by the United States Weather Bureau, 1870–1935." *Imago Mundi* 40 (1988): 15–31.

Moore, R. Laurence. *In Search of White Crows: Spiritualism, Parapsychology, and American Culture*. New York: Oxford University Press, 1977.

_____. "Spiritualism and Science: Reflections on the First Decade of the Spirit Rappings." *American Quarterly* 24 (1972): 474–500.

Morozov, Nikolai. *D. I. Mendeleev i znachenie ego periodicheskoi sistemy dlia khimii budushchago*. Moscow: I. D. Sytin, 1908.

Morozov, P. "Iz istorii nashei Akademii nauk." *Strana*, 4 December 1880, #95: 2–3.

Morrissey, Susan K. *Heralds of Revolution: Russian Students and the Mythologies of Radicalism*. Oxford: Oxford University Press, 1998.

Mosse, W. E. *Alexander II and the Modernization of Russia*. London: I. B. Tauris & Co., 1992 [1958].

Münsterberg, Hugo. "My Friends, the Spiritualists: Some Theories and Conclusions Concerning Eusapia Palladino." *The Metropolitan Magazine* 31, no. 5 (February 1910): 559–572.

Munting, Roger. "Economic Change and the Russian Gentry, 1861–1914." In Edmondson and Waldron, eds., *Economy and Society in Russia and the Soviet Union* (1992): 24–43.

Musabekov, Iu. S. "D. I. Mendeleev i konstantinovskii zavod v Iaroslavle." *VIET*, no. 4 (29) (1969): 132–133.

_____. "Novye materialy o V. V. Markovnikove (K 50-letiiu so dnia smerti)." *TIIEiT* 6 (1955): 307–317.

_____. "Novye materialy ob A. M. Butlerove." *TIIEiT* 6 (1955): 229–242.

_____. "Pervyi russkii khimicheskii zhurnal i ego osnovateli." In Figurovskii et al., eds., *Materialy po istorii otechestvennoi khimii* (1953): 288–302.

Musabekov, Iu. S., A. A. Makarenia, and V. A. Pozdysheva. "Vladimir Il'ich Lenin o trudakh i ideiakh D. I. Mendeleeva." *Zhurnal Vsesoiuznogo khimicheskogo obshchestva im. D. I. Mendeleeva* 20, no. 6 (1970): 704–705.

Naimark, Norman M. *Terrorists and Social Democrats: The Russian Revolutionary Movement under Alexander III*. Cambridge, Mass.: Harvard University Press, 1983.

Nametkin, S. S. "Trudy D. I. Mendeleev v oblasti izucheniia nefti i neftianoi promyshlennosti." In *Trudy iubileinogo mendeleevskogo s"ezda*, v. 1 (1936): 639–649.

"[Nasha Akademiia nauk i trebovaniia ee ustava]." *S.-Peterburgskie Vedomosti*, 17 (29) December 1880, #347: 1.

Nebeker, Frederik. *Calculating the Weather: Meteorology in the 20th Century*. New York: Academic Press, 1995.

Nelson, Geoffrey K. *Spiritualism and Society*. London: Routledge and Kegan Paul, 1969.

Nemchinov, V. S. "Ekonomicheskie vzgliady D. I. Mendeleeva." In Vol'fkovich, ed., *Dmitrii Ivanovich Mendeleev* (1957): 101–114.

"Nemtsy pobediteli!" *Novoe Vremia*, 12 (24) November 1880, #1692: 1.

Nevedomskii, M. P., and I. E. Repin. *A. I. Kuindzhi*. St. Petersburg: Izd. Obshchestva imeni A. I. Kuindzhi, 1913.

Neyman-Spallart, F. Ks. *Proteksionizm i vsemirnoe khoziaistvo (Kritika sistemy pokrovitel'stvennykh poshlin)*, tr. L. S. Zak. Odessa: Vysochaishe utverzhdennyi iuzhno-Russk. O-va Pechatnoe delo, 1896.

Nezdiurov, D. F. *Ocherki razvitiia meteorologicheskikh nabliudenii v Rossii*. Leningrad: Gidrometeoizdat, 1969.

Neznakomets. "Nedel'nye ocherki i kartinki." *Birzhevye Vedomosti*, 28 December 1875, #357: 1–2.

Nikitin, N. I., ed. *Materialy po istorii otechestvennoi khimii: Doklady, zaslushannye na zasedaniiakh Leningradskogo filiala komissii po istorii khimii*. Moscow: Izd. AN SSSR, 1954.

Nilson, L. F. "Om Scandium, en ny jordmetall." *Ofversigt af Kongl. Vetenskaps-Akademiens Förhandlingar*, no. 3 (1879): 47–51.

————. "Ueber das Atomgewicht und einige charakteristische Verbindungen des Scandiums." *Berichte der Deutschen Chemischen Gesellschaft* 13 (1880): 1439–1450.

Noakes, Richard J. "Telegraphy Is an Occult Art: Cromwell Fleetwood Varley and the Diffusion of Electricity to the Other World." *British Journal of the History of Science* 32 (1999): 421–459.

Noble, David F. *A World without Women: The Christian Clerical Culture of Western Science*. Oxford: Oxford University Press, 1992.

"Novaia vylazka akademicheskikh nemtsev." *Novoe Vremia*, 28 December 1880 (9 January 1881), #1736: 1.

"Novye vybory v Akademii." *Novoe Vremia*, 26 November (9 December) 1880, #1706: 1.

"Novyi tamozhennyi tarif." *Novoe Vremia*, 21 June (3 July) 1891, #5498: 1.

Nye, Mary Jo. "Berthelot's Anti-Atomism: A 'Matter of Taste'?" *Annals of Science* 38 (1981): 585–590.

————. *From Chemical Philosophy to Theoretical Chemistry: Dynamics of Matter and Dynamics of Disciplines, 1800–1950*. Berkeley: University of California Press, 1993.

————. "The Nineteenth Century Atomic Debates and the Dilemma of an 'Indifferent Hypothesis.'" *Studies in History and Philosophy of Science* 7 (1976): 245–268.

Nye, Robert A. *Masculinity and Male Codes of Honor in Modern France*. Berkeley: University of California Press, 1993.

O'Connell, Joseph. "Metrology: The Creation of Universality by the Circulation of Particulars." *Social Studies of Science* 23 (1993): 129–173.

Ol'khovskii, E. R. "Tainyi arest akademika Famintsyna. (Iz istorii izucheniia v 1878–1879 gg. prichin volnenii sredi uchashcheisia molodezhi)." *Ocherki po istorii Sankt-Peterburgskogo universiteta* 7 (1998): 132–145.

Olesko, Kathryn M. "The Meaning of Precision: The Exact Sensibility in Early Nineteenth-Century Germany." In Wise, ed., *The Values of Precision* (1995): 103–134.

Oppenheim, Janet. *The Other World: Spiritualism and Psychical Research in England, 1850–1914*. Cambridge: Cambridge University Press, 1985.

Orlov, Vladimir. *Hamayun: The Life of Alexander Blok*. Moscow: Progress, 1980.

Orlovsky, Daniel T. "High Officials in the Ministry of Internal Affairs, 1855–1881." In Pintner and Rowney, eds., *Russian Officialdom* (1980): 250–282.

_____. *The Limits of Reform: The Ministry of Internal Affairs in Imperial Russia, 1802–1881.* Cambridge, Mass.: Harvard University Press, 1981.

_____. "Recent Studies on the Russian Bureaucracy." *Russian Review* 35 (1976): 448–467.

"Ot Komisii dlia izsledovaniia mediumicheskikh iavlenii." *Golos*, 25 March 1876, #85.

Owen, Alex. *The Darkened Room: Women, Power and Spiritualism in Late Victorian England.* Philadelphia: University of Pennsylvania Press, 1990.

Owen, Thomas C. *Capitalism and Politics in Russia: A Social History of the Moscow Merchants, 1855–1905.* Cambridge: Cambridge University Press, 1981.

_____. "Entrepreneurship and the Structure of Enterprise in Russia, 1800–1880." In Guroff and Carstensen, eds., *Entrepreneurship in Imperial Russia and the Soviet Union* (1983): 59–83.

_____. "Impediments to a Bourgeois Consciousness in Russia, 1880–1905: The Estate Structure, Ethnic Diversity, and Economic Regionalism." In Clowes, Kassow, and West, eds., *Between Tsar and People* (1991): 75–89.

_____. *Russian Corporate Capitalism from Peter the Great to Perestroika.* Oxford: Oxford University Press, 1995.

_____. "The Russian Industrial Society and Tsarist Economic Policy, 1867–1905." *Journal of Economic History* 45 (1985): 587–606.

Ozarovskaia, O. E. *D. I. Mendeleev po vospominaniiam O. E. Ozarovskoi.* Moscow: Federatsiia, 1929.

Palfreman, Jon. "Between Scepticism and Credulity: A Study of Victorian Scientific Attitudes to Modern Spiritualism." *Sociological Review Monographs* 27 (1979): 201–236.

_____. "William Crookes: Spiritualism and Science." *Ethics in Science and Medicine* 3 (1976): 211–227.

Pang, Alex Soojung-Kim. "The Social Event of the Season: Solar Eclipse Expeditions and Victorian Culture." *Isis* 84 (1993): 252–277.

Parkhomenko, V. E. *D. I. Mendeleev i russkoe neftianoe delo.* Moscow: Izd. AN SSSR, 1957.

Parmenov, K. Ia. *Khimiia kak uchebnyi predmet v dorevoliutsionnoi i sovetskoi shkole.* Moscow: Akademiia pedagogicheskikh nauk RSFSR, 1963.

Pasetskii, V. M. *Adol'f Iakovlevich Kupfer, 1799–1865.* Moscow: Nauka, 1984.

Passek, T. P. "Vospominaniia T. P. Passek [Chapter 32]." *Russkaia Starina* 20 (1877): 277–300.

Pearson, Thomas S. *Russian Officialdom in Crisis: Autocracy and Local Self-Government, 1861–1900.* Cambridge: Cambridge University Press, 1989.

Petrov, L. P. "L. Errera i ego rol' v razvitii ucheniia o periodichnosti." *VIET*, no. 3–4 (1977): 74–76.

_____. "Prognozirovanie i razmeshchenie inertnykh elementov v periodicheskoi sisteme." In Trifonov, ed., *Uchenie o periodichnosti* (1981): 37–77.

Petrushevskii, F. "Postupok Akademii nauk. (Pis'mo k redaktsii)." *Golos*, 31 December 1880 (12 January 1881): 1.

Piccini, A. "Das periodische System der Elemente von Mendelejeff und die neuen Bestandteile der atmosphärischen Luft." *Zeitschrift für Anorganische Chemie* 19 (1899): 295–305.

Pikulev, I. *Ivan Ivanovich Shishkin, 1832–1898.* Moscow: Iskusstvo, 1955.

Pintner, Walter M. "The Evolution of Civil Officialdom, 1755–1855." In Pintner and Rowney, eds., *Russian Officialdom* (1980): 190–226.

_____. "The Russian Higher Civil Service on the Eve of the 'Great Reforms.'" *Journal of Social History* 8 (Spring 1975): 55–68.

_____. "The Social Characteristics of the Early Nineteenth-Century Russian Bureaucracy." *Slavic Review* 29 (1970): 429–443.

Pintner, Walter McKenzie, and Don Karl Rowney, eds. *Russian Officialdom: The Bureaucratization of Russian Society from the Seventeenth to the Twentieth Century*. Chapel Hill: University of North Carolina Press, 1980.

Pipes, Richard, ed. *Karamzin's Memoir on Ancient and Modern Russia: A Translation and Analysis*. New York: Atheneum, 1981.

————. "Russian Conservatism in the Second Half of the Nineteenth Century." *Slavic Review* 30 (1971): 121–128.

Pisarev, D. I. "Nasha universitetskaia nauka." *Russkoe Slovo*, no. 7 (1863): 1–75; no. 8 (1863): 1–54.

Pisarzhevsky, O. N. *Dmitry Ivanovich Mendeleyev: His Life and Work*. Moscow: Foreign Languages Publishing House, 1954.

Plate, A. F. "Novye materialy k biografii V. V. Markovnikova." In Figurovskii et al., eds., *Materialy po istorii otechestvennoi khimii* (1953): 70–80.

Pletner, Iu. V. *Dedushka russkoi khimii*. Kalinin: Kalininskoe knizhnoe izd., 1959.

Podmore, Frank. *Mediums of the 19th Century*, 2 v. New Hyde Park, N.Y.: University Books, 1963 [1902].

————. *Spiritizm: Istoricheskoe i kriticheskoe issledovanie*. St. Petersburg: Slovo, 1904–1905.

Pogodin, S. A. "Akademiia nauk i puti razvitiia khimii v dorevoliutsionnoi Rossii." In Khlopin, Balandin, and Pogodin, *Ocherki po istorii akademii nauk* (1945): 5–37.

————. "Bor'ba D. I. Mendeleeva za razvitie otechestvennoi khimicheskoi promyshlennosti." *Zhurnal Prikladnoi Khimii* 20, no. 6 (1947): 474–485.

————. "Otkrytie periodicheskogo zakona D. I. Mendeleevym i ego bor'ba za pervenstvo russkoi nauki." *Nauka i Zhizn'*, no. 3 (1949): 37–40.

Pogorelskin, Alexis. "'The Messenger of Europe.'" In Martinsen, ed., *Literary Journals in Imperial Russia* (1997): 129–149.

Poirier, Jean-Pierre. *Lavoisier: Chemist, Biologist, Economist*, tr. Rebecca Balinski. Philadelphia: University of Pennsylvania Press, 1993.

Pokhlebkin, Vil'iam. "Mendeleev i vodka." *Ogonek*, no. 50 (December 1997): 4–5.

Porter, Theodore M. "Objectivity as Standardization: The Rhetoric of Impersonality in Measurement, Statistics, and Cost–Benefit Analysis." *Annals of Scholarship* 9 (1992): 19–59.

Posin, Daniel Q. *Mendeleyev: The Story of a Great Scientist*. New York: Whittlesey House, 1948.

"Predlozhenie i balotirovanie professora F. F. Beil'shteina v ordinarnye akademiki po tekhnologii i khimii, prisposoblennoi k iskusstvam i remeslam." *Zapiski Imperatorskoi Akademii nauk* 41, no. 1 (1882): 84–167.

Pribytkov, Viktor. *Mediumicheskie iavleniia pered sudom vrachei*. St. Petersburg: Rebus, 1885.

————. "O fotografiiakh, predstavlennykh professorom N. P. Vagnerom v Tekhnicheskoe obshchestvo." *Rebus*, no. 13 (1894): 131.

————. "Professor Mendeleev priznaet mediumicheskiia iavleniia." *Rebus*, no. 1 (1894): 3–4.

Pushkin, Michael. "*Raznochintsy* in the University: Government Policy and Social Change in Nineteenth-Century Russia." *International Review of Social History* 26 (1981): 25–65.

Putnam, Allen. *Agassiz and Spiritualism: Involving the Investigation of Harvard College Professors in 1857*. Boston: Colby & Rich, [1874].

Putnam, George F. *Russian Alternatives to Marxism: Christian Socialism and Idealistic Liberalism in Twentieth-Century Russia*. Knoxville: University of Tennessee Press, 1977.

Pyman, Avril. *The Life of Aleksandr Blok. Volume I: The Distant Thunder, 1880–1908*. Oxford: Oxford University Press, 1979.

————. *The Life of Aleksandr Blok. Volume II: The Release of Harmony, 1908–1921*. Oxford: Oxford University Press, 1980.

Pynchon, Thomas. *V.* New York: Harper & Row, 1989 [1963].

Rabinowitz, Stanley J. "'Northern Herald': From Traditional Thick Journal to Forerunner of the Avant-Garde." In Martinsen, ed., *Literary Journals in Imperial Russia* (1997): 207–227.

Rachinskii, S. "Po povodu spiriticheskikh soobshchenii g. Vagnera." *Russkii Vestnik* (May 1875): 380–399.

Radovskii, M. I. "K uchastiiu russkikh uchenykh v mezhdunarodnykh soglasheniiakh o edinstve mer i vesov." *Istoricheskii Arkhiv*, no. 2 (1958): 120–133.

Raeff, Marc. "The Bureaucratic Phenomena of Imperial Russia, 1700–1905." *American Historical Review* 84 (1979): 399–411.

_____. "The Russian Autocracy and Its Officials." *Harvard Slavic Studies* 4 (1957): 77–91.

_____. *The Well-Ordered Police State: Social and Institutional Change through Law in the Germanies and in Russia*. New Haven: Yale University Press, 1983.

Raikov, B. E., ed. "Iz vospominanii zoologa Aleksandra Mikhailovicha Nikol'skogo." *Iz Istorii Biologicheskikh Nauk* (1966): 79–104.

Ramsay, William. "The Electron as Element." *Journal of the Chemical Society* 93 (1908): 774–788.

Rawson, Don C. "Mendeleev and the Scientific Claims of Spiritualism." *Proceedings of the American Philosophical Society* 122 (1978): 1–8.

_____. "The Process of Discovery: Mendeleev and the Periodic Law." *Annals of Science* 31 (1974): 181–204.

Reichman, Henry. *Railwaymen and Revolution: Russia, 1905*. Berkeley: University of California Press, 1987.

Remsen, Ira. Review of Mendeleev's *An Attempt towards a Chemical Conception of the Ether*. *American Chemical Journal* 33 (1905): 517–519.

Reynaud, Marie-Hélène. *Les Freres Montgolfier et leurs étonnantes machines*. Vals-les-Bains: De Plein Vent, 1982.

Rice, Richard E. "Mendeleev's Public Opposition to Spiritualism." *Ambix* 45 (1998): 85–95.

Richter, Friedrich. "F. K. Beilstein, sein Werk und seine Zeit: Zur Erinnerung an die 100. Wiederkehr seines Geburtstages." *Berichte der Deutschen Chemischen Gesellschaft* 71A (1938): 35–71.

Rieber, Alfred J. "Alexander II: A Revisionist View." *Journal of Modern History* 43 (1971): 42–58.

_____. "Bureaucratic Politics in Imperial Russia." *Social Science History* 2 (1978): 399–413.

_____. "Interest-Group Politics in the Era of the Great Reforms." In Eklof, Bushnell, and Zakharova, eds., *Russia's Great Reforms* (1994): 58–83.

_____. *Merchants and Entrepreneurs in Imperial Russia*. Chapel Hill: University of North Carolina Press, 1982.

_____, ed. *The Politics of Autocracy: Letters of Alexander II to Prince A. I. Bariatinskii, 1857–1864*. Paris: Mouton & Co., 1966.

_____. "The Rise of Engineers in Russia." *Cahiers du Monde Russe et Soviétique* 31 (1990): 539–568.

_____. "The Sedimentary Society." In Clowes, Kassow, and West, eds., *Between Tsar and People* (1991): 343–366.

Rimlinger, Gaston V. "Autocracy and the Factory Order in Early Russian Industrialization." *Journal of Economic History* 20 (1960): 67–92.

Riskin, Jessica. "Rival Idioms for a Revolutionized Science and a Republican Citizenry." *Isis* 89 (1998): 203–232.

Roberts, Gerrylynn K. "'A Plea for Pure Science': The Ascendancy of Academia in the Making of the English Chemist, 1841–1914." In Knight and Kragh, eds., *The Making of the Chemist* (1998): 107–119.

Rocke, Alan J. *Chemical Atomism in the Nineteenth Century: From Dalton to Cannizzaro.* Columbus: Ohio State University Press, 1984.

_____. "Gay-Lussac and Dumas: Adherents of the Avogadro–Ampère Hypothesis?" *Isis* 69 (1978): 595–600.

_____. "Kekulé, Butlerov, and the Historiography of Chemical Structure." *British Journal for the History of Science* 14 (1981): 27–57.

_____. *Nationalizing Science: Adolphe Wurtz and the Battle for French Chemistry.* Cambridge, Mass.: MIT Press, 2001.

_____. *The Quiet Revolution: Hermann Kolbe and the Science of Organic Chemistry.* Berkeley: University of California Press, 1993.

Roginskii, S. Z. "D. I. Mendeleev o neizbezhnosti izmenenii massy pri protsessakh pre-vrashcheniia elementov." *Uspekhi Khimii* 22, no. 3 (1951): 270–272.

Rogovsky, E. "On the Temperature and Composition of the Atmospheres of the Planets and the Sun." *Astrophysical Journal* 14 (1901): 234–260.

Romanovich-Slavatinskii, A. V. "Moia zhizn' i akademicheskaia deiatel'nost', 1832–1884 gg." *Vestnik Evropy* 38, no. 4 (1903): 527–566.

Romanovskii, S. I. *Nauka pod gnetom rossiiskoi istorii.* St. Petersburg: Izd. St. Petersburgskogo Universiteta, 1999.

Root, John D. "Science, Religion, and Psychical Research: The Monistic Thought of Sir Oliver Lodge." *Harvard Theological Review* 71 (1978): 245–263.

Roper, Michael, and John Tosh. *Manful Assertions: Masculinities in Britain since 1800.* London: Routledge, 1991.

Rosenthal, Bernice Glatzer, ed. *The Occult in Russian and Soviet Culture.* Ithaca: Cornell University Press, 1997.

_____. "The Search for a Russian Orthodox Work Ethic." In Clowes, Kassow, and West, eds., *Between Tsar and People* (1991): 57–74.

Rowney, Don Karl, and Walter M. Pintner. "Officialdom and Bureaucratization: An Introduction." In Pintner and Rowney, eds., *Russian Officialdom* (1980): 3–18.

Rozhdestvenskii, S. V. *Istoricheskii obzor deiatel'nosti Ministerstva narodnogo prosveshcheniia, 1802–1902.* St. Petersburg: Izd. Ministerstva narodnogo prosveshcheniia, 1902.

Ruane, Christine. *Gender, Class, and the Professionalization of Russian City Teachers, 1860–1914.* Pittsburgh: University of Pittsburgh Press, 1994.

Russell, Colin A. *Edward Frankland: Chemistry, Controversy and Conspiracy in Victorian England.* Cambridge: Cambridge University Press, 1996.

Russkoe khimicheskoe obshchestvo. XXV (1868–1893). Otdelenie khimii Russkago fiziko-khimich-eskago obshchestva. St. Petersburg: V. Demakov, 1894.

Ruud, Charles A. "The Russian Empire's New Censorship Law of 1865." *Canadian Slavic Studies* 3 (1969): 235–245.

Rydell, Robert W. *All the World's a Fair: Visions of Empire at American International Exhibitions, 1876–1916.* Chicago: University of Chicago Press, 1984.

Rykachev, M. A. *Istoricheskii ocherk Glavnoi fizicheskoi observatorii za 50 let eia deiatel'nosti, 1849–1899*, vol. 1. St. Petersburg: Tip. Imp. Akademii nauk, 1899.

_____. *Podniatie na vozdushnom share, v S. Peterburge, 20 maia/1 iiunia 1873 goda. Zapiski Imperatorskogo Russkogo geograficheskogo obshchestva po obshchei geografii* 6, no. 2 (1882).

Sacks, Oliver. *Uncle Tungsten: Memories of a Chemical Boyhood.* New York: Alfred A. Knopf, 2001.

Salleron, M. "Sur la nouvelle balance de M. Mendeleef." *Comptes Rendus* 80 (1875): 378–380.

Saltykov-Shchedrin, M. E. *Polnoe sobranie sochinenii*, vol. 10. Leningrad: Khudozhestvennaia literatura, 1936.

Sargent, Rose-Mary. *The Diffident Naturalist: Robert Boyle and the Philosophy of Experiment.* Chicago: University of Chicago Press, 1995.

Scerri, Eric R. "The Evolution of the Periodic System." *Scientific American* (September 1998): 78–83.

_____. "The Periodic Table and the Electron." *American Scientist* 85 (1997): 546–553.

Scerri, Eric R., and John Worrall. "Prediction and the Periodic Table." *Studies in History and Philosophy of Science* 32 (2001): 407–452.

Schaffer, Simon. "Late Victorian Metrology and Its Instrumentation: A Manufactory of Ohms." In Robert Bud and Susan E. Cozzens, eds., *Invisible Connections: Instruments, Institutions, and Science* (Bellingham, Wash.: SPIE Optical Engineering Press, 1991): 23–56.

_____. "Self Evidence." In James Chandler, Arnold I. Davidson, and Harry Harootunian, eds., *Questions of Evidence: Proof, Practice, and Persuasion across the Disciplines* (Chicago: University of Chicago Press, 1994): 56–91.

Schaffner, Kenneth F. *Nineteenth-Century Aether Theories.* Oxford: Pergamon Press, 1972.

Schapiro, Leonard. *Rationalism and Nationalism in Russian Nineteenth-Century Political Thought.* New Haven: Yale University Press, 1967.

Schinz, Ch. *Essai d'une Nouvelle Théorie Chimique.* Lausanne: Imprimerie-Librairie de Marc Ducloux, 1841.

Schulze, Ludmilla. "The Russification of the St. Petersburg Academy of Sciences in the Eighteenth Century." *British Journal for the History of Science* 18 (1985): 305–335.

Schütt, Hans-Werner. *Eilhard Mitscherlich: Prince of Prussian Chemistry,* tr. William E. Russey. Philadelphia: American Chemical Society and the Chemical Heritage Foundation, 1992.

Sechenov, I. M. *Avtobiograficheskie zapiski Ivana Mikhailovicha Sechenova.* Moscow: Izd. AN SSSR, 1945.

Sekanina, Zdenek. "Encke, the Comet." *Journal of the Royal Astronomical Society of Canada* 85 (1991): 324–376.

Semenov, N. N., ed. *Sto let periodicheskogo zakona khimicheskikh elementov: Doklady na plenarnykh zasedaniiakh.* Moscow: Nauka, 1971.

Semishin, V. I. "Inertnye gazy i periodicheskii zakon D. I. Mendeleeva." *VIET,* no. 1 (26) (1969): 49–53.

_____. *Literatura po periodicheskomu zakonu D. I. Mendeleeva (1869–1969).* Moscow: Vysshaia shkola, 1969.

_____. *Periodicheskii zakon i periodicheskaia sistema khimicheskikh elementov D. I. Mendeleeva v rabotakh russkikh uchenykh.* Moscow: Moskovskii institut khimicheskogo mashinostroeniia, 1959.

Sergeev, S. *Voprosy russkoi promyshlennosti.* Odessa: Vysochaishe utverzhdennyi iuzhno-Russk. O-va Pechatnoe delo, 1896.

Servos, John W. *Physical Chemistry from Ostwald to Pauling: The Making of a Science in America.* Princeton: Princeton University Press, 1990.

Shapin, Steven. *A Social History of Truth: Civility and Science in Seventeenth-Century England.* Chicago: University of Chicago Press, 1994.

Shapin, Steven, and Simon Schaffer. *Leviathan and the Air-Pump: Hobbes, Boyle and the Experimental Life.* Princeton: Princeton University Press, 1985.

Shapiro, Barbara J. *A Culture of Fact: England, 1550–1720.* Ithaca: Cornell University Press, 2000.

_____. *Probability and Certainty in Seventeenth-Century England: A Study of the Relationships between Natural Science, Religion, History, Law, and Literature.* Princeton: Princeton University Press, 1983.

Shchukarev, A. N. "Zakony Prirody i zakony Obshchestva." In *Sbornik po filosofii estestvoznaniia* (Moscow: Tvorcheskaia mysl', 1906): 86–103.

Shchukarev, S. A. "D. I. Mendeleev i Leningradskii gosudarstvennyi universitet." *Vestnik Leningradskogo universiteta*, no. 6 (1947): 148–154.

———. "Uchenie ob opredelennykh i neopredelennykh soedineniiakh v trudakh russkikh uchenykh." *Vestnik Leningradskogo universiteta*, no. 5 (1947): 5–25.

Shchukarev, S. A., and R. B. Dobrotin. "Pervye nauchnye raboty D. I. Mendeleeva kak etap na puti k otkrytiiu periodicheskogo zakona." *Vestnik Leningradskogo universiteta*, no. 2 (1954): 165–177.

Shchukarev, S. A., and S. N. Valk, eds. *Arkhiv D. I. Mendeleeva, t. 1: Avtobiograficheskie materialy, sbornik dokumentov*. Leningrad: Izd. Leningradskogo gosudarstvennogo universiteta imeni A. A. Zhdanova, 1951.

Shelley, Mary. *Frankenstein*, ed. J. Paul Hunter. New York: W. W. Norton and Company, 1996.

Sheynin, Oscar. "Mendeleev and the Mathematical Treatment of Observations in Natural Science." *Historia Mathematica* 23 (1996): 54–67.

Shkliarevskii, A. "Chto dumat' o spiritizme?: Po povodu pis'ma prof. Vagnera." *Vestnik Evropy* (June 1875): 906–918; (July 1875): 409–418.

———. "Kritiki togo berega." *Russkii Vestnik* 121 (January 1876): 470–495.

Shmulevich, L. A., and Iu. S. Musabekov. *Fedor Fedorovich Beil'shtein, 1838–1906*. Moscow: Nauka, 1971.

Shost'in, N. A. *D. I. Mendeleev i problemy izmereniia*. Moscow: Komitet po delam mer i izmeritel'nykh priborov pri Sovete ministrov SSSR, 1947.

Shpitser, S. M., ed. "D. I. Mendeleev po vospominaniiam V. I. Kovalevskogo." *VIET*, no. 13 (1962): 103–105.

Siegel, Daniel M. "Thomson, Maxwell, and the Universal Ether in Victorian Physics." In Cantor and Hodge, eds., *Conceptions of Ether* (1981): 239–268.

Siegfried, Robert. "The Chemical Basis for Prout's Hypothesis." *Journal of Chemical Education* 33 (1956): 263–266.

Sinclair, S. B. "J. J. Thomson and the Chemical Atom: From Ether Vortex to Atomic Decay." *Ambix* 34 (1987): 89–116.

Sinel, Allen. *The Classroom and the Chancellery: State Educational Reform in Russia under Count Dmitry Tolstoi*. Cambridge, Mass.: Harvard University Press, 1973.

Skinder, A., and Lamanskii, S. "Materialy dlia sostavleniia instruktsii o vyverke torgovykh mer i vesov." *VGPMV* 1 (1894): 102–123.

Skvortsov, A. V. "O priemakh rasshifrovki rukopisei D. I. Mendeleeva." *VIET*, no. 3 (1957): 33–38.

———. "Uchenyi titul D. I. Mendeleeva." In Arbuzov, ed., *Materialy po istorii otechestvennoi khimii* (1950): 116–121.

Smirnov, German. *Mendeleev*. Moscow: Molodaia Gvardiia, 1974.

Smith, Crosbie, and M. Norton Wise. *Energy and Empire: A Biographical Study of Lord Kelvin*. Cambridge: Cambridge University Press, 1988.

Smith, Douglas. *Working the Rough Stone: Freemasonry and Society in Eighteenth-Century Russia*. DeKalb: Northern Illinois University Press, 1999.

Smith, Pamela H. *The Business of Alchemy: Science and Culture in the Holy Roman Empire*. Princeton: Princeton University Press, 1994.

Smoluchowski de Smolan, M. "Etherion, a New Gas?" *Nature* 59 (5 January 1899): 223–224.

Soboleva, E. V. *Bor'ba za reorganizatsiiu peterburgskoi Akademii nauk v seredine XIX veka*. Leningrad: Nauka, 1971.

———. *Organizatsiia nauki v poreformennoi Rossii*. Leningrad: Nauka, 1983.

Solov'ev, Iu. I. *Istoriia khimii v Rossii: Nauchnye tsentry i osnovnye napravleniia issledovanii*. Moscow: Nauka, 1985.

———, ed. *Ocherki po istorii khimii*. Moscow: Izd. AN SSSR, 1963.

_____. "Prognoz i otkrytie inertnykh gazov." In Kedrov and Trifonov, eds., *Prognozirovanie v uchenii o periodichnosti* (1976): 71–78.

"Son na iavu." *Strekoza*, 7 December 1880, #49: 1.

"Spiriticheskiia podvigi." *Novoe Vremia*, 1 March 1876, #2: 2.

Stackenwalt, Francis Michael. "Dmitrii Ivanovich Mendeleev and the Emergence of the Modern Russian Petroleum Industry, 1863–1877." *Ambix* 45 (1998): 67–84.

_____. "The Economic Thought and Work of Dmitrii Ivanovich Mendeleev." Ph.D. Dissertation, University of Illinois at Urbana-Champaign, 1976.

Stalin, I. V. *Sochineniia*. Moscow: Gos. izd. politicheskoi literatury, 1954.

Stein, Howard. "'Subtler Forms of Matter' in the Period Following Maxwell." In Cantor and Hodge, eds., *Conceptions of Ether* (1981): 309–340.

Stephens, Holly DeNio. "The Occult in Russia Today." In Rosenthal, ed., *The Occult in Russian and Soviet Culture* (1997): 357–376.

Stites, Richard. *The Women's Liberation Movement in Russia: Feminism, Nihilism, and Bolshevism, 1860–1930*. Princeton: Princeton University Press, 1978.

Stock, Alfred. *Der internationale Chemiker-Kongreß Karlsruhe 3.–5. September 1860 vor und hinter den Kulissen*. Berlin: Verlag Chemie, 1933.

Stoney, G. Johnstone. "Argon—a Suggestion." *Chemical News* 71 (1895): 67–68.

_____. "On Atmospheres upon Planets and Satellites." *Astrophysical Journal* 7 (1898): 25–55.

_____. "On the Presence of Helium in the Earth's Atmosphere, and on Its Relation to the Kinetic Theory of Gas." *Astrophysical Journal* 8 (1898): 316–317.

Storonkin, A. V., and R. B. Dobrotin. "Kratkii ocherk ucheniia D. I. Mendeleeva o rastvorakh." *Vestnik Leningradskogo universiteta*, no. 2 (1955): 157–171.

_____. "Ob osnovnom soderzhanii ucheniia D. I. Mendeleeva o rastvorakh." *VIET*, no. 3 (1957): 14–23.

Strakhov, N. N. "Eshche pis'mo o spiritizme." *Novoe Vremia*, 1 (13) February 1884, #2848: 2–3.

_____. "Fizicheskaia teoriia spiritizma. (Pis'mo k redaktsii)." *Novoe Vremia*, 26 February (10 March) 1885, #3232: 2–3.

_____. "Tri pis'ma o spiritizme. Pis'mo II. Za neposviashchennykh." *Grazhdanin*, 22 November 1876, #43: 1015–1018.

_____. "Tri pis'ma o spiritizme. Pis'mo III. Granitsy vozmozhnago." *Grazhdanin*, 29 November 1876, #44: 1056–1059.

_____. "Tri pis'ma o spiritizme. Pis'mo pervoe. Idoly." *Grazhdanin*, 15 November 1876, #41–42: 981–983.

_____. "Zakonomernost' stikhii i poniatii. (Otkrytoe pis'mo k A. M. Butlerovu)." *Novoe Vremia*, 11 (23) November 1885, #3487: 2–3; 26 November (8 December) 1885, #3502: 2.

Stranges, Anthony N. *Electrons and Valence: Development of the Theory, 1900–1925*. College Station: Texas A&M University Press, 1982.

Suny, Ronald Grigor. "Rehabilitating Tsarism: The Imperial Russian State and Its Historians." *Comparative Studies in Society and History* 31 (1989): 168–179.

Sutton, Geoffrey. "Electric Medicine and Mesmerism." *Isis* 72 (1981): 375–392.

Swedenborg, Emanuel. *Concerning Heaven and Its Wonders, and Concerning Hell from Things Heard and Seen*. New York: New Jerusalem, 1865.

Szporluk, Roman. *Communism and Nationalism: Karl Marx versus Friedrich List*. Oxford: Oxford University Press, 1988.

Tait, P. G. *Svoistva materii*, tr. I. M. Sechenov. St. Petersburg: L. F. Panteleev, 1887.

Taranovski, Theodore. "Alexander III and His Bureaucracy: The Limitations on Autocratic Power." *Canadian Slavonic Papers* 26 (1984): 207–219.

Taylor, Norman W. "Adam Smith's First Russian Disciple." *Slavonic and East European Review* 45 (1967): 425–438.

Tchitchérine, B. "Le système des éléments chimiques." *Bulletin de la Société Impériale des Naturalistes de Moscou*, N.S., v. 4 (1890): 42–81.

Thackray, Arnold. *Atoms and Powers: An Essay on Newtonian Matter-Theory and the Development of Chemistry*. Cambridge, Mass.: Harvard University Press, 1970.

Thaden, Edward C. *Conservative Nationalism in Nineteenth-Century Russia*. Seattle: University of Washington Press, 1964.

———, ed. *Russification in the Baltic Provinces and Finland, 1855–1914*. Princeton: Princeton University Press, 1981.

Thaden, Edward C., with Marianna Foster Thaden. *Russia's Western Borderlands, 1710–1870*. Princeton: Princeton University Press, 1984.

Thomsen, Julius. "Über die mutmaßliche Gruppe inaktiver Elemente." *Zeitschrift für Anorganische Chemie* 9 (1895): 283–288.

Tishchenko, V. E. "A. M. Butlerov i D. I. Mendeleev v ikh vzaimnoi kharakteristike." *ZhRFKhO* 61, no. 4, ch. khim. (1929): 641–651.

———. "D. I. Mendeleev i Russkoe khimicheskoe obshchestvo." *Zhurnal Prikladnoi Khimii* 7, no. 8 (1934): 1527–1534.

———. "Dmitrii Ivanovich Mendeleev. Kratkii biograficheskii ocherk." In Tishchenko, ed., *Trudy pervago mendeleevskogo s"ezda* (1909): 8–32.

———, ed. *Trudy pervago mendeleevskogo s"ezda po obshchei i prikladnoi khimii, sostoiavshagosia v S.-Peterburge 20-go po 30-go dekabria 1907 g*. St. Petersburg: M. P. Frolova, 1909.

———. "Vospominaniia o D. I. Mendeleeve." *Priroda* (March 1937): 126–136.

Tishchenko, V. E., and M. N. Mladentsev. *Dmitrii Ivanovich Mendeleev, ego zhizn' i deiatel'nost: Universitetskii period, 1861–1890 gg*. Moscow: Nauka, 1993. Published as *Nauchnoe Nasledstvo*, v. 21.

Tishkin, G. A. "Peterburgskii universitet i nachalo vysshego zhenskogo obrazovaniia v Rossii." *Ocherki po istorii Leningradskogo universiteta* 4 (1982): 15–32.

Todes, Daniel P. *Darwin without Malthus: The Struggle for Existence in Russian Evolutionary Thought*. Oxford: Oxford University Press, 1989.

———. *Pavlov's Physiology Factory: Experiment, Interpretation, Laboratory Enterprise*. Baltimore: The Johns Hopkins University Press, 2001.

Tolf, Robert W. *The Russian Rockefellers: The Saga of the Nobel Family and the Russian Oil Industry*. Stanford: Hoover Institution Press, 1976.

Tolz, Vera. *Russian Academicians and the Revolution: Combining Professionalism and Politics*. New York: St. Martin's Press, 1997.

Torke, Hans J. "Continuity and Change in the Relations between Bureaucracy and Society in Russia, 1613–1861." *Canadian Slavic Studies* 4 (1971): 457–476.

Trifonov, D. N. "K istorii voprosa ob analiticheskom vyrazhenii periodicheskogo zakona." *VIET*, no. 4 (29) (1969): 83–90.

———. *O kolichestvennoi interpretatsii periodichnosti*. Moscow: Nauka, 1971.

———. "O matematicheskom modelirovanii periodicheskoi sistemy elementov." In Markov et al., eds., *Filosofiia i estestvoznanie* (1974): 238–256.

———. *Problema redkikh zemel'*. Moscow: Gosatomizdat, 1962.

———. *Redkozemel'nye elementy*. Moscow: Izd. AN SSSR, 1960.

———. *Redkozemel'nye elementy i ikh mesto v periodicheskoi sisteme*. Moscow: Nauka, 1966.

———, ed. *Uchenie o periodichnosti: Istoriia i sovremennost'*. Moscow: Nauka, 1981.

———. "Versiia-2. (K istorii otkrytiia periodicheskogo zakona D. I. Mendeleevym)." *VIET*, no. 2 (1990): 24–36; no. 3 (1990): 20–32.

Trifonov, D. N., and I. S. Dmitriev. "O kolichestvennoi interpretatsii periodicheskoi sistemy." In Trifonov, ed., *Uchenie o periodichnosti* (1981): 221–253.

Trirogova-Mendeleeva, O. D. *Mendeleev i ego sem'ia.* Moscow: Izd. AN SSSR, 1947.

Trotsky, Leon. "Dialectical Materialism and Science." In Leon Trotsky, *Problems of Everyday Life: And Other Writings on Culture and Science* (New York: Monad Press, 1973): 206–226.

[Tsvet, G.]. *Spiritizm i spirity: Znakomstvo, vyvody, obshchestvennyia i psikhicheskiia usloviia, istoricheskaia pochva.* St. Petersburg: Obshchestvennaia Pol'za, 1876.

Tucker, Jennifer. "Voyages of Discovery on Oceans of Air: Scientific Observation and the Image of Science in an Age of 'Balloonacy.'" *Osiris* 11 (1996): 144–176.

Turgenev, I. S. *Pis'ma*, v. 2. Moscow: Nauka, 1987.

Turner, Frank Miller. *Between Science and Religion: The Reaction to Scientific Naturalism in Late Victorian England.* New Haven: Yale University Press, 1974.

Ulam, Adam B. *Prophets and Conspirators in Prerevolutionary Russia*, 2d. ed. New Brunswick, N.J.: Transaction Publishers, 1998 [1977].

Ulanovskaia, M. A. "N. N. Beketov o periodicheskom zakone D. I. Mendeleeva." *TIIEiT* 39 (1962): 272–282.

Urbain, G. "L'œuvre de Lecoq de Boisbaudran." *Revue générale des sciences pures et appliquées* 23, no. 17 (1912): 657–664.

Ustavy Akademii nauk SSSR. Moscow: Nauka, 1974.

V. N. "Neskol'ko myslei po povodu pis'ma prof. Vagnera o spiritizme." *Vestnik Evropy* (May 1875): 458–464.

V. P. "D. I. Mendeleev i Akademiia." *Novoe Vremia*, 19 November (1 December) 1800, #1699: 3–4.

V. Zh. "Novyi podvig Akademii nauk." *Golos*, 15 (27) November 1880, #316: 1.

Vagner, N. P. "Eshche dva slova o kartine Kuindzhi." *Novoe Vremia*, 13 (25) November 1880, #1693: 3.

_____. "Eshche po voprosu o mediumizme." *Golos*, 9 (21) June 1875, #158: 3.

_____. "[Letter to the editor]." *S.-Peterburgskie Vedomosti*, 3 (15) December 1875, #325: 3.

_____. "Mediumizm." *Russkii Vestnik* 119 (October 1875): 866–951.

_____. "Peregorodochnaia filosofiia i nauka. (Otkrytoe pis'mo k g. Strakhovu)." *Novoe Vremia*, 13 (25) July 1883, #2647: 2; 20 July (1 August) 1883, #2654: 2.

_____. "Pis'mo k redaktoru: Po povodu spiritizma." *Vestnik Evropy* (April 1875): 855–875.

_____. "Razdvoennaia filosofiia: Otkrytyi otvet na pis'mo N. N. Strakhova." *Novoe Vremia*, 3 (15) April 1884, #2909: 2–3.

_____. "Sine ira et studio. (Po povodu mediumicheskikh fotografii)." *Rebus*, no. 15 (1894): 151–152; no. 16 (1894): 164–166.

_____. "Za i protiv: Otvet na prigovor spiriticheskoi komisii universetetskago fizicheskago obshchestva, pomeshchennoi v No. 85-m 'Golosa.'" *Golos*, 12 (24) April 1876, #101: 5.

Valkenier, Elizabeth. *Russian Realist Art: The State and Society: The Peredvizhniki and Their Tradition.* New York: Columbia University Press, 1989.

Van Spronsen, J. W. *The Periodic System of Chemical Elements: A History of the First Hundred Years.* Amsterdam: Elsevier, 1969.

Vasetskii, G. S. "Mirovozzrenie D. I. Mendeleeva." *Sovetskaia Nauka*, no. 3 (1938): 99–106.

_____. "Nekotorye cherty mirovozzreniia D. I. Mendeleeva." In *Periodicheskii zakon D. I. Mendeleeva i ego filosofskoe znachenie* (1947): 209–233.

Vavilov, S. I. "Fizika v nauchnom tvorchestve D. I. Mendeleeva." In *Trudy iubileinogo mendeleevskogo s"ezda*, vol. 2 (Moscow: Izd. AN SSSR, 1937): 3–11.

Vdovenko, V. M., and R. B. Dobrotin. "D. I. Mendeleev i voprosy radioaktivnosti (po materialam arkhiva D. I. Mendeleeva)." *VIET*, no. 5 (1957): 175–177.

Veinberg, V. P. "D. I. Mendeleev (k stoletiiu so dnia rozhdeniia)." *Sotsialisticheskaia Rekonstruktsiia i Nauka*, no. 4 (1933): 106–115.

_____. *Iz vospominanii o Dmitrii Ivanoviche Mendeleeve kak lektor.* Tomsk: Tip. Gubernskago Upravleniia, 1910.

_____. "Khimik ili fizik Mendeleev?" *Nauchnoe Slovo*, no. 6 (1928): 63–79.

Venturi, Franco. *Roots of Revolution: A History of the Populist and Socialist Movements in Nineteenth-Century Russia,* tr. Francis Haskell. Chicago: University of Chicago Press, 1960.

Verner, Andrew M. *The Crisis of Russian Autocracy: Nicholas II and the 1905 Revolution.* Princeton: Princeton University Press, 1990.

Verzhbolovich, M. O. *Spiritizm pred sudom nauki i khristianstva.* Moscow: Universitetskaia tip., 1900.

Veselvoskii, K. *O klimate Rossii.* St. Petersburg: Tip. Imp. Akademiia Nauk, 1857.

Visokovatov, Vladimir. "Iz vospominanii o D. I. Mendeleeve." *S.-Peterburgskie Vedomosti*, 17 February (2 March) 1907, #38: 2.

Vlasov, A. G. "Istoricheskaia spravka po vvedeniiu metricheskoi sistemy v SSSR." *Metrologiia i Poverochnoe Delo*, no. 3 (July 1938): 8–12.

"Vnutrennee obozrenie." *Otechestvennye Zapiski* 225 (April 1876): 254–280.

Vol'fkovich, S. I. "D. I. Mendeleev i otechestvennoi khimii." In Semenov, ed., *Sto let periodicheskogo zakona khimicheskikh elementov* (1971): 115–132.

_____, ed. *Dmitrii Ivanovich Mendeleev: Zhizn' i trudy.* [Moscow]: Izd. AN SSSR, 1957.

Vol'fkovich, S. I., and V. S. Kiselev, eds. *75 let periodicheskogo zakona D. I. Mendeleeva i Russkogo khimicheskogo obshchestva.* Moscow: Izd. AN SSSR, 1947.

Volgin, I. L., and V. L. Rabinovich. "Dostoevskii i Mendeleev: Antispiriticheskii dialog." *Voprosy Filosofii*, no. 11 (1971): 103–115.

_____. "Dostoevsky and Mendeleev: An Antispiritist Dialogue." *Soviet Studies in Philosophy* 11 (1972): 170–194.

Volk, S. S. "Revoliutsionnye izdaniia narodovol'cheskikh kruzhkov Peterburgskogo universiteta v 1882 g." *Ocherki po istorii Leningradskogo universiteta* 2 (1968): 52–62.

Volkov, Solomon. *St. Petersburg: A Cultural History,* tr. Antonina W. Bouis. New York: Free Press, 1995.

Volkov, V. A. "Novye dokumenty o literaturnom nasledstve D. I. Mendeleeva i o ego sem'e." *VIET*, no. 4 (29) (1969): 138–141.

Volkova, T. V. "A. M. Butlerov i Peterburgskii universitet." *Vestnik Leningradskogo universiteta*, no. 6 (1952): 149–160.

_____. "D. I. Mendeleev i kniga: Biblioteka D. I. Mendeleeva." *Sotsialisticheskaia Rekonstruktsiia i Nauka*, no. 9 (1934): 99–103.

_____. "Materialy k deiatel'nosti A. M. Butlerova v Akademii nauk." *Zhurnal Prikladnoi Khimii* 21, no. 12 (1948): 1300–1305.

_____. "Materialy k deiatel'nosti A. M. Butlerova v Peterburge (1869–1886)." In Nikitin, ed., *Materialy po istorii otechestvennoi khimii* (1954): 5–22.

_____. "Osnovy khimii i periodicheskii zakon (k 75-letiiu periodicheskogo zakona D. I. Mendeleeva)." *Priroda*, no. 2 (1944): 74–78.

_____. "Perepiska D. I. Mendeleeva s inostrannymi uchenymi (pis'ma Vil'iamsona, Viurtsa i Kannitsaro k D. I. Mendeleevu." *Uspekhi Khimii* 10, no. 6 (1941): 734–742.

_____. "Perepiska I. M. Sechenova s D. I. Mendeleevym." *Priroda*, no. 2 (1940): 86–92.

_____. "Pis'ma A. P. Borodina k D. I. Mendeleevu." *Uspekhi Khimii* 9, no. 9 (1940): 1060–1071.

_____. "Russkoe fiziko-khimicheskoe obshchestvo i Peterburgskii-Leningradskii universitet." *Vestnik Leningradskogo universiteta*, no. 5 (1950): 120–125.

_____. "Ukrepiteli periodicheskogo zakona (pis'ma Lekok de Buabodrana, Vinklera, Nil'sona i Braunera D. I. Mendeleevu)." *Uspekhi Khimii* 13, no. 4 (1944): 317–327.

Von Asten, E. "Über die Existenz eines widerstehenden Mittels im Weltenraume." *Bulletin de l'Académie Imperiale des Sciences de St.-Pétersbourg* 20 (1875): 187–197.

Von Laue, Theodore H. "Factory Inspection under the Witte System: 1892–1903." *American Slavic and East European Review* 19 (1960): 347–362.

_____. *Sergei Witte and the Industrialization of Russia.* New York: Columbia University Press, 1963.

Vorob'ev, V. N. "D. I. Mendeleev i vozdukhoplavanie." *Sovetskaia Nauka*, no. 8 (1939): 125–140.

_____. *Genezis russkoi vozdukhoplavatel'noi mysli v trudakh D. I. Mendeleeva.* Moscow: Izd. AN SSSR, 1965.

"Voskresnye nabroski." *Golos*, 16 (28) November 1880, #317: 1–2.

Vozdvizhenskii, G. S. *Stranitsy iz istorii kazanskoi khimicheskoi shkoly.* Kazan: Kazanskii khimiko-tekhnologicheskii institut imeni S. M. Kirova, 1960.

"Vremennaia (1898 g.) instruktsiia No. 1 sostavlennaia Glavnoiu Palatoiu mer i vesov, dlia rukovodstva pri primenenii obraztsovykh mer i vesov v mestnykh poverochnykh uchrezhdeniiakh." *VGPMV* 4 (1899): 46–49.

"Vremennaia (1898 g.) instruktsiia No. 2 sostavlennaia Glavnoiu Palatoiu mer i vesov, dlia rukovodstva pri poverke i kleimenii torgovykh mer i vesov v mestnykh poverochnykh uchrezhdeniiakh." *VGPMV* 4 (1899): 50–56.

"Vremennyia pravila dlia ispytaniia i poverki elektricheskikh izmeritel'nykh priborov, predstavliaemykh v Glavnuiu Palatu mer i vesov." *VGPMV* 6 (1903): 109–113.

Vucinich, Alexander. *Empire of Knowledge: The Academy of Sciences of the USSR (1917–1970).* Berkeley: University of California Press, 1984.

_____. "Mendeleev's Views on Science and Society." *Isis* 58 (1967): 342–351.

_____. *Science in Russian Culture, 1861–1917.* Stanford: Stanford University Press, 1970.

_____. *Social Thought in Tsarist Russia: The Quest for a General Science of Society, 1861–1917.* Chicago: University of Chicago Press, 1976.

Wagner, William G. *Marriage, Property, and Law in Late Imperial Russia.* Oxford: Clarendon Press, 1994.

_____. "The Trojan Mare: Women's Rights and Civil Rights in Late Imperial Russia." In Crisp and Edmondson, eds., *Civil Rights in Imperial Russia* (1989): 65–84.

Walden, Paul. "Dmitri Iwanowitsch Mendelejeff." *Berichte der Deutschen Chemischen Gesellschaft* 41 (1908): 4719–4800.

_____. "O trudakh D. I. Mendeleeva po voprosu o rastvorakh." In Tishchenko, ed., *Trudy pervago mendeleevskogo s"ezda* (1909): 58–88.

Walicki, Andrzej. *The Controversy over Capitalism: Studies in the Social Philosophy of the Russian Populists.* Notre Dame, Ind.: University of Notre Dame Press, 1969.

_____. *Legal Philosophies of Russian Liberalism.* Notre Dame, Ind.: University of Notre Dame Press, 1992.

Wallace, Alfred Russel. *Miracles and Modern Spiritualism.* London: George Redway, 1896.

Weeks, Theodore R. *Nation and State in Late Imperial Russia: Nationalism and Russification on the Western Frontier, 1863–1914.* DeKalb: Northern Illinois University Press, 1996.

Wheatcroft, S. G. "The 1891–92 Famine in Russia: Towards a More Detailed Analysis of Its Scale and Demographic Significance." In Edmondson and Waldron, eds., *Economy and Society in Russia and the Soviet Union* (1992): 44–64.

Whelan, Heide W. *Alexander III and the State Council: Bureaucracy and Counter-Reform in Late Imperial Russia.* New Brunswick, N.J.: Rutgers University Press, 1982.

Whittaker, Cynthia H. *The Origins of Modern Russian Education: An Intellectual Biography of Count Sergei Uvarov, 1786–1855.* DeKalb: Northern Illinois University Press, 1984.

Whittaker, Edmund T. "The Aether: Past and Present." *Endeavour* 2 (1943): 117–120.

———. *A History of the Theories of Aether and Electricity. I. The Classical Theories. II. The Modern Theories.* College Park, Md.: American Institute of Physics, 1987 [1951–1953].

Wilson, David B. "The Thought of Late Victorian Physicists: Oliver Lodge's Ethereal Body." *Victorian Studies* 15 (1971): 29–48.

Winkler, Clemens. "Germanium, Ge, ein neues, nichtmetallisches Element." *Berichte der Deutschen Chemischen Gesellschaft* 19 (1886): 210–211.

———. "Ueber die Entdeckung neuer Elemente im Verlaufe der letzten fünfundzwanzig Jahre und damit zusammenhängende Fragen." *Berichte der Deutschen Chemischen Gesellschaft* 30 (1897): 6–21.

———. "Zur Entdeckung des Germaniums." *Berichte der Deutschen Chemischen Gesellschaft* 32 (1899): 307–308.

Winter, Alison. *Mesmerized: Powers of Mind in Victorian Britain.* Chicago: University of Chicago Press, 1998.

Wirtschafter, Elise Kimerling. *Social Identity in Imperial Russia.* DeKalb: Northern Illinois University Press, 1997.

———. *Structures of Society: Imperial Russia's "People of Various Ranks."* DeKalb: Northern Illinois University Press, 1994.

Wise, M. Norton. "German Concepts of Force, Energy, and the Electromagnetic Ether: 1845–1880." In Cantor and Hodge, eds., *Conceptions of Ether* (1981): 269–307.

———, ed. *The Values of Precision.* Princeton: Princeton University Press, 1995.

Witte, Sergei. *The Memoirs of Count Witte,* tr. and ed. Sidney Harcave. Armonk, N.Y.: M. E. Sharpe, 1990.

Wood, DeVolson. *The Luminiferous Aether.* New York: D. Van Nostrand, 1886.

Wortman, Richard S. *The Crisis of Russian Populism.* Cambridge: Cambridge University Press, 1967.

———. *The Development of a Russian Legal Consciousness.* Chicago: University of Chicago Press, 1976.

———. "Property Rights, Populism, and Russian Political Culture." In Crisp and Edmondson, eds., *Civil Rights in Imperial Russia* (1989): 13–32.

———. *Scenarios of Power: Myth and Ceremony in the Russian Monarchy. Volume 2: From Alexander II to the Abdication of Nicholas II.* Princeton: Princeton University Press, 2000.

Wynne, Brian. "Physics and Psychics: Science, Symbolic Action, and Social Control in Late Victorian England." In Barry Barnes and Steven Shapin, eds., *Natural Order: Historical Studies of Scientific Culture* (Beverly Hills, Calif.: Sage, 1979): 167–186.

Wyrouboff, G. "On the Periodic Classification of the Elements." *Chemical News* 74 (1896): 31.

Yaney, George L. *The Systematization of Russian Government: Social Evolution in the Domestic Administration of Imperial Russia, 1711–1905.* Urbana: University of Illinois Press, 1973.

Zabrodskii, G. A. *Mirovozzrenie D. I. Mendeleeva: K piatidesiatiletiiu so dnia smerti (1907–1957).* Moscow: Gos. Izd. politicheskoi literatury, 1957.

Zaionchkovsky, Peter A. *The Russian Autocracy in Crisis, 1878–1882,* ed. and tr. Gary M. Hamburg. Gulf Breeze, Fla.: Academic International Press, 1979.

Zaitseva, L. L. "Nekotorye neopublikovannye materialy, otnosiashchiesia k istorii ucheniia o radioaktivnosti." *TIIEiT* 35 (1961): 149–166.

Zaitseva, L. L., and N. A. Figurovskii. *Issledovaniia iavlenii radioaktivnosti v dorevoliutsionnoi Rossii.* Moscow: Izd. AN SSSR, 1961.

Zakharova, Larissa. "Autocracy and the Reforms of 1861–1874 in Russia: Choosing Paths of Development," tr. Daniel Field. In Eklof, Bushnell, and Zakharova, eds., *Russia's Great Reforms* (1994): 19–39.

Zamecki, Stefan. "Mendeleev's First Periodic Table in Its Methodological Aspect." *Organon* 25 (1995): 107–126.

Zelnik, Reginald E. *Labor and Society in Tsarist Russia: The Factory Workers of St. Petersburg, 1855–1870*. Stanford: Stanford University Press, 1971.

Zhukova, R. G. "Revoliutsionnye studencheskie kruzhki S.-Peterburgskogo universiteta: 80-kh godov XIX veka." *Ocherki po istorii Leningradskogo Universiteta* 2 (1968): 41–51.

Zhukovskii, N. E. "O rabotakh Dmitriia Ivanovicha Mendeleeva po soprotivleniiu zhidkostei i vozdukhoplavaniiu." In Tishchenko, ed., *Trudy pervago mendeleevskogo s"ezda* (1909): 116–124.

"Zur Nichtwahl Mendelejew's." *St. Petersburger Zeitung*, 24 December 1880, #359: 2.

INDEX